Biopharmaceutics

ADVANCES IN PHARMACEUTICAL TECHNOLOGY

A Wiley Book Series

Series Editors:
Dennis Douroumis, University of Greenwich, UK
Alfred Fahr, Friedrich–Schiller University of Jena, Germany
Jurgen Siepmann, University of Lille, France
Martin Snowden, University of Greenwich, UK
Vladimir Torchilin, Northeastern University, USA

Titles in the Series

Hot-Melt Extrusion: Pharmaceutical Applications
Edited by Dionysios Douroumis

Drug Delivery Strategies for Poorly Water-Soluble Drugs
Edited by Dionysios Douroumis and Alfred Fahr

Computational Pharmaceutics: Application of Molecular Modeling in Drug Delivery
Edited by Defang Ouyang and Sean C. Smith

Pulmonary Drug Delivery: Advances and Challenges
Edited by Ali Nokhodchi and Gary P. Martin

Novel Delivery Systems for Transdermal and Intradermal Drug Delivery
Edited by Ryan Donnelly and Raj Singh

Drug Delivery Systems for Tuberculosis Prevention and Treatment
Edited by Anthony J. Hickey

Continuous Manufacturing of Pharmaceuticals
Edited by Peter Kleinebudde, Johannes Khinast, and Jukka Rantanen

Pharmaceutical Quality by Design
Edited by Walkiria S Schlindwein and Mark Gibson

***In Vitro* Drug Release Testing of Special Dosage Forms**
Edited by Nikoletta Fotaki and Sandra Klein

Characterization of Pharmaceutical Nano- and Microsystems
Edited by Leena Peltonen

Biopharmaceutics: From Fundamentals to Industrial Practice
Edited by Hannah Batchelor

Forthcoming Titles:

Process Analytics for Pharmaceuticals
Edited by Jukka Rantanen, Clare Strachan and Thomas De Beer

Mucosal Drug Delivery
Edited by Rene Holm

Biopharmaceutics

From Fundamentals to Industrial Practice

Edited by

HANNAH BATCHELOR
Strathclyde Institute of Pharmacy and Biomedical Sciences
University of Strathclyde
Glasgow, United Kingdom

Registered Office(s)
John Wiley & Sons Ltd, The Atrium, Southern Gate, Chichester, West Sussex, PO19 8SQ, UK

Editorial Office
Boschstr. 12, 69469 Weinheim, Germany

For details of our global editorial offices, customer services, and more information about Wiley products visit us at www.wiley.com.

Wiley also publishes its books in a variety of electronic formats and by print-on-demand. Some content that appears in standard print versions of this book may not be available in other formats.

Library of Congress Cataloging-in-Publication Data

Names: Batchelor, Hannah, editor.
Title: Biopharmaceutics : from fundamentals to industrial practice / edited by Hannah Batchelor.
Other titles: Biopharmaceutics (Batchelor) | Advances in pharmaceutical technology.
Description: First edition. | Chichester, UK ; Hoboken, NJ : John Wiley & Sons, 2022. | Series: Advances in pharmaceutical technology | Includes bibliographical references and index.
Identifiers: LCCN 2021050446 (print) | LCCN 2021050447 (ebook) | ISBN 9781119678281 (cloth) | ISBN 9781119678274 (adobe pdf) | ISBN 9781119678373 (epub)
Subjects: MESH: Biopharmaceutics–methods | Drug Design
Classification: LCC RM301.5 (print) | LCC RM301.5 (ebook) | NLM QV 35 | DDC 615.7–dc23/eng/20211101
LC record available at https://lccn.loc.gov/2021050446
LC ebook record available at https://lccn.loc.gov/2021050447

Cover Design: Wiley
Cover Image: © SCIEPRO/Getty Images

Set in 10/12pt Times by Straive, Pondicherry, India
Printed and bound by CPI Group (UK) Ltd, Croydon, CR0 4YY

C9781119678281_021221

Contents

14 Inhalation Biopharmaceutics   **239**

Precious Akhuemokhan, Magda Swedrowska, and Ben Forbes

15 Biopharmaceutics of Injectable Formulations   **253**

Wang Wang Lee and Claire M. Patterson

List of Contributors

Precious Akhuemokhan Institute of Pharmaceutical Sciences, King's College London, London, United Kingdom

Abdul W. Basit Department of Pharmaceutics, UCL School of Pharmacy, University College London, London, United Kingdom

Hannah Batchelor Strathclyde Institute of Pharmacy and Biomedical Sciences, University of Strathclyde, Glasgow, United Kingdom

Shanoo Budhdeo Seda Pharmaceutical Development Services, Alderley Edge, Alderley Park, Cheshire, United Kingdom

James Butler Biopharmaceutics, Product Development and Supply, GlaxoSmithKline R&D, Ware, United Kingdom

Paul A. Dickinson Seda Pharmaceutical Development Services, Alderley Edge, Alderley Park, Cheshire, United Kingdom

Talia Flanagan UCB Pharma S.A., Avenue de l'industrie, 1420Braine l'Alleud, Belgium

Ben Forbes Institute of Pharmaceutical Sciences, King's College London, London, United Kingdom

Nikoletta Fotaki Department of Pharmacy and Pharmacology, University of Bath, Bath, United Kingdom

Simon Gaisford Department of Pharmaceutics, UCL School of Pharmacy, University College London, London, United Kingdom

Francesca K. H. Gavins Department of Pharmaceutics, UCL School of Pharmacy, University College London, London, United Kingdom

Pavel Gershkovich School of Pharmacy, Centre for Biomolecular Sciences, The University of Nottingham, Nottingham, United Kingdom

Wang Wang Lee Seda Pharmaceutical Development Services, Alderley Edge, Alderley Park, Cheshire, United Kingdom

Christine M. Madla Department of Pharmaceutics, UCL School of Pharmacy, University College London, London, United Kingdom

Mark McAllister Pfizer Drug Product Design, Sandwich, United Kingdom

Laura E. McCoubrey Department of Pharmaceutics, UCL School of Pharmacy, University College London, London, United Kingdom

Hamid A. Merchant Department of Pharmacy, School of Applied Sciences, University of Huddersfield, Huddersfield, United Kingdom

Mine Orlu Department of Pharmaceutics, UCL School of Pharmacy, University College London, London, United Kingdom

Claire M. Patterson Seda Pharmaceutical Development Services, Alderley Edge, Alderley Park, Cheshire, United Kingdom

Chris Roe Quotient Sciences, Mere Way, Ruddington, Nottingham, United Kingdom

Linette Ruston Advanced Drug Delivery, Pharmaceutical Sciences, R&D AstraZeneca, Macclesfield, United Kingdom

Konstantinos Stamatopoulos Biopharmaceutics, Pharmaceutical Development. PDS, MST, RD Platform Technology and Science, GSK, Ware, Hertfordshire, United Kingdom

Magda Swedrowska Institute of Pharmaceutical Sciences, King's College London, London, United Kingdom

Sarah J. Trenfield Department of Pharmaceutics, UCL School of Pharmacy, University College London, London, United Kingdom

Vanessa Zann Quotient Sciences, Mere Way, Ruddington, Nottingham, United Kingdom

Panagiota Zarmpi Department of Pharmacy and Pharmacology, University of Bath, Bath, United Kingdom

Foreword

The term biopharmaceutics causes confusion, particularly with the advent of biopharmaceutical drug products. However, the origins of the word come from the combination of the prefix 'bio' from the Greek, 'relating to living organisms and tissues' where the patient is the organism (or owns the tissues) and pharmaceutics defined as the science relating to the preparation of medicines. Biopharmaceutics encompasses the physical/chemical properties of the drug, the dosage form (drug product) in which the drug is given, and the route of administration to better understand the rate and extent of systemic drug absorption.

Since its introduction in 1970, biopharmaceutics knowledge and testing has enabled scientists to predict drug absorption using a range of *in vitro* and *in silico* tools ensuring that the development of new pharmaceutical products is efficient by refining or reducing the burden of clinical testing. Technological advances during this period have resulted in changes in testing based both on scale, where understanding is now sought at the molecular level and also by the use of dynamic testing systems rather than static apparatus.

The application of biopharmaceutics has reformed the pharmaceutical development process, most notably allowing for clinically relevant risk assessments against formulation and processes changes. Recent advances have witnessed the extension of biopharmaceutics tools towards the earliest and latest stages in product development. Most major pharmaceutical companies have dedicated biopharmaceutics team, invested in improving the predictive power of the suite of tools available to minimise risks of clinical impact of formulation-based changes. As the pharmaceutical industry seeks to accelerate the timelines for drug discovery and development, there is a need for efficient and robust formulations and manufacturing processes to meet the needs of the clinical programme. The biopharmaceutics teams must therefore rise to the challenge of ensuring that formulations used in the clinical programme provide the necessary exposure of drug and are adequately risk assessed.

The Academy of Pharmaceutical Sciences (APS) is a UK-based professional membership body for pharmaceutical scientists. The biopharmaceutics focus group within the APS has a mission to promote scientific education and training in the field of biopharmaceutics to meet the needs of scientists working in both industrial and academic sectors. This book arose as a result of this mission where the members of the focus group recognised the need to provide training for those wanting to better understand biopharmaceutics and serves as a handbook to introduce the tools used within biopharmaceutics as applied to drug development.

The first six chapters cover the basics of biopharmaceutics and provide the context that underpins the later chapters. Chapters 7 and 8 specifically deal with how biopharmaceutics is integrated into product development within the pharmaceutical industry. Chapters 9 and 10 provide information on the regulatory aspects of biopharmaceutics. Chapter 11 highlights the impact of physiology and anatomy and how this can affect the rate and extent of drug absorption. Physiologically based modelling is a valuable asset in the biopharmaceutics toolkit, and this is introduced in Chapter 12. Chapter 13 provides information on a more advanced topic, the application of biopharmaceutics to special populations. Although oral administration remains the mainstay of drug therapy, it is not the exclusive route of administration, Chapters 14–16 cover alternative routes of administration. A newly emerging topic that considers the impact of the microbiome on the rate and extent of drug absorption is included as the final chapter in this book.

The chapters are written by members of the APS Biopharmaceutics focus group and includes academic and industrial authors with a diverse range of experience from those currently undertaking a PhD to those with more than 20 years of experience in this field.

It has been a real joy and privilege to bring this book together, and I thank all of the authors and the editorial team for their dedication in producing this book. I truly hope that this book is useful for those who want to explore biopharmaceutics in the future.

Hannah Batchelor
Strathclyde Institute of Pharmacy and Biomedical Sciences,
Glasgow, United Kingdom

1

An Introduction to Biopharmaceutics

Hannah Batchelor

Strathclyde Institute of Pharmacy and Biomedical Sciences,
University of Strathclyde, Glasgow, United Kingdom

1.1 Introduction

The aim of this chapter is to introduce biopharmaceutics and to define some key terms used within biopharmaceutics. It will also briefly introduce where biopharmaceutics sits in the drug development process.

1.2 History of Biopharmaceutics

The term **biopharmaceutics** was introduced in the 1960s by Levy [1]. The word originates from the combination of bio- from the Greek meaning relating to living organisms or tissue and pharmaceutics defined as the science of pharmaceutical formulations; in this case the living organism is the person (or animal being treated). In modern parlance, the term biopharmaceutics encompasses the science associated with the physical/chemical properties of the drug product (including all components therein) and the interactions of this product with parameters linked to the route of administration that affect the rate and extent of drug uptake or presence at the site for local action. It combines knowledge of materials science; physiology; anatomy and physical sciences.

In more simple terms it is everything that controls the availability of the drug: that is how the drug exits the dosage form and travels to the systemic circulation (for systemically

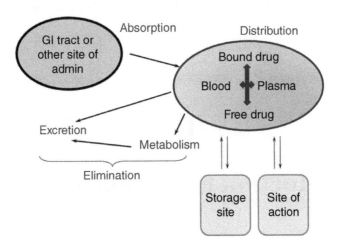

Figure 1.1   *Schematic of the fate of drugs once administered orally; biopharmaceutics relates to the absorption aspect of this image.*

acting drugs) or to the local site of action for locally acting agents. It provides a link between the formulation and the clinical performance of a drug; a mechanistic understanding of biopharmaceutics ensures that the formulation is optimised in terms of exposure. This is shown schematically in Figure 1.1 where biopharmaceutics is focussed on absorption.

The term biopharmaceutics can cause confusion; particularly with the advent of biopharmaceutical drug products. There is evidence in confusion in terminology back in the 1970s where efforts were made to standardise the terminology used [2]; these efforts defined biopharmaceutics in several ways according to the experts at the time of publication. The most widely used definition is, 'The study of the influence of formulation on the therapeutic activity of a drug product. Alternatively, it may be defined as a study of the relationship of the physical and chemical properties of the drug and its dosage form to the biological effects observed following the administration of the drug in its various dosage forms' [3].

An analysis of new drug approvals in 2019 (US, EU and Japan) showed that oral products represented the majority of approvals (50%) with tablets and capsules as the dominant oral dosage forms [4]. Thus biopharmaceutics has tended to focus on oral more than alternative routes of administration.

Historically biopharmaceutics was part of **clinical pharmacology** and **pharmaceutical chemistry**, only becoming its own scientific discipline in the 1970s. In scientific terms, the MeSH definition (MeSH [Medical Subject Headings] is the United States National Library of Medicine controlled vocabulary thesaurus used for indexing articles for PubMed) of biopharmaceutics (introduced in 1970) is, 'The study of the physical and chemical properties of a drug and its dosage form as related to the onset, duration and intensity of its action'. The MeSH term 'biopharmaceutics' being introduced in the 1970s provides an

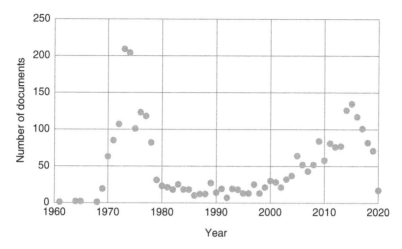

Figure 1.2 *Frequency of biopharmaceutics as a MESH terms in publications versus time. Source: Data from Pubmed.gov, November 2020.*

insight into the history of the topic; the scientific discipline existed long before but was previously listed in scientific data based under a bigger heading of pharmacology as:

- Chemistry, Pharmaceutical (1966–1969)
- Drug Compounding (1966–1969)
- Drugs (1966–1969)
- Pharmacology (1966–1969)

A search in PubMed of 'Biopharmaceutics' [Mesh] conducted in November 2020 resulted in 2725 retrieved documents with a peak in the early 1970s as the science of biopharmaceutics developed. There has also been a general trend of increased use of the term biopharmaceutics since the year 2000. This is shown in Figure 1.2.

There have been a number of key events in the history of biopharmaceutics and these are highlighted in Figure 1.3.

1.3   Key Concepts and Definitions Used Within Biopharmaceutics

There is a strong link between biopharmaceutics and **pharmacokinetics**. Pharmacokinetics measures the concentration of drug at a site in the body versus time. Understanding the biopharmaceutics will influence the pharmacokinetic profile observed. In particular, biopharmaceutics has a focus on the absorption phase of a drug as this is the phase where the dosage form design has influence over the pharmacokinetic profile. The metabolism and subsequent elimination and excretion are driven by the drug properties rather than those of the formulation used to administer the drug.

Pharmacokinetic studies provide information on drug concentrations (typically in plasma or blood) versus time; these studies can be used to demonstrate safety and efficacy of a drug as well as compare the relative performance of alternative dosage forms (for further

Figure 1.3   *Overview of the biopharmaceutics timeline of key events.*

details see Chapter 2). This performance can be by design, for example, to develop a sustained release product to alter dosing frequency. Generation of statistically similar pharmacokinetic profiles for alternative drug products provides reassurance that these medicines can be interchanged with limited effects on clinical efficacy. These statistically similar pharmacokinetic profiles show **bioequivalence** between drug products, this bioequivalence is discussed more in the chapter on regulatory biopharmaceutics (Chapter 10). This is of great importance for generic medicine development to ensure that medicines can be interchanged with not clinical impact to the patient.

Pharmacokinetic data can be analysed to demonstrate what fraction of the drug administered orally was measured within the system; this fraction is termed the **bioavailable dose**. It is recognised that not all drug administered will reach the site of measurement as some will be lost due to: localised degradation; failure to permeate membranes to reach the site of measurement; metabolism between site of absorption and site of measurement. Calculation of the **bioavailability** of a drug is important in dosage form design as it will influence the dose to be administered as well as the likelihood of reaching the target concentration at the site of measurement (and site of action). This can also be termed the **bioperformance** of a product.

The processes that influence the bioavailable dose are key to the science of biopharmaceutics. There is emphasis on the fraction of drug absorbed as this relates to the inherent drug properties and how they link with the dosage form as well as the site where absorption occurs. Formulation scientists can design dosage forms for a range of sites for administration and understanding how the **fraction absorbed** varies by site of administration is important for systemically acting drugs. **Absorption** can be complex and is not a single-step process; there are often several membranes or other barriers that lie between the site of administration and the site of measurement (or action) for a drug. The **permeability** (Chapter 5) of the drug across each of these barriers will dictate the fraction that can traverse the membrane. Measuring the fraction absorbed at each membrane is not possible and often there is a single point for administration and a single point for measurement which can complicate accurate determination of the fraction absorbed. This is exemplified in oral absorption of drugs. Drugs will enter the gastro-intestinal system where some of the drug will be solubilised and will traverse the intestinal membrane; however, there may be some metabolism at the intestinal wall meaning that not all the drug absorbed reaches the systemic circulation. Furthermore, the portal vein drains from the intestine directly into the liver where further metabolism is likely to occur again reducing the quantity of drug present in the systemic circulation. The site of measurement; typically a blood or plasma sample taken peripherally will only show the drug that successfully traversed the intestinal wall AND was not metabolised within the liver; therefore this is lower than the actual fraction of drug absorbed.

First pass metabolism is the term used to describe the fraction of drug lost between entering the portal vein directly from the intestine and existing the liver. This describes the fraction of drug lost during the first pass through the liver, prior to reaching the sampling site.

The oral route is the most common route of drug administration and as such much of this book will focus on oral biopharmaceutics; however there are chapters on alternative routes of administration (Chapter 14: Inhaled Biopharmaceutics; Chapter 15: Biopharmaceutics of Injectable Formulations and Chapter 16: Topical Bioavailability).

A key factor that influences the absorption of drug from the gastro-intestinal tract is the **solubility** (Chapter 4) of the drug within the intestinal fluids. The intestinal fluids are complex, affected by food and many other factors associated with ethnicity, disease and gender

(Chapter 13: Special Populations). Understanding the composition of intestinal fluids and replication of this for *in vitro* models is of huge interest to those working within biopharmaceutics. Due to the transit time within the intestinal tract, it is not just the solubility that is important but the rate of drug **dissolution** (Chapter 6) within the fluids present that will influence the rate and extent of drug absorption.

The **biopharmaceutics classification system (BCS)** (Chapter 9), introduced in 1995 by Gordon Amidon [5], sought to classify drugs based on their dissolution and permeability as these factors are fundamental in controlling the rate and extent of oral absorption. This system is still in use in regulatory science and has been extended to also look at the developability of drugs [6]. The BCS can also justify a **biowaiver**; this is a situation where the *in vitro* solubility and permeability data can negate the need for a clinical study to demonstrate bioequivalence, resulting in a large cost saving for those involved in development.

The major emphasis of research in biopharmaceutics is the development of *in vitro* and *in silico* model that predict how a drug will be absorbed *in vivo*. Thus the use of **biorelevant** models that replicate the physiology, anatomy and local environment within the gastro-intestinal tract (or other site of administration) are important. In particular, the use of **physiologically based pharmacokinetic (PBPK)** models (Chapter 12) that not only replicate the body but also provide indications on the population-based variability in drug absorption.

1.4   The Role of Biopharmaceutics in Drug Development

Drug development is a complex process that involves many scientists, a lot of money and at least 10 years. The process starts with target identification where chemicals are manufactured to 'fit' the receptor of interest and they are typically ranked by potency for that receptor. At this stage, there is little biopharmaceutics input. The next step is to evaluate the lead chemicals using **preclinical** models; this can be cell lines or animal models to determine whether the chemical is as potent *in vivo*. At this stage, some biopharmaceutics input is crucial as the drug may need to be formulated for administration to the animal model and may even be administered orally so the fraction absorbed can be measured. This often relates to the 'drugability' of the lead candidates; defined as the technical evaluation of whether a compound will be a commercially successful drug. Drugability here relates to the likelihood for sufficient and non-variable pharmacokinetic exposure.

Success in preclinical models will trigger **clinical evaluation** in humans. There are three phases of clinical trial prior to launch of a product: **phase 1** will measure safety and efficacy of a compound in healthy volunteers where possible; at this stage the bioavailable dose will be assessed. **Phase 2** studies explore the safety and efficacy of the drug in patients with the disease of interest. The product used for phases 1 and 2 is often different to the final commercial product as the dose is still to be defined. Thus there may be differences in the bioavailable fraction of each formulation administered that needs to be accounted for when interpreting the data and determining the dose. The term **bridging** is used to describe how any differences between formulations used in preclinical and clinical testing are managed during the clinical testing. **Phase 3** studies evaluate the efficacy and safety in a large patient population. Where possible the final commercial formulation will be used in phase 3 studies as these are **pivotal** to underpinning the evidence to justify the introduction of a new product. Biopharmaceutics is integral to the phases of clinical testing as predictive models to understand absorption and consequences of bridging are critical to the success of the interpretation of clinical data.

In parallel to the clinical evaluation (phase 1, 2 and 3 studies) work will be ongoing to ensure that the **chemistry manufacturing and controls (CMC)** activities are on track. These CMC activities ensure that the product and manufacturing process meet the stringent regulatory requirements ensuring that a safe and high-quality product is available to the patient population. Any changes to the product or manufacturing process need to be understood, particularly if there are likely to be consequences to the patient; thus biorelevant predictive tests are of value in de-risking the development process. In addition to biorelevant tests, often **discriminatory dissolution testing** is required; this is a method that links to clinical data and shows where changes in the product (as a result of composition or manufacturing changes) are likely to have an effect on the clinical performance. These discriminatory dissolution tests are generated by links to *in vivo* clinical data; either using an *in vitro in vivo* **relationship (IVIVR)** or using the principles of **quality by design (QbD)**.

Regulatory approval of new products is essential. The International Council for Harmonisation of Technical Requirements for Pharmaceuticals for Human Use (**ICH**) brings together the regulatory authorities and pharmaceutical industry to discuss scientific and technical aspects of drug registration. ICH guidelines include information on biopharmaceutics that are essential for the approval of medicines. Two guidelines are focussed on biopharmaceutics specifically: ICH M9 Biopharmaceutics classification system based biowaivers and M13 Bioequivalence for Immediate release solid oral dosage forms. Within the US the major regulatory agency is the **FDA** (Food and Drug Administration); the FDA have a Biopharmaceutics council within the centre for drug evaluation and research. This office is responsible for the generation, implementation and review of biopharmaceutics-related guidance, policies and practices. There are several biopharmaceutics specific FDA regulatory guidance papers issues that are critical to the approval of new drugs. Similar to the USA there are many global regulatory bodies where biopharmaceutics guidance has been issued including the **EMA (European Medicines Agency)** and the **Japanese Food and Drug Administration**. Recently the ICH M9 guidance has sought to align these where possible for the BCS classification.

Biopharmaceutics interfaces with several other scientific disciplines, this book aims to provide a background to biopharmaceutics and to showcase how knowledge can be applied to the efficient development of drug products. The level of detail in terms of biopharmaceutics knowledge of a drug and a drug product will increase during the drug development process. This is shown schematically in Figure 1.4.

Biopharmaceutics is an important scientific discipline, particularly for those developing new drugs. An understanding of biopharmaceutics aids in the design of appropriate drug candidates (Chapter 7) as well as optimised drug products (Chapter 8) to ensure that the drug is well absorbed from the site of administration. Clinical testing of drugs, from phase 1 to phase 4 clinical trials is expensive and time-consuming. Biopharmaceutics tests and knowledge are critical to de-risk changes in the clinical performance as a result of minor changes in the product and process used to manufacture the drug product used within these clinical trials. There is a strong relationship between biopharmaceutics and regulatory science during the development of drug products.

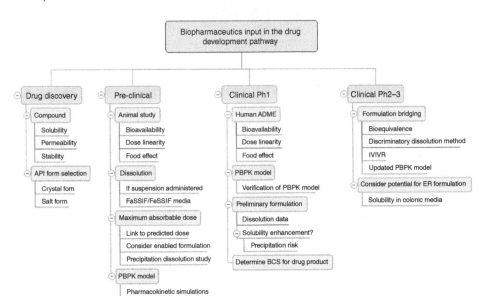

Figure 1.4   Overview of biopharmaceutics input in the drug development pathway.

1.5   Conclusions

Biopharmaceutics is a relatively new science that brings together knowledge on anatomy and physiology to understand the biological environment where drugs are absorbed with materials science to appreciate the drug and excipient related effects on these processes. This book brings together the knowledge required to better understand biopharmaceutics and to apply this knowledge in the development of drug products.

References

[1]   Levy, G. and Nelson, E. (1961). Pharmaceutical formulation and therapeutic efficacy. *JAMA* **177** (10): 689–691.

[2]   Zathurecký, L. (1977). Progress in developing a standard terminology in biopharmaceutics and pharmacokinetics. *Drug. Intell. Clin. Pharm.* **11** (5): 281–296.

[3]   Wagner, J.G. (1961). Biopharmaceutics: absorption aspects. *J. Pharm. Sci.* **50**: 359–387.

[4]   Sedo, T.K. (2020). 2019 Global drug delivery & formulation report part 1 a review of 2019 product approvals. *Drug Develop. Deliv.* **20** (2): 18–23.

[5]   Amidon, G.L., Lennernäs, H., Shah, V.P., and Crison, J.R. (1995). A theoretical basis for a biopharmaceutic drug classification: the correlation of *in vitro* drug product dissolution and *in vivo* bioavailability. *Pharm. Res.* **12** (3): 413–420.

[6]   Butler, J.M. and Dressman, J.B. (2010). The developability classification system: application of biopharmaceutics concepts to formulation development. *J. Pharm. Sci.* **99** (12): 4940–4954.

2

Basic Pharmacokinetics

Hamid A. Merchant

*Department of Pharmacy, School of Applied Sciences, University of Huddersfield,
Huddersfield, United Kingdom*

2.1 Introduction

This chapter aims to introduce basic pharmacokinetic terminologies and principles underpinning a drug's life cycle in the body, from administration to elimination. The chapter will
help readers to understand a typical pharmacokinetic profile following drug administration
via absorptive versus non-absorptive routes.

The understanding of the critical role of absorption processes in pharmacokinetics will
help in understanding how biopharmaceutics principles and strategies can modulate the
pharmacokinetic profile of a drug and in turn affect the therapeutics.

2.2 What is 'Pharmacokinetics'?

The term pharmacokinetics refers to the principles underpinning the *absorption, distribution, metabolism* and *elimination* of a drug following its administration into the body. The
drug can be administered via a range of routes; many require absorption of the drug from the
site of administration to get into the blood circulation, whereas drug can also be administered directly into the blood circulation (intravenously) bypassing the absorption process.

The clinical efficacy of a drug relies on the delivery of a drug to the receptor site at the
required concentration to produce a therapeutic effect. A drug, therefore, must bypass the

Biopharmaceutics: From Fundamentals to Industrial Practice, First Edition. Edited by Hannah Batchelor.
© 2022 John Wiley & Sons Ltd. Published 2022 by John Wiley & Sons Ltd.

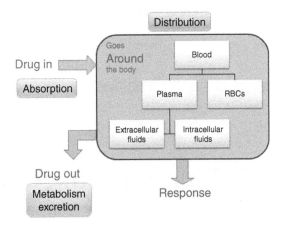

Figure 2.1   *Illustration showing a typical life cycle of a drug in the body (RBCs = red blood cells).*

barriers at the absorptive site and the metabolic challenges for successful delivery to the target site. The life cycle of a drug typically means that it goes '*into*' the body, goes '*around*' the body, exerts its pharmacological effects and comes '*out*' of the body (Figure 2.1).

When a drug is administered into the body, it is *absorbed* from the tissues where it was administered (e.g., muscles, gastrointestinal tract under the skin, etc.) to get into blood circulation. It is then *distributed* across the body once it is in the systemic circulation where a fraction of the drug arrives at the target sites, binds the receptors and produces a therapeutic effect. The drug is also *metabolised* by various tissues (mainly liver and gut) into other forms (mainly inactive) during its voyage into the body and then it is *excreted* out of the body, mainly via urine. The 'pharmacokinetics' is the study of the processes governing the 'absorption', 'distribution', 'metabolism' and 'excretion' of a drug, often referred to as *ADME*.

2.3   Pharmacokinetic Profile

A typical pharmacokinetic profile represents a drug concentration in blood (or plasma or serum) over a period of time usually measured post drug administration until most of the drug is eliminated from the body. The shape of the plasma drug concentration–time profile depends on the route of administration of the drug. If the drug was administered intravenously, it becomes available in the systemic circulation instantly and the drug concentration is hypothetically at its maximum following its administration; this will gradually decline with time as the drug is eliminated from the body (Figure 2.2).

When the drug is administered into the tissues, for example, muscles (intramuscular injection) or taken orally (tablets, capsules, etc.), it has to go through a complex *absorption* process before it appears in the blood. Hence, there is a significant lag time to see the therapeutic effect of the drug compared to intravenous administration. This is one reason when the intravenous route of administration is preferred over the oral route in accidents or medical emergencies when a prompt response of a lifesaving drug is desired.

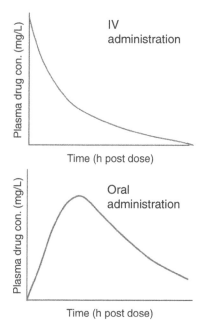

Figure 2.2 *A typical pharmacokinetic profile (plasma drug concentration–time profile) of a drug following intravenous (IV) or oral administration.*

Following oral administration of a drug, once the drug starts to be absorbed and appears in the blood circulation, its plasma concentration increases as a function of time. As soon as the drug starts to circulate in the blood, it leads to drug clearance (by metabolism and excretion). However, during the initial phase (termed as the *absorption phase*), the drug absorption surpasses the elimination leading to an increase in plasma drug concentration over time. A therapeutic response from the drug (for example a pain relief following analgesic drug) is observed when the drug concentration reaches the minimum thresholds required to produce the therapeutic effect, known as *minimum effective concentration* (MEC). The time taken for the plasma drug concentration to get to the MEC post administration is referred to as the *onset time*.

Following administration to sites other than intravenous, the plasma concentration reaches a plateau, termed as C_{max} or *maximum plasma concentration* of the drug; after this point, the elimination becomes dominant compared to absorption. The t_{max} refers to the time taken to get to the C_{max} from when the dose was administered. Following this point, the elimination surpasses the absorption leading to a net loss of the drug from the body and plasma drug concentration starts to decrease with time; this refers to the *elimination phase*. For drugs which are not absorbed efficiently from the gut, formulation strategies underpinning principles of biopharmaceutics can improve the drug absorption. Conversely, often a slower rate of absorption is beneficial in sustaining the drug effect for an extended period, for instance, modified release dosage forms (e.g., sustained-release, slow-release or controlled-release). This often helps to reduce the dosage frequency of a drug and improves patient compliance. The pharmacokinetic profiles of such formulation exhibit a 'flip-flop' model where there is significant drug absorption still taking place beyond C_{max}.

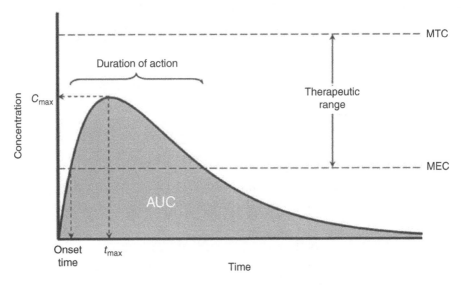

Figure 2.3   *A typical pharmacokinetic profile (plasma concentration–time profile) following oral drug administration illustrating common pharmacokinetic parameters.*

Pharmacokinetic studies help to determine the safe and effective dose of a drug by optimising the dose to keep the drug concentrations well above MEC but well below the *minimum toxic concentration* (MTC) that is also known as the *maximum safe concentration* (MSC). The concentration difference from MEC to MSC is referred to as the *therapeutic window* or the *therapeutic index*, shown in Figure 2.3. The wider the window, the safer is the dosage regimen. Where there is a narrow window, a much tighter dosage regimen is needed to ensure that fluctuations in the plasma drug concentration do not exceed the MTC or fall below the MEC. Drugs where the therapeutic window is very narrow are often referred to as NTI (*narrow therapeutic index*) or *potent drugs* (pharmacologically active at a very small dose) and require close monitoring (refer to Section 2.12) and often need personalised adjustment of dosages to prevent toxic adverse effects.

The term *duration of action* refers to the time period when drug concentration stays above the MEC threshold; hence, it represents the time period during which the drug remains therapeutically effective. The term *AUC* (*area under the curve*) refers to the total area under the plasma concentration–time profile and represents the total circulatory concentration of a drug over a period of time (Figure 2.3).

2.4   Bioavailability

The absorption of a drug at the administration site is a complex interplay between the *physicochemical* characteristics of the drug and the *physiology* of the surrounding tissues.

The drug absorption through the gastrointestinal tract, for instance, is a function of drug's *solubility* (Chapter 4), its *dissolution* (Chapter 6) within the gastrointestinal milieu and drug's *permeability* across the gastrointestinal mucosa (Chapter 5) taking into account the complex *gastrointestinal anatomy and physiology* (Chapter 11).

The fraction of the administered dose of a drug that reaches the systemic circulation is termed the *Bioavailability* (*F*) which is a function of the fraction that permeates through the gastrointestinal mucosa (f_a), the fraction that survives gut-wall metabolism during absorption (f_g) and the fraction which escapes the hepatic metabolism (f_h) when it passes through the liver. Figure 2.4 summarises various factors associated with drug absorption through the gastrointestinal tract. The fraction absorbed (f_a) from the gastro-intestinal tract is an interplay between physicochemical properties of the drug, its solubility and dissolution rate within the gastrointestinal luminal fluids and its stability in the gastrointestinal milieu (acidic/basic pH, the digestive enzymes, and the interaction with food and its digestive products). The fraction absorbed also depends on a drug's ability to cross the gastrointestinal absorptive barrier which in turn also depends on its physicochemical characteristics, lipophilicity, the partition coefficient and its suscepti-bility to various *influx* or *efflux transporters* present within gastrointestinal absorptive cells. The absorption of most of the drugs administered into the gastrointestinal tract, is, therefore, a complex phenomenon; hence, the oral bioavailability (*F*) is rarely equal to 1 (or 100%). In contrast, if the drug was administered into the blood circulation directly, such as intravenously (injected into the veins), it bypasses the absorption bar-riers and entirety of the administered dose is available in the systemic circulation, hence bioavailability (*F*) is 1.

The science of biopharmaceutics is invaluable to understand the complex interplay between physicochemical properties of the drug and gastrointestinal physiology to develop innovative formulations and drug delivery systems (Chapter 8) to improve the drug absorp-tion and bioavailability through various routes of administration.

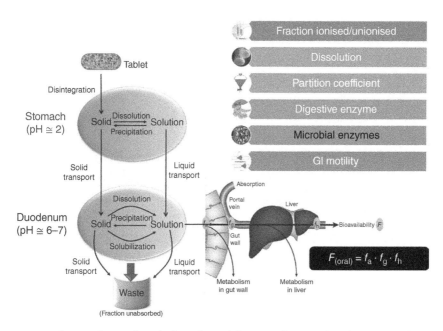

Figure 2.4   *Physicochemical and physiological factors affecting the absorption of the drug from the gastrointestinal tract following oral administration.*

2.5   Drug Distribution

The blood circulates *'around the body'* (termed *distribution*) to various tissues and provides oxygen and nutrients, and also removes metabolic waste from the tissues; the drug is also distributed to the body tissues through the blood circulation. Therefore, measuring drug concentration in blood gives a reflection of the drug concentration in various body tissues. A blood sample can be more easily obtained from a human volunteer or the patients as compared to the other body tissues. Blood is a complex matrix containing various cellular components (blood cells) and dissolved constituents in the liquid part of the blood. Refer to the schematic in Figure 2.5 to understand how *plasma* and *serum* differ from the whole *blood*.

Typically, when a drug concentration is measured in a whole blood sample, the concentration may not be the same as if measured in serum, or plasma. This is due to the changes in the volume that occur when plasma or serum is separated from the blood as well as potential binding of the drug to blood components, refer the illustration in Box 2.1 for an example to understand this phenomenon.

Often, drugs have a tendency to bind to plasma proteins (like albumin) or the red blood cells (RBCs); in this case, it is necessary to differentiate between the 'fraction bound' and the free fraction of drug that is present. The drug concentration measured in plasma or serum may, therefore, specifically refer to the 'fraction unbound' of the drug and will not be the same as the total drug concentration in the whole blood. The 'fraction unbound' is also termed the *'free fraction'* of the drug. It is this free fraction that is distributed to the body tissues, exerts its pharmacological effects and is also excreted via urine. The 'fraction bound' is usually confined to the blood circulation, is not distributed to the tissues and is often 'unavailable' for elimination via the kidneys. The 'fraction bound' and 'fraction unbound' are maintained in equilibrium, and the fraction bound becomes unbound with time due to a continuous reduction of the unbound fraction following drug elimination.

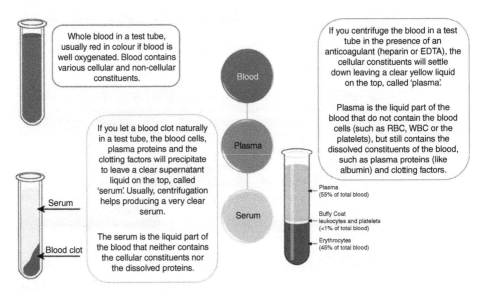

Figure 2.5   *Illustration explaining how plasma and serum differ from blood.*

Box 2.1   Drug Concentration in *blood* is not the same as the drug concentration in *plasma*!

A drug concentration in a whole blood was measured as 1 mg/L. If the plasma was then separated from the whole blood, then the 1 mg of the drug that was present in the whole blood will now be present in the plasma provided that the drug does not bind to the RBCs (red blood cells).

   In a typical healthy male, plasma represents ~55% of the total blood volume. This will mean that the 1 mg of the drug will now be present in 0.55 L of the plasma instead of 1 L of the whole blood.

   This represents a plasma concentration of 1.8 mg/L, which is higher than the 1 mg/L – the concentration measured in the whole blood sample.

Box 2.2   Why measure drug concentrations?

The intensity of pharmacological effect (or toxic effect) is often related to drug concentration at the receptor site (in tissue). Measuring plasma drug concentration may help to adjust the dose to optimise the drug response or prevent serious toxic effects in an individual patient.

   Measuring drug concentration in blood during bioequivalence studies also helps to establish therapeutic equivalences between two different products of a drug (such as brand vs. generic) or different formulations of a drug (such as tablet vs. capsule).

The fraction bound can also be subject to a potential drug interaction; if a co-administered drug had a higher affinity for the binding site at the plasma protein or the RBC, it could displace the other drug. This may result in a significant increase in the free fraction of a drug which can result in increased therapeutic effect (or toxicity) and increased elimination.

The potency of drug response (the pharmacological or toxic effect) depends on the drug concentration at the target site in the tissues; measuring drug concentration in the blood (or in serum or plasma) helps to estimate the intensity of the drug response at a given dose.

2.6   Volume of Distribution

The volume of distribution (V) of a drug is a pharmacokinetic term that represents the hypothetical volume of total body tissues where a drug is distributed following administration. It is denoted in appropriate units of volume, usually litres (L) or litres per kilogram body weight (L/kg). Water represents most of the body weight, with total body water estimated approximately 42 L in a 70 kg healthy male that accounts for the 60% of the total body weight (Box 2.3). The total body water is represented by the blood volume (~5 L), the interstitial fluids (~11 L) and the intracellular fluids (~28 L).

Box 2.3   Total body water.

Plasma: ~3 L (~4% body weight)
Interstitial fluids: ~11 L (~16% body weight)
Extracellular fluids (ECF) ~20% body weight
Intracellular fluids: ~28 L (~40% body weight)
Total body water, ~42 L ~60% body weight

Based on an average 70 kg healthy male.

A drug with a very low volume of distribution may mean that the drug is mainly distributed to the extracellular fluids, like blood and/or interstitial fluids.

The rate and extent of drug distribution into the body tissues is an interplay between physicochemical properties of the drug (e.g., molecular weight, solubility, lipophilicity, ionisation constant and partition coefficient) and its interaction with biomolecules (e.g., ability to bind to the blood cells or the plasma proteins). The rate and extent of drug distribution to a particular organ or a tissue also depends on the blood flow. The heart pumps about five litres of blood every minute (*the cardiac output*) of which about 30% passes through the liver, ~25% to the kidneys and about 14% to the brain. The drug is usually distributed to the highly perfused organs first (such as brain, heart, lung, kidney and liver) before it is distributed to other tissues with poor blood flow (Figure 2.6). In pharmacokinetics, a very large volume of distribution suggests that the drug is distributed deeper down the body tissues, for instance, the muscles and the adipose tissue.

2.7 Elimination

The drug is removed out of the body through *metabolism* and direct elimination via *excretion*. The process is often referred to as *clearance* in pharmacokinetics. The metabolism involves structural changes by biotransformation of the drug into different chemical entities,

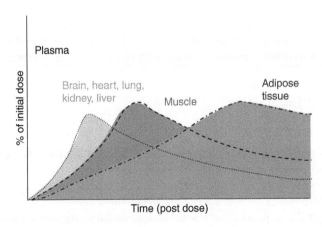

Figure 2.6   *A typical drug distribution into the body following an intravenous dose.*

usually referred to as *metabolites* which can still be pharmacologically active (produce a therapeutic effect or are responsible for adverse reactions). These metabolites along with the unmetabolized drug are then subjected to *excretion* via the kidneys.

2.7.1   Metabolism

Drugs absorbed from the gastrointestinal tract can undergo gut-wall metabolism by various mucosal enzymes expressed in the intestinal mucosa, such as the cytochrome P450 enzyme system. The fraction absorbed that survives the gut-wall metabolism then passes through the liver via the hepatic portal vein (Figure 2.4) where it gets metabolized, also referred to as the *'first-pass effect'*; some drugs are subjected to a significant first-pass metabolism by the liver. The drug can also be metabolised by various other tissues during its distribution in the body. Some drugs are also subjected to *hepatobiliary excretion*, where a fraction of the absorbed drug is excreted back into the small intestine through the bile, which either gets reabsorbed or excreted via *faeces*.

The *liver* is responsible for most of the metabolism of the drug circulating in the body, irrespective of the route of administration. The drug metabolism in the liver mainly encompasses *Phase-I* chemical reactions to make minor structural changes, for instance, oxidation, reduction or hydrolysis, that helps in excretion of the drug through the kidneys. Some drugs also undergo a *Phase-II* (conjugation) reactions, where the drug is coupled with large biomolecules, such as glucuronic acid, glutathione or amino acids. These conjugated drugs are then eliminated in the urine by the kidneys.

The degree of drug metabolism can be affected by the liver function of the individual. Damage to the liver can reduce the metabolic capacity of the liver resulting in increased bioavailability of certain drugs. The hepatic enzymes can also be inhibited by some co-administered drugs and a potential interaction between two drugs may also lead to an increased drug fraction escaping metabolism, therefore, may require dosage adjustments. Conversely, certain foods or co-administered drugs can also act as enzyme inducers and can lead to an increased hepatic clearance of the drug and may reduce the bioavailability of certain drugs.

2.7.2   Excretion

Most of the drug (unchanged and its metabolites), along with other metabolic wastes that are produced in the body as part of the daily routine, are excreted in urine via the kidneys. *Renal excretion*, therefore, is considered as the prime route of elimination for many drugs. The kidney receives about a quarter of the cardiac output; ~1.25 L blood flows through kidneys every minute. The *renal tubule* is the unit structure in the kidney that is responsible for filtering the drug, metabolites and other wastes, and is also referred to as *nephron*. There are hundreds of thousands of nephrons present in each kidney. The drug and metabolites along with other waste products are filtered at the *glomerulus* that is a bunch of blood capillaries twisted into Bowman's capsule in each nephron (Figure 2.7).

The drug and its metabolites are chiefly filtered passively at the glomerulus. The plasma flow to the glomerulus is ~120 mL/min, also known as the *Glomerulus filtration rate* (GFR). The drugs and metabolites that cannot be passively filtered (ionised, large molecular structure, conjugates, etc.) rely on transport proteins to actively secrete these molecules in the *proximal tubule* by a process known as the *active tubular secretion* (ATS), for

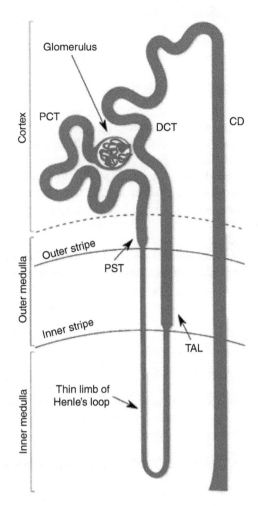

Figure 2.7 *An illustration of the renal tubule (nephron) and its cortex and the medullary regions. CD, collecting duct; DCT, distal convoluted tubule; PCT, proximal convoluted tubule; PST, proximal straight tubule and TAL, thick ascending limb. Source: From Kumaran and Hanukoglu [1] / John Wiley & Sons / CC BY 4.0.*

example, penicillins. Often unionised drugs (permeable) are reabsorbed back into the body from the *distal tubule*. The changes in urine pH (acidification or alkalinisation) can therefore affect the urinary excretion of many ionisable (acidic or basic drugs), such as aspirin.

The rate and extent of drug elimination by the body are therefore significantly affected by the renal function. The renal function is related to age, sex, body weight, hydration state, pregnancy, oedema, altered protein binding and other factors. The renal function can also be compromised by co-administered drugs or toxins or due to a pre-existing pathological condition, such as chronic kidney disease. The dosages for drugs that are chiefly cleared by the kidney are, therefore, adjusted according to the patient's renal function. The renal function in a patient can be estimated by the *creatinine clearance*.

Creatinine is an endogenous waste produced as a result of muscle metabolism, that is filtered at the glomerulus and eliminated via the urine. Normal creatinine clearance can, therefore, indicate a healthy renal function. If renal function is compromised, renal excretion of creatinine is reduced and the accumulation of creatinine results in increased serum concentration of creatinine. Creatinine clearance can be accurately measured in a patient by measuring the serum concentration of creatinine (requires a blood sample) and a 24-hour urine collection to measure creatinine excretion rate in the urine. The creatinine clearance can also be estimated in a patient using *Cockcroft and Gault* method (refer British National Formulary) which only requires a single-point serum creatinine concentration and uses patients body weight and age to estimate the creatinine clearance, and therefore is quicker and easier method for the routine clinical practice. However, this method is subject to significant estimation errors in either subjects with a very lean muscle mass (very low body mass index) or morbidly obese individuals, therefore, requires a careful clinical interpretation.

2.8   Elimination Half-Life ($t_{1/2}$)

Elimination half-life is denoted as $t_{1/2}$ and reported in a unit of time (such as minutes or hours). It is an important pharmacokinetic parameter that helps to understand the rate of drug elimination from the body. Half-life can be defined as *'the time it takes for the plasma (or blood or serum) drug concentration to reduce by half'*. Drug elimination from the body is non-linear and follows first-order kinetics for most drugs; therefore, the elimination phase of the pharmacokinetic curve can be explained by the drug's half-life. Half-life is a concentration-independent property; therefore, it can be determined at any point in the elimination phase of the plasma drug concentration–time profile.

Figure 2.8 shows a first-order pharmacokinetic profile following intravenous drug administration on a log-linear scale. The profile shows that the drug concentration reduces to half every two hours; hence, the drug's elimination half-life ($t_{1/2}$) is two hours.

Half-life, can be used to calculate how long it will take for a drug to be completely removed from the body following a dose, often referred to as the *washout period*. Typically, it takes three to five half-lives for most of the drug to be eliminated from the body.

This would mean that a drug with $t_{1/2} = 2$ h will take about 6 to 10 hours to eliminate from the body (Box 2.4). The drug concentration in the body after five half-lives does not reach a mathematical zero but is so little that five half-lives principle is usually acceptable in pharmacokinetic studies. The wash-out period is important in clinical research and drug development to plan studies and to generate the cleanest possible data set.

2.9   Elimination Rate Constant

The elimination rate constant (k) (also written as k_e or k_{el}) represents the proportion of the drug in the body that is eliminated in a given time. The k is expressed as the inverse of time, for example, h^{-1} or min^{-1}. For instance, a drug with $k = 0.1\,h^{-1}$ will mean that 10% of the drug is being eliminated out of the body each hour. The elimination rate constant is closely related to the elimination half-life and it complements the understanding of drug

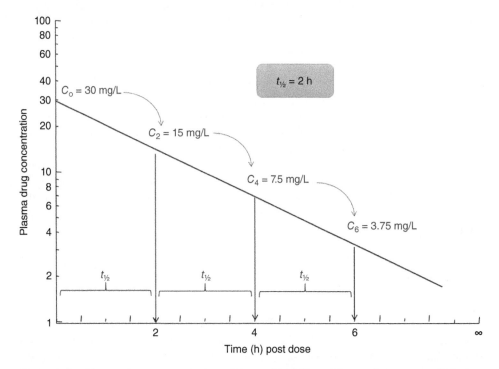

Figure 2.8 *Plasma drug concentration–time profile followed by an intravenous (IV) dose exhibiting first-order kinetics. Figure shows the estimation of drug half-live at three different time points.*

Box 2.4   How long it will take for the drug to eliminate completely from the body?

Number of $t_{1/2}$ elapsed	Percentage of drug eliminated	Drug	Half-life (h)	Time for 90% elimination
1	50	Gentamicin	2–3	7–10 h
2	75			
3	87.5	Theophylline	6–8	24 h
3.3	90			
4	94			
4.3	95			
5	97	Lithium	15–30	2–4 days
6	98.4			
6.6	99			
7	99.2	Digoxin	36–51	5–7 days

Typically, it takes three to five half-lives for most of the drug to be eliminated from the body.

elimination. The elimination rate constant of a drug can be calculated from the elimination half-life of the drug from the equation: $k = \dfrac{0.693}{t}$.

For a drug with elimination rate constant, $k = 0.1\ h^{-1}$, the half-life will be ~7 hours. If 100 mg of the drug was present in the body, it will take about seven hours to reduce this to 50 mg by elimination, i.e., to the half; this is also represented by the amount eliminated every hour using elimination rate constant (Box 2.5).

2.9.1   Clearance

In pharmacokinetics, *clearance (Cl)* is defined as the '*volume of body fluid cleared of the drug per unit time*' and is expressed in units of volume per time, e.g., $L\,h^{-1}$ or $mL\,min^{-1}$. Clearance can be determined using *elimination rate constant (k)* and the *volume of distribution (V)*, i.e., $Cl = k \times V$

The *total body clearance (Cl)* is a function of *renal (Cl_R)* and *non-renal (Cl_{NR})* clearance, and can be expressed as $Cl = Cl_R + Cl_{NR}$. The renal clearance is dependent on kidney function as explained earlier in section 2.7.2. The drug metabolism by the liver, termed as *hepatic clearance*, represents most of the non-renal clearance. Total body clearance is, therefore, affected by the renal or the hepatic function or both, as the case may be. For a drug, primarily eliminated unchanged in the urine, any changes in renal function will significantly impact on its clearance. Impaired clearance of the drug may lead to accumulation of the drug in the body and may require adjustment in subsequent dosages to avoid potential toxic effects. The dosage adjustment decisions should not be solely based on renal clearance and must consider total body clearance to ensure the hepatic clearance is taken into consideration (Figure 2.9). Many patients with compromised renal function, such as in chronic kidney disease (CKD) or those with acute kidney injury (AKI) and taking drugs that are primarily eliminated by the kidneys, may require dosage adjustments. The total body clearance of the drug in a patient

---

Box 2.5   The elimination rate constant, k and the elimination half-life ($t_{1/2}$)

If the $k = 0.1\ h^{-1}$ it will mean that ~10% of the drug left in the body will be eliminated in the next hour.

Time	Amount eliminated (mg)	Amount in body (mg)
0	0	100
1	10	90
2	9	81
3	8.1	72.9
4	7.3	65.6
5	6.6	59.0
6	5.9	53.1
7	5.3	47.8

The amount of the drug left in the body at seven hours is reduced to half of the original amount at time zero, correlating to the drug's half-life, i.e. $t_{1/2} = 7\,h$.

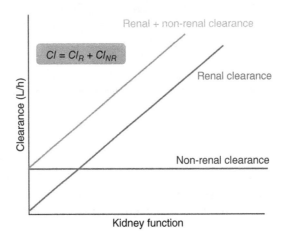

Figure 2.9 *The relationship of total body clearance, renal and non-renal clearance and kidney function.*

in association with renal function (as estimated by the creatinine clearance) helps to make personalised dosage adjustments in a patient (section 2.7.2).

2.10   Area Under the Curve (AUC)

The term AUC refers to the total area under the drug's plasma concentration–time curve and represents the total circulatory concentration of a drug over a period of time, often also referred to as *total exposure*. It is usually expressed as either AUC_{0-t} (from the administration of dosage to the last point of measurement in pharmacokinetic profile) or $AUC_{0-\infty}$ (from the administration of dosage to an infinite time when drug concentration reaches near zero).

AUC is an important pharmacokinetic parameter which represents the extent of absorption of a drug from a dosage form and, therefore, helps in determining the *bioequivalence* of two dosage forms of the same drug or comparing an innovators' product (brand) with a generic.

AUC can be estimated by geometrically calculating the area under the curve of a plasma drug concentration–time profile. This is usually done by dividing the curve into various small geometries such as trapezoids or rectangles and triangles to enable accurate measurements. This is shown in Figure 2.10. If the drug concentration is expressed as $mg\,L^{-1}$ against time in hours in a pharmacokinetic profile, then AUC is expressed as $mg\,h\,L^{-1}$

2.11   Bioequivalence

The pharmacokinetic profiles of two products containing the same drug are often compared. Two formulations which give 'essentially' equivalent circulating concentrations of a drug at each point in time in a pharmacokinetic profile are likely to elicit equivalent therapeutic effects and therefore can be regarded as *bioequivalent*.

Bioequivalence studies are helpful to assess the pharmacokinetics of two drug products, for instance: originally formulated product versus a reformulated or modified drug product,

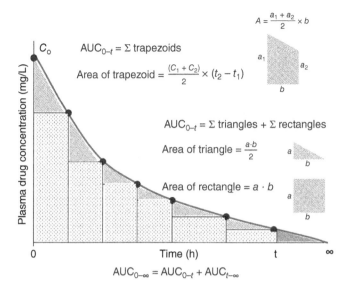

Figure 2.10   *An illustration showing the calculation of AUC in a plasma drug concentration–time profile.*

two different dosage forms of the same product e.g., tablets versus oral liquid or comparison of the generic with innovator's original product. Biopharmaceutical considerations in early development help to achieve optimum absorption and bioavailability of a drug product or developing targeted delivery systems. The biopharmaceutical approaches in formulation design are discussed in detail in Chapter 8.

Bioequivalence studies usually compare the C_{max}, t_{max} and AUC of the two products and if found within tolerances as set by the regulatory agencies (e.g., MHRA, FDA and EMA), the products may be declared as bioequivalent. Chapter 10 provides a detailed account of regulatory perspectives in bioavailability and bioequivalence studies.

2.12   Steady State

Medicines are often not taken as just a single dose and repeat administration of dosages is often required to achieve remission. In chronic conditions, such as diabetes or hypertension, medicines are administered regularly to maintain the plasma concentration of drugs constantly within the therapeutic window to ensure maintenance of the therapeutic response. After multiple dosage administrations, when a pharmacokinetic equilibrium is achieved, the plasma concentration becomes constant (fluctuates in a constant pattern within the therapeutic window), termed as the *'steady state'*.

At the steady state, the rate of drug input into the systemic circulation becomes equal to the rate of drug output (elimination); hence, a pharmacokinetic equilibrium keeps a fairly constant level of drug concentration in the body. It takes about three to five half-lives for a drug to get to the steady state (Figure 2.11). This may mean that a drug with $t_{1/2} = 8$ h if taken twice a day will achieve a steady-state plasma concentration in 40 hours (Box 2.6). For some drugs, this steady-state plasma concentration needs very close monitoring during treatment for a patient, termed as *therapeutic drug monitoring* (TDM).

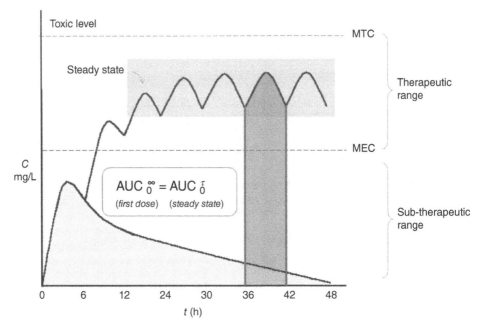

Figure 2.11 *Repeat dose pharmacokinetic profile showing steady state compared to the first dose.*

Box 2.6   How long it takes to get to the steady state?

It takes about five half-lives to get to the steady-state concentration after administering multiple dosages of a drug.

Drug	Dosage interval, τ	Halt-life, $t_{1/2}$	Time to steady state, $t_{C_{ss}}$
Lignocaine	1 h	90 min	7.5 h
Theophylline	12 h (MR*)	8 h	40 h
Digoxin	24 h	48 h	10 days
Amiodarone	24 h	50 days	250 days

*modified release dosage form.

Therapeutic drug monitoring (TDM) involves frequent checking of the plasma drug concentration in the patient and consequent adjustment of the dose to ensure plasma concentrations are maintained within the desired therapeutic range. TDM is usually performed for drugs that are known to exhibit significant pharmacokinetic variability, those with a very narrow therapeutic range such as potent drugs or those where the pharmacological response strongly depends on maintaining a target drug concentration in the body (Box 2.7). Often such variability in pharmacokinetics is due to intra- and inter-subject variability in drug

Box 2.7 Therapeutic drug monitoring (TDM).

Drug	Therapeutic range
Lithium	0.4–0.8 mmol L^{-1}
Phenytoin	10–20 mg L^{-1}
Amiodarone	1.0–2.5 mg L^{-1}
Digoxin	0.5–2.0 mcg L^{-1}
Theophylline	10–20 mg L^{-1}

absorption, metabolism or elimination processes. The differences in rate and extend of drug absorption (often due to physiological or formulations related factors) may have implications in steady-state pharmacokinetics and can therefore affect therapeutic efficacy.

2.13 Compartmental Concepts in Pharmacokinetics

The compartmental concepts in pharmacokinetics are used to ensure accurate estimation of pharmacokinetic parameters, such as elimination half-life, the volume of distribution and elimination rate constant of a drug. The knowledge and understanding of compartmental concepts enable fitting of the pharmacokinetic profile into appropriate models to ensure the dosage regimens are predicted correctly in clinical settings.

The compartmental behaviour of a drug strongly depends on its distribution characteristics in the body and how quickly a drug achieves a distribution equilibrium. Many drugs follow a simple one-compartmental model, where the whole body is treated as *one single compartment* and the distribution equilibrium is achieved instantaneously, see Figure 2.12. Achieving distribution equilibrium implies that the rate of transfer of drug from the blood to all body tissues and the rate of drug transfer from the body tissues back to the blood become equal instantaneously.

For some drugs, the distribution equilibrium takes time (several minutes to hours) depending upon the physicochemical properties of the drug. Often, the distribution of these drugs to the body tissues is very slow and the drug is distributed to highly perfused organs first, such as brain, heart, lung, liver and kidneys, which together with systemic circulation forms the *first or the central compartment*. The drug distribution to other tissues, such as muscles, bones and adipose tissues, is slower. These tissues can be grouped and referred to as the *second compartment*, also known as peripheral or the tissue compartment. For such drugs, the distribution equilibrium between the *two compartments* can take several minutes to hours to establish. Therefore, it requires the fitting of plasma concentration–time profile data into appropriate models for accurate estimation of pharmacokinetic parameters that could enable clinically relevant predictions of the dosage regimens. This is shown in Figure 2.13.

Advanced *in silico* tools that can also predict drug absorption, metabolism, distribution and elimination in the body using inputs on physicochemical properties and ADME parameters and they are described in detail in Chapter 12.

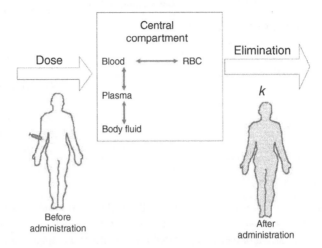

Figure 2.12   *One-compartmental pharmacokinetic model, distribution equilibrium is achieved instantaneously.*

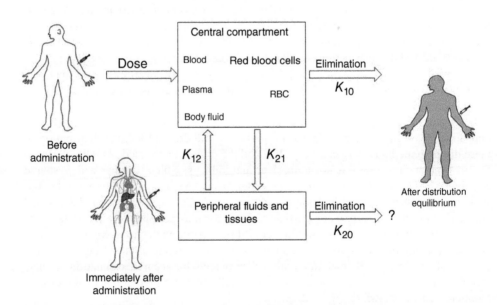

Figure 2.13   *Two-compartmental pharmacokinetic model, distribution equilibrium is slow and takes time.*

2.14 Concept of Linearity in Pharmacokinetics

For drugs that follow the first-order kinetics, the pharmacokinetic parameters such as the plasma concentration at the steady state (C_{ss}) and the area under the curve (AUC) of a plasma concentration–time profile is linearly related to the administered dose. An increase in the dose, therefore, can result in a linear increase in the C_{ss} or the AUC (Figure 2.14). This linear relationship helps in simple dosage adjustment to achieve desired blood concentrations in the body.

Often for some drugs, one of the pharmacokinetic processes, i.e., $t_{1/2}$, k, V, Cl is not governed by the simple first-order kinetics, instead it follows *non-linear pharmacokinetics*. This is when one of the absorption, distribution, metabolism or excretion processes in the body are saturable on increasing the dose. Most drugs still follow first-order pharmacokinetics in practice as the saturation in ADME processes is attained beyond clinical dose ranges. However, this can be a problem for some drugs where saturation is observed at a clinically relevant dose, for instance, *phenytoin* (Figure 2.15). This non-linear behaviour in phenytoin pharmacokinetics is explained by the saturable hepatic metabolism of the drug. The pharmacokinetics of such drugs can be explained by the *Michaelis–Menten* model which can be used to calculate dosages required to achieve a desired steady state for a patient.

The saturable processes in drug distribution, for instance, plasma protein binding, can also lead to non-linear pharmacokinetics for certain drugs. For instance, the volume of distribution of *disopyramide* increases with the dose due to increased free fraction of the

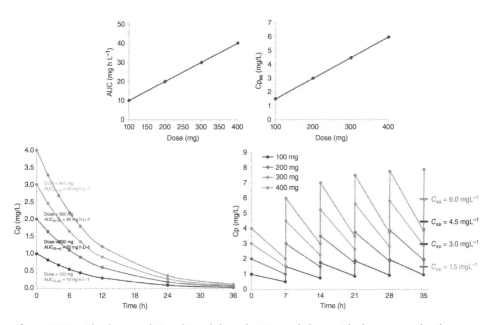

Figure 2.14 *The linear relationship of the administered dose with the area-under-the-curve (AUC) and plasma concentration of a drug. Increase in the dose proportionally increases the steady-state concentration and the AUC.*

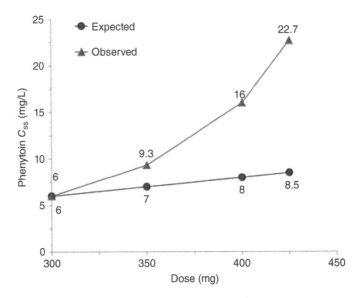

Figure 2.15   *Illustration showing non-linear increase in steady state concentration of phenytoin on increasing the dose.*

drug once plasma protein binding sites are saturated at higher dosages. For drugs that are actively secreted into the renal tubule by active transport, for instance, *dicloxacillin*, the renal clearance decreases when transport proteins at the renal tubules are saturated on increasing the dose. The decreased clearance leads to disproportional increase in circulatory concentration of the drug when the dose is increased.

Dose-dependent saturation in drug absorption can also be responsible for the non-linear pharmacokinetics for some drugs. For example, *amoxicillin* relies on transporters in the gut for its absorption (*influx or active transport*) that can be saturated on increasing the dose. Therefore, drug absorption does not increase proportionally on increasing the dose from a point when absorption transporters are saturated. The drugs exhibiting saturable pharmacokinetics are often prone to more drug–drug or drug–food interactions when co-administered drug or food competes for the similar molecular pathway involved in its absorption, distribution, metabolism or elimination. Some excipients and formulation strategies can manipulate this interaction and can affect drug's binding to the transport proteins or enzymes at the gut, liver or kidney and therefore can manipulate drug's pharmacokinetics. The underpinning biopharmaceutical principles are therefore key considerations in the dosage form design.

2.15   Conclusions

Pharmacokinetic studies are useful to study the absorption, metabolism, distribution and elimination of drugs. For biopharmaceutics, they are used predominantly to better understand absorption; in particular, how formulation or physiological properties affect the absorption of a particular drug and implications in bioavailability, biowaivers and establishing bioequivalence. The correlation between *in vivo* pharmacokinetic data and *in vitro* or *in silico* data is critical in the development of drug products. Typically, pharmacokinetic studies are conducted in animals during the preclinical development phase prior to the 'first in-human' trials or the pharmacokinetic assessment during Phase I clinical trials.

Further Reading

[1] Batchelor, H.K. and Marriott, J.F. (2015). Paediatric pharmacokinetics: key considerations. *British Journal of Clinical Pharmacology* **79** (3): 395–404.

[2] Jambhekar, S.S. and Breen, P.J. (2012). *Basic Pharmacokinetics*. Pharmaceutical Press. ISBN: ISBN: 978 0 85369 772 5.

[3] McConnell, E.L., Fadda, H.M., and Basit, A.W. (2008). Gut instincts: Explorations in intestinal physiology and drug delivery. *International Journal of Pharmaceutics* **364** (2): 213–226.

[4] Varum, F.J.O., Hatton, G.B., and Basit, A.W. (2013). Food, physiology and drug delivery. *International Journal of Pharmaceutics* **457** (2): 446–460.

[5] Varum, F.J.O., Merchant, H.A., and Basit, A.W. (2010). Oral modified-release formulations in motion: The relationship between gastrointestinal transit and drug absorption. *International Journal of Pharmaceutics* **395** (1): 26–36.

[6] Madla, C.M., Gavins, F.K.H., Merchant, H., Orlu, M., Murdan, S., and Basit, A.W. (2021). Let's Talk About Sex: Differences in Drug Therapy in Males and Females. Advanced Drug Delivery Reviews. https://doi.org/10.1016/j.addr.2021.05.014

[7] Winter, M.E. (2009). *Basic Clinical Pharmacokinetics*. Wolters Kluwer Health/Lippincott, Williams & Wilkins. ISBN: ISBN-10: 0-7817-7903-0.

3

Introduction to Biopharmaceutics Measures

Hannah Batchelor[1] and Pavel Gershkovich[2]

[1]*Strathclyde Institute of Pharmacy and Biomedical Sciences, University of Strathclyde, Glasgow, United Kingdom*
[2]*School of Pharmacy, Centre for Biomolecular Sciences, The University of Nottingham, Nottingham, United Kingdom*

3.1 Introduction

Biopharmaceutics is centred on certain key measurements; this chapter aims to define and distinguish the following terms:

- Solubility
- Permeability
- Dissolution

The chapter will also explain why these three measures are important in biopharmaceutics assessment; further chapters will explore these concepts in additional detail in the context of the drug development process.

3.2 Solubility

The International Union of Pure and Applied Chemistry (IUPAC) definition of solubility is, 'The analytical composition of a saturated solution, expressed in terms of the proportion of a designated solute in a designated solvent, is the solubility of that solute' [1].

Biopharmaceutics: From Fundamentals to Industrial Practice, First Edition. Edited by Hannah Batchelor.
© 2022 John Wiley & Sons Ltd. Published 2022 by John Wiley & Sons Ltd.

In terms of biopharmaceutics, it is the drug that is the solute of interest and the solvent differs depending on where the drug is during the process of absorption or disposition. Thus, solubility will affect absorption, distribution within the body as well as elimination. A drug needs to be in solution to undergo pharmacological processes *in vitro* (e.g., permeation, metabolism, bioassay target binding) and *in vivo* (e.g., intestinal absorption). The solubility of a drug at the site of absorption is especially important to its uptake.

Poor solubility has been linked to the following outcomes that are of interest in biopharmaceutics:

1. Insufficient aqueous solubility to formulate an intravenous product
2. Inability to achieve sufficiently high blood/plasma concentrations to attain a therapeutic effect due to saturation of intestinal solubility limiting exposure
   a. Inability to generate data from an ascending dose study or toxicity study due to dose-limited exposure
   b. Poor or variable bioavailability due to insufficient solubility within the GI environment
   c. High variability in pharmacokinetic data due to erratic dissolution and potential for precipitation
   d. Flattened pharmacokinetic curves as the dissolution of the poorly soluble drug does not allow for a rapid T_{max} or high C_{max} to be achieved.

The solubility of a list of commercial drugs was reported by Lipinski et al. [2]. It was identified that 87% of drugs on the market had a solubility of $\geq 65\,\mu g/mL$ thus this value has since been used as a target solubility for compounds during drug discovery and development.

The solubility of a solute in a solvent can be expressed using a wide range of terminologies and units. This can range from units of moles per litre; grams per litre; parts per million; a percentage weight per volume and a range of others.

Pharmacopeial sources (British Pharmacopoeia, European Pharmacopoeia and US Pharmacopoeia) use broad descriptive terms ranging from very soluble to practically insoluble as listed in Table 3.1.

Analytically the solubility is based on a saturated solution; saturation depends on an excess solute being present and there being sufficient agitation for the saturated

Table 3.1   *Solubility criteria as listed in the BP and USP.*

Descriptive term	Parts of the solvent required per part of solute	Expressed as mass per volume (g/mL)
Very soluble	<1	>1
Freely soluble	From 1 to 10	0.1–1
Soluble	From 10 to 30	0.033–0.1
Sparingly soluble	From 30 to 100	0.01–0.033
Slightly soluble	From 100 to 1 000	0.001–0.01
Very slightly soluble	From 1 000 to 10 000	0.0001–0.001
Practically insoluble	10 000 and over	<0.0001

concentration to be achieved; this is a situation where the soluble and insoluble fractions are in equilibrium within the solvent. However, the time to reach equilibrium can be long, up to days for some compounds. The equilibrium solubility value achieved will depend on the affinity of the solvent molecules to the solute. These interactions can be formed via ionic, van der Waals, dispersion and hydrogen bonds.

Chemical solubility uses well-defined conditions that are maintained over the duration of the experiments. However, the process of drug absorption is dynamic with changes in the volume, composition and agitation of the solute in biological fluids during the time-course of absorption. Therefore, it is of interest to measure the solubility of a drug in a range of relevant liquids and to consider the time spent within each fluid as well as the level of agitation present. When considering the solubility of a drug it is essential to understand the factors that are limiting solubility within the absorptive environment. There are many strategies that formulation scientists can apply to improve the solubility of a drug; there are also aspects that medicinal chemists can adapt to provide optimised drug candidates. However, there are also instances where the solubility needs to be limited; specifically in taste masking of drugs to limit the exposure with taste buds on the tongue or in modified release compounds where the rate of drug release is controlled. Further details are presented in Chapter 4.

3.3 Dissolution

Dissolution is defined as the transfer rate of individual drug molecules from the solid particles (usually crystalline) into solution as individual free drug molecules. It is often the dissolution rate that is of greater interest than solubility due to the dynamic nature of the absorption process.

Whilst solubility assessment is typically focussed on the drug substance; dissolution is a parameter more relevant for a formulated drug product (e.g. tablet or capsule). In fact, dissolution testing is commonly used as an analytical technique to measure the rate of drug release from a pharmaceutical product. Although it is the drug product under test during dissolution it is the rate of drug substance release that is measured within the test. For immediate release drug products, rapid and complete dissolution is the goal; whereas for extended release a slower and controlled rate is desirable. Further details on dissolution testing apparatus, including images, is presented in Chapter 6.

Within pharmaceutical product development dissolution testing is used in two distinct ways:

- In a 'quality' environment dissolution testing is used to ensure reproducibility in products and to provide assurance on batch to batch variability. When used for quality control purposes the dissolution testing conditions are designed to evaluate the robustness of the formulation and manufacturing process. Thus the test must be sufficiently sensitive to discriminate between formulations that are different based on their content or process for manufacture. This level of discrimination will depend upon the therapeutic window of the product under test and is further explored in Chapter 8.

- The second application of dissolution is to mimic the physiological conditions at the site of absorption to provide a biorelevant environment. In this format, the equipment will be adapted to reflect the relevant composition, volume and agitation. There are also efforts to mimic the timeframe within each environment and reflect the dynamic conditions within the body. Correlations between dissolution testing and *in vivo* data are frequently sought so that dissolution testing could be used as a surrogate for some *in vivo* preclinical and clinical testing.

There is growing evidence to combine these approaches to have a dissolution methods that are biorelevant and also provide information on the product robustness.

The primary goal of a dissolution test is to develop a discriminating, robust and reproducible dissolution method that can highlight significant clinically relevant changes in product performance due to changes in the formulation or manufacturing process.

The dissolution method(s) will change during the development of a new product. In the early stages, the dissolution is used to understand the factors that govern drug release and to ensure that there are no interactions within the dosage form that adversely affect drug release. The next phase of testing would seek to correlate the *in vitro* dissolution release to *in vivo* release, once this data is available. Further details are provided in Chapter 6.

3.4 Permeability

For drug to pass from the exterior of the body to the interior it needs to permeate at least one membrane. Moreover, membrane permeability is also needed for drug distribution and elimination following absorption. Permeability of the drug across the relevant membrane is a key measurement within biopharmaceutics. Measurement of permeability will help understand the rate-limiting processes associated with absorption and manage risks during the development process.

Permeability is reported as the mass of drug that is transported across a unit area of membrane over a given time. In most cases, the drug needs to be solubilised at the membrane surface in order to permeate the membrane. Yet cellular membranes are formed of lipid bilayers where the core is lipophilic, which can be a barrier to very hydrophilic compounds. This lipid bilayer favours transport of unionised and non-polar compounds. In permeability modelling the lipid bilayer can be considered to be a homogenous organic layer thus lipidic molecules are preferred.

Figure 3.1 provides an overview of the transport pathways across cells as an example for permeability.

Passive membrane permeation is a concentration gradient driven process that typically follows a first-order kinetics. Passive transport can be further divided into passive transcellular where the drug is transported through the cell and passive paracellular where the drug permeates the tight junctions between epithelial cells. In passive transcellular permeability there is a correlation between a drug's $\log P$ value between $\log P$ of -2 to $+4$ and the rate of transport; where the higher the $\log P$ value the greater the permeability [3]. Whilst cationic small molecules are able to permeate the paracellular pathway larger anionic molecules

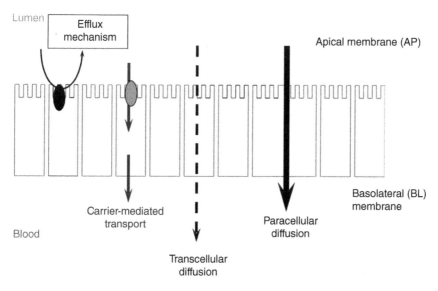

Figure 3.1   *Image of transport pathways across a cellular membrane.*

cannot [4]. There is much data on the most suitable animal or cell-based models to mimic the transcellular permeation observed in humans that is described in Chapter 5.

Active transport (also termed carrier-mediated transport) across a membrane is facilitated by transporters present at either side of the membrane; this can result in net uptake or net efflux of a compound depending on the affinity to the transporter and the relative density of transporters present at the site of absorption. Transporters can also be subject to competitive binding that can lead to drug–drug or drug–nutrient interactions.

3.5   Absorptive Flux

The overall net absorption or absorptive flux can be determined using the following equation, Fick's Law.

$$\text{Absorptive flux}\left(J\right) = -P_e \times \text{SA} \times \Delta C$$

where P_e is the effective permeability across a membrane; SA is the total surface area available for absorption and ΔC is the concentration gradient. The absorptive flux is written as a negative value as it is measuring the concentration from the outside to the inside of a membrane thus the directionality means that the flux is negative (drug is lost from the original site due to the transfer across the membrane).

Using the simple biopharmaceutics measures the permeability and likely concentration gradient (identified via the solubility) can be predicted thus the absorptive flux estimated. It is often used to consider relative values of absorptive flux to compare new drugs to existing and well characterised compounds.

3.6   Lipinsky's Rule of 5

A key article published in 1997 reported the Lipinsky rule of 5 [2]; this paper reviewed commercial drugs to highlight key biopharmaceutics properties (specifically solubility and permeability) of compounds within the drug discovery setting. The purpose of this work was to identify a library of compounds that possess favourable absorption properties which could then be used to risk assess future lead compounds.

An overview of the findings is presented as this can also support key biopharmaceutics properties:

3.6.1   Molecular Weight

A higher molecular weight results in a larger molecule that will have poorer permeability due to its size/volume. Larger compounds can also display poorer solubility due to the large number of solvent molecules required to solvate them. Thus, the target molecular weight was <500.

3.6.2   Lipophilicity

The ratio of solubility of a drug in octanol:aqueous phase is reported as its $\log P$ value. This is also known was the partition coefficient between octanol and water. This value captures the relative hydrophilic to lipophilic property of a particular compound, in biopharmaceutics this is the API. Following oral ingestion of a drug product, the solubility is measured within an aqueous system which drives absorption, yet the membrane permeability favours an unionised and lipophilic compound thus a balance of hydrophilic and lipophilic property is desirable for a drug. A $\log P$ value of <5 was reported as preferred.

It should be noted that human physiology is not as simple as a ratio between octanol and water. The aqueous phase within the gastrointestinal tract for example can vary from an acidic gastric media to a neutral pH within the small intestine. Therefore it is often more relevant to consider the partition coefficient between an aqueous buffer of the appropriate pH and octanol. Particular pH values of interest include the following: pH 7.4 (the pH of blood/serum); pH 6.5–6.8 (the pH within the small intestine) and pH 1–2 (the pH within the stomach).

Drug compounds that are not ionisable will show the same value of partition ratio across a full pH range. Many drugs are ionisable and thus the partition between octanol and an aqueous phase will differ according to their ionisation status. Ionised compounds are more polar and will thus favour the aqueous phase. This is shown schematically in Figure 3.2.

The distribution coefficient (D) (usually expressed as $\log D$) is the effective lipophilicity of a compound at a given pH and is a function of both the lipophilicity of the unionised compound and the degree of ionisation.

Consider a weak acid; this compound will be unionised at low pH and thus will partition into octanol to a greater extent at lower pH values. An example of this is shown in Figure 3.3 for indomethacin which has a pK_a value of 4.5. Note that the $\log P$ value is the partition coefficient when the compounds under test are fully unionised. A $\log D$ value should always be reported with a pH value.

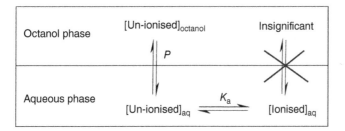

Figure 3.2 *Schematic diagram to demonstrate the impact of ionisation on the partition of a compound between octanol and an aqueous phase.*

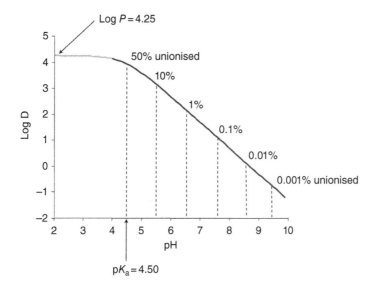

Figure 3.3 *Impact of log D vs pH for a compound with a pK$_a$ value of 4.5.*

3.6.3 Hydrogen Bond Donors/Acceptors

A high number of hydrogen bond donors can impair membrane permeability due to the polarity of the compound. Compounds with fewer than five hydrogen bond donor groups were preferred. Similarly, hydrogen bond acceptors indicate polarity and fewer than 10 were correlated to better absorption characteristics.

References

[1] IUPAC (1997). *Compendium of Chemical Terminology (the "Gold Book")*, 2e. Compiled by A. D. McNaught and A. Wilkinson. Oxford: Blackwell Scientific Publications. ISBN: 0-9678550-9-8. https://doi.org/10.1351/goldbook. Online version (2019) created by S.J. Chalk (online version created 2019).

[2] Lipinski, C.A., Beryl, F.L., Dominy, W., and Feeney, P.J. (1997). Experimental and computational approaches to estimate solubility and permeability in drug discovery and development settings. *Advanced Drug Delivery Reviews* **23** (1): 3–25.

[3] Camenisch, G., Alsenz, J., van de Waterbeemd, H., and Folkers, G. (1998). Estimation of permeability by passive diffusion through Caco-2 cell monolayers using the drugs' lipophilicity and molecular weight. *European Journal of Pharmaceutical Sciences* **6** (4): 313–319.
[4] Adson, A., Raub, T.J., Burton, P.S. et al. (1994). Quantitative approaches to delineate paracellular diffusion in cultured epithelial cell monolayers. *Journal of Pharmaceutical Sciences* **83** (11): 1529–1536.

4

Solubility

Hannah Batchelor

Strathclyde Institute of Pharmacy and Biomedical Sciences,
University of Strathclyde, Glasgow, United Kingdom

4.1   Definition of Solubility

Solubility is a measurement of the amount of a substance that can stay in a solvent without precipitation. Solubility may be expressed in units of concentration, molality, mole fraction, mole ratio and other units. For the purposes of biopharmaceutics, the substance of interest is usually the API and the solvent will vary depending upon the physiological region of interest; for example the stomach, small intestine or pulmonary fluid.

A drug's structure determines its solubility as the chemical structure will also determine lipophilicity, hydrogen bonding, molecular volume, crystal energy and ionisability. Thus during lead compound optimisation, there is scope to balance potency with solubility [1].

The solubility of a substance can change when pressure, temperature and/or the composition of the solvent changes thus it is important to accurately detail solubility data to provide sufficient information.

4.2   The Importance of Solubility in Biopharmaceutics

Poor aqueous solubility of a drug can lead to issues during the preclinical phase of drug development and beyond. Insufficient solubility within the gastrointestinal lumen can limit absorption and subsequent exposure to the drug in question. Thus, the solubility of

Biopharmaceutics: From Fundamentals to Industrial Practice, First Edition. Edited by Hannah Batchelor.
© 2022 John Wiley & Sons Ltd. Published 2022 by John Wiley & Sons Ltd.

a drug candidate will impact upon decisions and risk assessments undertaken during development.

There has been an increase in the proportion of poorly water-soluble drugs with estimates of up to 75% of candidates in development being classified as low aqueous solubility [2]. This trend towards increasing proportions of low aqueous solubility drugs is linked to the drive towards potent and selective drugs where candidate optimisation often adds lipophilic groups to enhance binding, yet the lipophilicity of the resulting molecule increases. However, poorly water-soluble candidate drugs are associated with higher rates of attrition as well as higher costs during drug development [3]. Furthermore, poorly water-soluble drugs are associated with greater inter-individual pharmacokinetic variability as well as being susceptible to food effects [3].

Solubility data will determine the need for an enabling formulations (e.g., wetting agents, micronisation, solubilising agents, solid solutions, emulsions and nanoparticles).

The importance of solubility is recognised within the biopharmaceutics classification system, where further details are provided in Chapter 9.

4.3   What Level of Solubility Is Required?

It can be complicated to define the minimum solubility required for any compound as this depends on the permeability and dose. A review of published high throughput solubility screening tests reported by the pharmaceutical industry reported that the target solubility ranged from 100 to 1000 μM in media containing a mix of DMSO and buffers at pH 6.5–7.4 [3] or greater than 65 μg/mL [1].

For a compound with average permeability, a lower aqueous solubility would be acceptable whereas a higher solubility is required for a poorly permeable compound.

The equation for the maximum absorbable dose links drug solubility with permeability and the intestinal physiology.

$$\text{Maximum absorbable dose} = S \times K_a \times \text{SIWV} \times \text{SITT}$$

where S = aqueous solubility (mg/mL, at pH 6.5); K_a = intestinal absorption rate constant (min^{-1}) (permeability in rat intestinal perfusion experiment, quantitatively similar to human K_a); SIWV = small intestine water volume (~250 mL); SITT = small intestine transit time (~270 min).

The aqueous solubility is the simplest of these inputs to adjust thus there are some estimates for the minimum solubility to achieve the maximum absorbable dose for given values of the permeability. An example is shown in Table 4.1.

Table 4.1 shows that as the dose increases the solubility must also increase; this highlights the importance of potent molecules to achieve sufficient exposure. Formulation design can aid in transient increases in the solubility to minimise precipitation of compounds to ensure that sufficient amount of the drug is absorbed.

There are clinical examples of very poorly soluble drugs that are marketed, for example candesartan cilexetil, an antihypertensive drug in use since 1997, has a water solubility of approximately 0.1 μg/mL.

Table 4.1   *Impact of dose and permeability on the target solubility for a candidate drug.*

Dose (mg)	Permeability (K_a) High = 0.03		Minimum acceptable aqueous solubility (mg/L)	
1	Low = 0.003	High = 0.03	0.0493	0.494
10	Low = 0.003	High = 0.03	0.494	4.94
100	Low = 0.003	High = 0.03	4.94	49.4
1000	Low = 0.003	High = 0.03	49.4	494

4.4   Solubility-Limited Absorption

Drug solubility within intestinal fluids can limit the overall absorption, as described in the maximum absorbable dose equation. This can become evident in early clinical testing where a single ascending dose study is planned. This study is used to determine the pharmacokinetic profile of a drug following a series of ascending oral doses as well as to explore the pharmacodynamic effects of an increasing dose. It is used within the safety assessment of a drug. If the solubility limits the exposure then work is required to ensure that the formulation used within the ascending dose study achieves the appropriate exposure to ensure linear pharmacokinetics. This is shown in Figure 4.1.

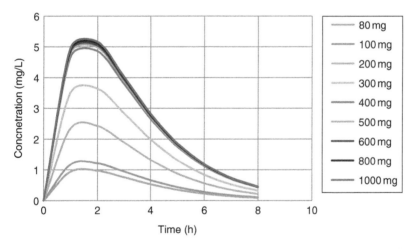

Figure 4.1   *Image showing plasma pharmacokinetics from a study in which an oral ranging from 80 to 1000 mg was administered. It can be seen that the solubility limitation is reached at 400 mg whereby further increases in dose do not show increased C_{max} or AUC.*

4.5   Methods to Assess Solubility

It is important to report the method used to assess solubility when reporting the solubility value. A range of methods are available, yet the most common method used for biopharmaceutics is the shake-flask method that is recommended by the regulatory agencies (FDA and EMA). In the shake-flask method excess drug is added to the solvent to form a saturated solution which is noted by the observation of undissolved material within the flask. The total content of the flask is then transferred to a shaker and agitated for a predetermined time to reach equilibrium solubility at a fixed temperature (usually 37 °C).

However, during the drug development process often high throughput or *in silico* methods are used during lead optimisation and candidate selection. *In silico* methods to predict solubility use structural parameters including the use of 2D and 3D chemical structures, log P and melting point models [4]. Within high throughput methods, solubility is measured based on a small amount of drug that was dissolved in DMSO; then precipitated prior to dissolution in buffers at pH 6.5 or 7.4. However, these high throughput methods have been demonstrated to overestimate solubility, most likely due to residual DMSO present within the solvent.

A second common method is the intrinsic dissolution rate (IDR) which is defined as the dissolution rate of the drug substance from a constant surface area and stirring speed in a solvent with defined pH and ionic strength. The IDR is calculated as the mass rate transferred from the solid surface to the solvent phase. The major differences between IDR and the shake-flask are that the shake-flask method provides an equilibrium solubility measurement whereas the IDR is a rate measurement. The notation of dissolution can cause confusion in this context. However, the intrinsic dissolution rate relates to the drug substance and how rapidly the drug substance achieves saturated solubility. Dissolution in the context of biopharmaceutics relates to the rate of solvation of the drug substance from the formulated drug product, dissolution is discussed in detail in Chapter 6.

It is important that solubility is assessed by the most suitable method at the appropriate stage during development, particularly as it is known that the crystalline form will affect the solubility recorded.

4.6   Brief Overview of Forces Involved in Solubility

The solubility of a solute depends upon its relative affinity to the solvent as well as other solute molecules; the nature of the affinity will dictate the type of bond or interaction involved.

4.6.1   van der Waals Interactions

Solute molecules that have a permanent dipole, as a result of the molecular structure, will result in a degree of polarity. Some molecules can be strongly polar whilst others are weakly polar. Van der Waals forces are a result of dipole interactions where compounds with a similar polarity will mix whereas those with differences in polarity will not mix. Attractive forces are the result of the positive head of one molecule positioning itself close to the negative head of another molecule. However, it is also important to consider physical, geometric constraints which may limit the efficiency of mixing.

4.6.2   Hydrogen Bonding

The hydrogen bonding capacity of a drug molecule can be calculated based on its chemical structure. Much of biopharmaceutics is related to aqueous media where hydrogen bonding is a significant factor in aqueous solubility. Thus hydrogen bonds are important for biopharmaceutics.

4.6.3   Ionic Interactions

Oppositely charged ions will interact. This is of relevance in biopharmaceutics as many drugs can be prepared as salt forms which will interact with aqueous media as it is polar. Pharmaceutical salts can often have a higher solubility compared to their free base form.

4.7   Solid-State Properties and Solubility

Chemicals that exist in more than one solid-state form are termed polymorphs; the classic example is carbon that can exist as diamond or graphite. Alternative polymorphic forms of a drug substance will have different crystalline forms which can affect their hydration and the time-frame to reach equilibrium solubility. A change in polymorph can affect oral bio-availability due to the change in solubility as well as dissolution from the drug product.

The crystal structure determines the free energy within the solid and this can influence the rate of solvation. In early preclinical phases drug is often already in solution for evaluation of the potency and other key factors so polymorphism is irrelevant as this only affects solid materials. However, once the drug is produced and administered as a solid it is essential to understand the polymorphic form and relative stability of each form. Polymorphic screening is undertaken by dissolving the drug into a range of solvents and evaporating the solvents to characterise the crystalline or amorphous material formed.

Drug substances are manufactured as solids; a polymorph screen will be undertaken to identify the polymorphs present. This screen is usually undertaken early during drug development to identify and characterise the range of polymorphs.

It is important to not only recognise and characterise all polymorphs (for patenting purposes amongst others) but also to control the form of polymorph present upon manufacture and throughout the shelf-life of the product. Solubility and dissolution can be used as surrogate measures of change in polymorph due to the measured difference in solubility and dissolution rate.

This phenomenon was significant for ritonavir in 1998 where the continued supply of the HIV drug was threatened due to the change in polymorph to a more stable and less soluble crystalline form. The original formulation contained ritonavir as an oral liquid or within semi-solid capsules such that the drug was solubilised in an ethanol/water mix. A solubilised drug does not have a crystal structure thus no polymorphic control was required. However, it was noted that several batches of the semi-solid capsules failed dissolution and upon investigation a second polymorph was identified which had much reduced solubility and subsequently reduced bioavailability which would limit exposure. As a result, a reformulation was required for both products [5].

4.8   pH and Drug Solubility

Many drugs are ionisable compounds and can exist as weak acids or bases. The solubility of a weak acid/base is highly dependent upon the pH of the solvent. As the pH changes

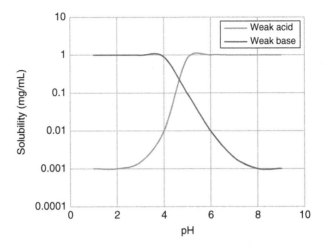

Figure 4.2   *Impact of pH on solubility of a weakly acidic and a weakly basic drug.*

along the gastrointestinal tract from the acidic conditions within the stomach to the neutral environment within the small intestine the impact of the pH of the solvent on the drug solubility needs to be understood. This is of greatest importance for weakly basic drugs as their solubility reduces within the small intestine which is the major site for drug absorption. Figure 4.2 shows the impact of pH on the solubility of a weakly acidic and a weakly basic compound.

Very small changes in pH can have a major impact on the solubility of an ionisable drug. Solubility measurements are usually undertaken in buffered solutions to prevent small changes in pH affecting the measurement obtained.

pH solubility measurements are usually performed early in the drug development process to understand risks and whether there are drug candidates with favourable pH solubility profiles (see Chapter 7).

The pK_a of the drug is another useful drug characteristic to understand the impact of pH on solubility as this is linked via the Henderson–Hasselbach equation.

As well as the influence of pK_a on solubility it is important to note that it is the undissociated form that more readily penetrates biological tissues to exert a therapeutic effect. As the equilibrium between the dissociated and undissociated fractions is constant, it can be assumed that the undissociated permeated material will be replenished to maintain equilibrium and thus absorption can occur for both weak acids and weak bases.

4.9   Solvents

Drug solubility in a range of solvents is of interest for both the prediction of biopharmaceutics parameters and also to aid in the formulation strategy. Drug solubility in a range of organic solvents can facilitate polymorph identification and process optimisation. For compounds that show poor aqueous solubility, additional solubility screening is performed with a range of solubilising vehicles to guide the formulation design and associated risk assessment.

4.9.1   Biorelevant Solubility

It has been demonstrated that replication of the composition of the gastrointestinal media improves the ability to predict *in vivo* performance of drug products. Therefore, much effort has been made to replicate *in vivo* gastrointestinal media to provide simulated fluids for solubility assessment. Four levels of simulation of luminal composition have been proposed ranging from simple to more complex [6]; these are outlined in Table 4.2.

During the drug development process, the level of complexity will increase to ensure that the measurements are appropriate to the risk assessment of the product. Often biorelevant media are first used just before the human clinical phase of development.

As well as the levels of media it is important to note that there are significant compositional changes from the stomach to small intestine to large intestine that also need to be addressed. Furthermore, the impact of food cannot be ignored thus several variations of simulated fluids are reported in the literature. Most data has come from matching composition to intestinal fluids taken from healthy adult volunteers. However, there is also a need to account for populations that may differ including those with diseases or at the extremes of age (see Chapter 13).

There are several papers that describe the characterisation of human intestinal fluids and translate these findings into simulated fluids for the fasted state [7–11] and for the fed state [10, 11].

The composition of commonly used media is presented in Table 4.3.

Table 4.2   *An overview of the four levels of biorelevant media that can simulate the gastrointestinal luminal environment.*

Level	Purpose
Level 0 media	Replicates pH using a simple buffer
Level 1 media	Replicates pH and buffer capacity
Level 2 media	Replicates pH, buffer capacity, osmolality and solubilisation capacity using bile salts, dietary lipids and digestion products
Level 3	Builds on level 2 by also incorporating proteins and enzymes In addition can be viscosity matched to *in vivo* fluid

Table 4.3   *Composition of commonly used simulated intestinal media: fasted state simulated gastric fluid (FaSSGF); fasted state simulated intestinal fluid (FaSSIF) and fed state simulated intestinal fluid (FeSSIF).*

	FaSSGF	FaSSIF	FeSSIF
pH	1.6	6.5	5.0
Bile salt (taurocholate) (mM)	0.08	3	15
Phospholipids (mM)	0.02	0.75	3.75
Sodium ions	34	148	319
Chloride ions	59	106	203
Phosphate ions		29	
Acetic acid			144

4.9.2   Buffer System – Phosphate vs Bicarbonate

In vivo, the gastrointestinal system is buffered by bicarbonate within the anaerobic conditions. Laboratory solubility and dissolution testing are technically difficult to conduct within anaerobic conditions which are required for bicarbonate buffers. Therefore, phosphate buffers are typically used instead. However, there is evidence that bicarbonate buffers are more representative and should be considered for certain drugs [12].

4.9.3   Solubilisation by Surfactants

Surfactants can improve the solubility of poorly soluble drugs. Surfactants are amphiphilic molecules that can form colloidal structures including micelles within an aqueous environment. Drug can then associate with these micelles, which improves the measured solubility. It is worth noting that the drug associated with micelles is not strictly dissolved but it is also no longer present as a solid. This distinction can be important for advanced modelling of drug solubilisation.

In biorelevant media, the bile salts and phospholipids form colloidal structures that include micelles that can enhance the measured solubility of the drug. Bile salt micelles will only form above the critical micelle concentration of the individual surfactant. Thus, small changes can have a large influence of the solubilisation capacity for a drug [13].

Studies have reported increased solubility of different drugs in FaSSIF compared to the same media without the bile salts [14]. This is shown in Figure 4.3.

Typically the solubility of drug in FaSSIF will be measured early in development as this can have implications for determining the most appropriate biopharmaceutics classification for the drug and to calculate the most relevant maximum absorbable dose.

Solubility increases are often also observed in FeSSIF compared to FaSSIF. However, the inherent variability associated with administration of drugs in the fed state makes reliance upon this solubility increase a risky strategy.

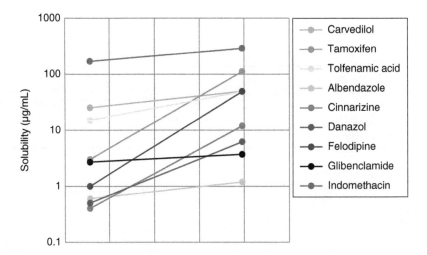

Figure 4.3 *Comparison of the solubility of a series of drugs in blank FaSSIFf (left data point) and FaSSIF containing bile salts (right data point). Source: Data from Fagerberg et al. [14].*

4.9.4   Solubilisation During Digestion

The gastrointestinal tract is maintained in equilibrium via secretions and its composition has been well characterised in the fasted state. In the fed state the secretions change and in addition there is digestion of the consumed food. Thus it is not only the effect of endogenous surfactants but also ingested nutrients and digestion products that can affect drug solubility. For example, when short-chain fatty acids are ingested, they are hydrolysed and then incorporated into the bile micelles to further solubilise certain drugs. Work has been conducted on simulated media that include lipolysis products within the media [15]. Oleic acid and monoolein are often used within the media to replicate the lipolysis products [16].

4.9.5   Excipients and Solubility

In cases where drug solubility is low and where it may limit exposure efforts are focussed on understanding the factors that limit solubility and identification of strategies to improve it. Typically different enabling formulations are used. Formulation solubilising vehicles including cosolvents, surfactants, complexation agents and oils/lipids are considered. In early stages of development, simple strategies are required to ensure that solubility does not limit the generation of data from an ascending dose study, thus a cosolvent approach can be a simple workaround at this stage yet would not be suitable for the commercial formulation. Typical cosolvents used include polyethylene glycol (PEG400), propylene glycol and ethanol.

For ionic drug compounds, commonly used approaches include pH adjustment and salt formulation. A salt screen can be performed to identify the most suitable counter ion. There are several examples of salt forms having large differences in solubility. For example, terfenadine salts formed with phosphoric acid, hydrochloric acid, methanesulphonic acid and lactic acid showed up to 10-fold differences, ranging from 0.5 to 5 mg/mL [17]. There are some concerns about the common-ion effect where the presence of ionic components, for example NaCl within the gastrointestinal media may affect the solubilisation rate (dissolution) where sodium or chloride was the counterion for the salt present.

Amorphous drug forms can be a useful strategy for drug compounds with a high melting temperature; the amorphous form disrupts the crystal packing and thus shows rapid dissolution, although there can be risks of precipitation. It is important that the amorphous form is stabilised during manufacture and storage. Often a solid dispersion, where the amorphous form of a drug is intimately mixed with a polymer provides a matrix to stabilise the amorphous form and prevent recrystallisation. There are many commercial examples including Kaletra which contains both lopinavir and ritonavir stabilised in PVPVA [18].

The use of lipid or oil-based formulations is a strategy used for lipophilic drug compounds; these formulation can also improve permeability and lymphatic transport. In these formulations, the drug is blended with lipidic excipients including long-chain triglyceride (such as sesame or other vegetable oils), long-chain monoglycerides and diglycerides (such as Peceol®) or medium-chain triglyceride (eg, Miglyol® 812), medium-chain monoglycerides & diglycerides (eg, Capmul® MCM), phospholipids (eg, Phosal® 53 MCT). The most commonly cited example is Neoral, a lipidic formulation of cyclosporin A.

Surfactant formulations are another strategy to improve solubility of drugs. In oral liquid formulations, surfactants can be added to the formulation to improve solubility, typically by the formation of micelles within the liquid that can contain the drug to solubilise it. Examples of surfactants used include the following: Polysorbate 80 (Tween 80), Polyoxyl-35 castor oil (Cremophor® EL/Kolliphor® EL), caprylocaproyl polyoxyl-8 glycerides (Labrasol) and sodium lauryl sulphate (SLS) [19].

Cyclodextrin formulations can be also used to improve drug solubility. These are also known as inclusion complex formulations. Cyclodextrins have an exterior that is water soluble yet they contain an internal hydrophobic cavity; drugs can sit within this cavity forming an inclusion complex. There are several marketed cyclodextrin-containing formulations [20].

4.10   Risk of Precipitation

Solubility occurs under dynamic equilibrium, which means that solubility results from the simultaneous and opposing processes of dissolution and phase joining (e.g., precipitation of solids). Equilibrium solubility is defined as the maximum amount of solid that can be dissolved in a solvent when the system reaches equilibrium. In some cases, it can take some time for the system to reach equilibrium thus it is important that the time course is reported. In some cases where solubility enhancing excipients are used there is a transient increase in the measured concentration that generates a solubility value higher than the equilibrium solubility. In these cases, a supersaturated solution has been formed and there is a risk of precipitation. Within the gastrointestinal tract, there is removal of the dissolved drug by absorption. If the rate of absorption is high then the risk of precipitation is low. Figure 4.4 shows some examples schematically.

As shown in Figure 4.4 there are three distinct solubility profiles. In terms of biopharmaceutics, it is important that the greatest solubility value is reached whilst the drug is at the major site of absorption, the small intestine. Thus for the solid line, it would be best if the time at equilibrium (highest solubility) was reached within the small intestine. For the

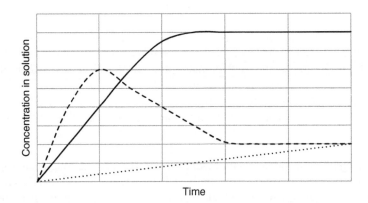

Figure 4.4 *Examples of solubility vs time plots for a rapidly dissolving solute reaching equilibrium (solid line); a slowly dissolving solute where equilibrium was not reached during the time course measured (dotted line) and a supersaturated solution where precipitation occurred and equilibrium was reached (dashed line).*

dotted line, it would be useful to include solubility enhancing excipients to improve the rate of solubilisation to maximise concentrations within the GI fluids. For the dashed line if the peak concentration obtained coincides with the small intestine for absorption then this would improve bioavailability. However, the risks around using a supersaturated formulation are high as it is not always possible to control the rate of precipitation for an orally administered formulation. Furthermore, it may be that the form that precipitates is of a different polymorph to that administered, which can further complicate the solubility and dissolution profile. The risks of precipitation are further described in Chapter 6.

4.11   Solubility and Link to Lipophilicity

The lipophilicity of a compound is described in terms of a partition coefficient, $\log P$, which is defined as the ratio of the concentration of the unionised compound, at equilibrium, between organic and aqueous phases. Since it is virtually impossible to determine $\log P$ in a realistic biological medium, octanol has been widely adopted as a model of the lipid phase. Generally, drug compounds with $\log P$ values between 1 and 3 show good bioavailability.

4.12   Conclusions

Drug solubility depends on the fluid composition, pH and temperature. Poor solubility within the intestinal environment can limit drug absorption; thus understanding solubility is key to prediction of pharmacokinetics for orally administered drugs. Various solubilisation techniques can be employed to improve drug solubility and these will be discussed in Chapters 8 and 9.

References

[1] Kerns, E.H., Di, L., and Carter, G.T. (2008). *In vitro* solubility assays in drug discovery. *Current Drug Metabolism* **9** (9): 879–885.
[2] Di, L., Kerns, E.H., and Carter, G.T. (2009). Drug-like property concepts in pharmaceutical design. *Current Pharmaceutical Design* **15** (19): 2184–2194.
[3] Di, L., Fish, P.V., and Mano, T. (2012). Bridging solubility between drug discovery and development. *Drug Discovery Today* **17** (9): 486–495.
[4] Bergström, C.A.S. and Larsson, P. (2018). Computational prediction of drug solubility in water-based systems: qualitative and quantitative approaches used in the current drug discovery and development setting. *International Journal of Pharmaceutics* **540** (1): 185–193.
[5] Bauer, J., Spanton, S., Henry, R. et al. (2001). Ritonavir: an extraordinary example of conformational polymorphism. *Pharmaceutical Research* **18** (6): 859–866.
[6] Markopoulos, C., Andreas, C.J., Vertzoni, M. et al. (2015). *In-vitro* simulation of luminal conditions for evaluation of performance of oral drug products: choosing the appropriate test media. *European Journal of Pharmaceutics and Biopharmaceutics* **93**: 173–182.
[7] Galia, E., Nicolaides, E., Hörter, D. et al. (1998). Evaluation of various dissolution media for predicting *in vivo* performance of class I and II drugs. *Pharmaceutical Research* **15** (5): 698–705.
[8] Vertzoni, M., Fotaki, N., Nicolaides, E. et al. (2004). Dissolution media simulating the intralumenal composition of the small intestine: physiological issues and practical aspects. *Journal of Pharmacy and Pharmacology* **56** (4): 453–462.

[9] Vertzoni, M., Dressman, J., Butler, J. et al. (2005). Simulation of fasting gastric conditions and its importance for the *in vivo* dissolution of lipophilic compounds. *European Journal of Pharmaceutics and Biopharmaceutics* **60** (3): 413–417.

[10] Jantratid, E., Janssen, N., Reppas, C., and Dressman, J.B. (2008). Dissolution media simulating conditions in the proximal human gastrointestinal tract: an update. *Pharmacuetical Research* **25** (7): 1663.

[11] Vertzoni, M., Diakidou, A., Chatzilias, M. et al. (2010). Biorelevant media to simulate fluids in the ascending colon of humans and their usefulness in predicting intracolonic drug solubility. *Pharmaceutical Research* **27** (10): 2187–2196.

[12] Amaral Silva, D., Al-Gousous, J., Davies, N.M. et al. (2019). Simulated, biorelevant, clinically relevant or physiologically relevant dissolution media: the hidden role of bicarbonate buffer. *European Journal of Pharmaceutics and Biopharmaceutics* **142**: 8–19.

[13] Wiedmann, T.S. and Kamel, L. (2002). Examination of the solubilization of drugs by bile salt micelles. *Journal of Pharmaceutical Sciences* **91** (8): 1743–1764.

[14] Fagerberg, J.H., Tsinman, O., Sun, N. et al. (2010). Dissolution rate and apparent solubility of poorly soluble drugs in biorelevant dissolution media. *Molecular Pharmaceutics* **7** (5): 1419–1430.

[15] Fatouros, D.G., Walrand, I., Bergenstahl, B., and Müllertz, A. (2009). Colloidal structures in media simulating intestinal fed state conditions with and without lipolysis products. *Pharmaceutical Research* **26** (2): 361.

[16] Kleberg, K., Jacobsen, F., Fatouros, D.G., and Müllertz, A. (2010). Biorelevant media simulating fed state intestinal fluids: colloid phase characterization and impact on solubilization capacity. *Journal of Pharmaceutical Sciences* **99** (8): 3522–3532.

[17] Streng, W.H., Hsi, S.K., Helms, P.E., and Tan, H.G.H. (1984). General treatment of pH-solubility profiles of weak acids and bases and the effects of different acids on the solubility of a weak base. *Journal of Pharmaceutical Sciences* **73** (12): 1679–1684.

[18] Newman, A. (2017). Rational design for amorphous solid dispersions. In: *Developing Solid Oral Dosage Forms*, 2e (eds. Y. Qiu, Y. Chen, G.G.Z. Zhang, et al.), 497–518. Boston: Academic Press.

[19] van der Vossen, A.C., van der Velde, I., Smeets, O.S.N.M. et al. (2017). Formulating a poorly water soluble drug into an oral solution suitable for paediatric patients; lorazepam as a model drug. *European Journal of Pharmaceutical Sciences* **100**: 205–210.

[20] Saokham, P., Muankaew, C., Jansook, P., and Loftsson, T. (2018). Solubility of cyclodextrins and drug/cyclodextrin complexes. *Molecules (Basel, Switzerland)* **23** (5): 1161. https://doi.org/10.3390/molecules23051161. PMID: 29751694; PMCID: PMC6099580.

5

Permeability

Chris Roe and Vanessa Zann

Quotient Sciences, Mere Way, Ruddington, Nottingham, United Kingdom

5.1 Introduction

This chapter will discuss the importance of permeability and the tools that can be used to predict, assess or observe oral absorption. Although this is limited to oral permeability, the main principles can be applied to a range of absorption sites (Chapters 14–16 cover non-oral sites).

In terms of gastrointestinal (GI) physiology, the main area of interest for permeability and absorption is the small intestine (see Chapter 11 on impact of anatomy and physiology for more details). Figure 5.1 shows the different ways compounds can permeate the intestinal epithelial barrier.

Intestinal permeability can be either a passive process, through either transcellular or paracellular routes, via movement down a concentration gradient, or an active process, which requires a transmembrane protein and energy and can be absorptive (influx) or exsorptive (efflux).

Measuring permeability can help define the rate-limiting process for absorption, drive formulation strategy and manage risk (see Chapters 7 and 8). It can also be a key input for modelling and simulation activities (see Chapter 12).

Biopharmaceutics: From Fundamentals to Industrial Practice, First Edition. Edited by Hannah Batchelor.
© 2022 John Wiley & Sons Ltd. Published 2022 by John Wiley & Sons Ltd.

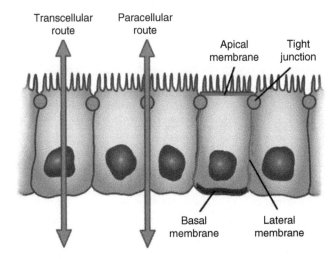

Figure 5.1 *How compounds cross the intestinal epithelial barrier and key features. Source: BallenaBlanca / Wikimedia Commons / CC BY-SA 4.0. https://commons.wikimedia.org/wiki/ File:Selective_permeability_routes_in_epithelium.png*

5.2 Enzymes, Gut Wall Metabolism, Tissue Permeability and Transporters

Many transporters and enzymes are present within the GI tract, both to aid digestive processes and to protect the body against potential ingestion of harmful toxins. Some transporters and enzymes are ubiquitous throughout the body and others are expressed preferentially in the certain anatomical regions. Expression gradients have been documented for a number of enzymes and transporters and their relevance to impacting drug absorption and oral bioavailability is molecule specific.

5.2.1 Enzymes

Enzymes in the GI tract arise from three principle GI sources, luminal (e.g. peptidases, esterases), bacterial, and mucosal/gut wall. Combined with hepatic metabolism, these are the key factors in drug biotransformation. This, together with absorption, will determine the bioavailability of orally administered drugs.

Many enzymes present in the liver (the major site for drug metabolism) are also expressed in intestinal epithelial cells, with the Cytochrome P450 (CYP) family being particularly important given its responsibility for the majority of Phase 1 liver reactions (e.g. oxidation) [1]. CYPs are membrane-associated haemoproteins located within the enterocytes (absorptive enterocytes, make up>80% of all small intestinal epithelial cells). CYPs are expressed and distributed (unequally) as various isoforms throughout the GI tract. The CYP3A subfamily is the most frequently expressed, representing 70–80% of total intestinal CYP content, with CYP3A4 being the most prevalent [2]. CYP2C is the second most prevalent isoenzyme subfamily at levels of approximately 18% of total CYP content [2]. Analysis from tissue biopsies has confirmed the low-level expression of many other CYP variants in the GI tract including CYP2D6, CYP2E1, and CYP2J2 [2, 3].

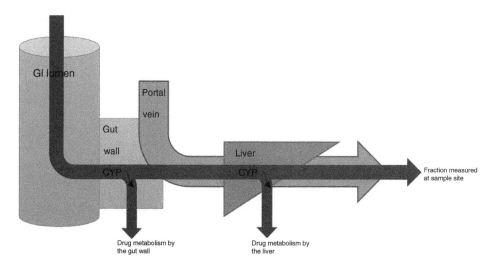

Figure 5.2   *Cartoon to show loss of drug at the gut wall and the liver due to CYP enzymes.*

Pre-systemic metabolism or first-pass metabolism (metabolism of the drug prior to it reaching the site of action) can significantly affect the bioavailability of many orally administered drugs and is a function of both gut wall and liver metabolic first-pass processes. The relative contribution of each to the bioavailability of a compound can vary and has been open to conjecture; however, the role of gut wall metabolism should not be underestimated. This is shown schematically in Figure 5.2.

Drugs that are CYP substrates, and hence liable to pre-systemic gut wall metabolism with reduced bioavailability in addition to first-pass liver metabolism, present several challenges for achieving effective oral drug delivery. They are likely to exhibit non-proportional exposure as dose increases due to saturation of CYP-mediated metabolism. As the CYP metabolism is saturated, a greater than proportional amount of drug will be absorbed at higher dose levels, providing a potential strategy to maximise exposure. In addition, CYP substrates can demonstrate high inter-subject variability as a result of high variability (up to 30 fold) in inter-individual CYP3A4 expression [4, 5] (further details are provided in Chapter 13). Higher dose levels, which saturate CYP metabolism, can decrease this intrasubject variability.

Several strategies have been evaluated to overcome these problems. For antiviral therapies, the co-administration of low-dose ritonavir, a potent albeit non-specific CYP3A4 inhibitor, was successfully found to inhibit metabolism and subsequently increase bioavailability of other protease inhibitors including saquinavir [6]. A >50-fold enhancement of plasma concentrations was observed in humans following a single co-dose of ritonavir (600 mg) and saquinavir (200 mg), which was believed to be due to inhibition of both gut wall and hepatic metabolism of saquinavir. Certain non-ionic and lipid-based excipients have also been demonstrated (*in vitro*) to be CYP inhibitors including polysorbate 80, D-α-tocopheryl polyethylene glycol (1000) succinate, sucrose laurate, Cremophor EL, and Cremophor RH40 [7].

An additional drug delivery strategy for CYP3A4 substrates arises from the observation that enzyme distribution increases in the more proximal regions of the small intestine as evidenced by both CYP3A protein level and the intrinsic clearance of midazolam, a known CYP substrate [4, 8]. As such, the potential to deliver drugs beyond this location, to the colon and more distal regions of the GI tract, offers the potential to increase relative bioavailability. This phenomenon has been reported for simvastatin, which exhibited a 3-fold increase in bioavailability following delivery to the lower GI tract [9]. While regional delivery offers potential success for the bolus delivery of drugs, this needs careful consideration in the design of modified release formulations.

In common with all CYPs, some drugs or dietary components can also act as inhibitors or inducers of the CYP3A4 enzyme, further complicating the drug delivery challenge and also raising the potential for drug-drug interactions (DDIs) and drug–food interactions, affecting the safety and efficacy of any medications, which may be metabolised by CYP3A4. Notable examples reviewed by Thelen and Dressman [1] included ketoconazole and grapefruit juice (both inhibitors) with a concomitant 2- to 3-fold increase in the bioavailability of co-administered drugs, and rifampicin (an inducer), which has been shown to increase tissue CYP3A4 mRNA levels by a factor of 5–8.

5.2.2   Drug Transporters

Many intestinal-based transporters have been shown to influence drug absorption (either through assisting or limiting the entry of substances into cells), and are also involved in DDIs [10]. Membrane transporters can influence drug absorption by either increasing transport into the intestinal enterocyte via influx transporters or limiting transport into the cell via efflux back into the intestinal lumen.

There are many transporter proteins expressed throughout the body, which are responsible for the transport of a wide range of substances. The transporters can be divided into two major classes: the adenosine triphosphate (ATP)-binding cassette family (ABC transporters) and the solute carrier family (SLC). Transporters can be divided into active transporters, i.e. requiring energy to affect the movement of the substrate, or passive ones such as facilitated diffusion carrier or channel proteins in the cell membrane that assist in the movement of molecules across a concentration gradient, whereby the driving force is diffusion down an electrochemical gradient without using energy [11].

Transporters can be further classified depending on the direction in which they transport substances, either to assist with nutrient and ion absorption (influx transporters) [12] or to expel harmful substances (efflux transporters) and prevent absorption into the systemic circulation [13, 14]. Influx transporters can increase absorption of drug substrates and have been exploited to increase drug exposure.

The SLC family is typically influx transporters and assists the movement of substances into the intracellular environment through cellular uptake. Movement of substances via influx transporters are powered via a number of mechanisms; diffusion down an electrochemical gradient or transport against a diffusion gradient by a secondary active transporter, whereby the energy is provided via a symport or antiport system transporting ions. However, certain SLC transporters can operate in a bidirectional manner and efflux substances out of the cell as well.

The ABC transporters are efflux transporters requiring energy from ATP to facilitate the movement of the substance against a concentration gradient. Efflux transporters will prevent absorption into a cell by transporting substances from the intracellular environment to the external environment and thus limit cellular entry. This is shown in Figure 5.3.

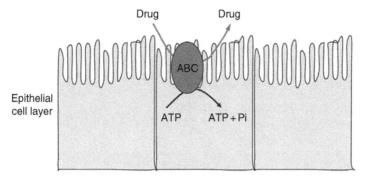

Figure 5.3   *Cartoon showing efflux of a drug due to an ABC transporter.*

Due to the nature of active transporter processes, they can be saturated at high drug concentrations, thus permeability, and hence absorption and drug exposure can be non-linear. Once an influx transporter is saturated, the passive permeability process will become the predominant mechanism as concentration increases, and this can result in a sub-proportional increase in exposure as the drug dose escalates.

For drugs subject to efflux transport, at low concentrations where the transporter is not saturated, absorption may be limited. However, as the dose and thus local concentration in the GI tract increases, the efflux transporter becomes saturated and can no longer limit drug permeability as the passive permeability process will become the dominant mechanism. Saturation of an efflux transporter in the intestine can result in a supra-proportional increase in exposure as the dose escalates.

The impact of the transporter contribution to intestinal absorption depends on the contribution of the active process compared to the passive permeability for the specific drug dose. *In vitro* cell lines expressing transporters of interest are often used to identify if a drug substance is a substrate for a transporter. If an active transport mechanism is identified using *in vitro* models, caution should be given to interpreting the *in vivo* significance of the finding. Teasing out the underlying causal mechanism for non-linear pharmacokinetics is complicated, potentially being impacted by solubility, metabolism, permeability and transporters, and many *in vitro* and *in vivo* experiments are required.

5.2.3   Efflux Transporters

Many drug substances are substrates for efflux transporters and, as such, the implication of the impact of efflux in limiting drug absorption is often assessed during drug discovery. P-gp/MDR-1 is by far the most researched efflux transporter to date. It has a broad substrate range, previously suggested to encompass up to 50% of marketed drugs [15] and can transport small molecules to polypeptides (e.g. 350–4000 Da). The crystal structure of the P-gp protein exhibits partially overlapping binding sites within the internal cavity of the protein and, thus, can explain the broad substrate specificity [16].

Identifying the true role of intestinal P-gp/MDR-1 transporter in drug absorption is often complicated due to a number of factors. CYP3A4 and P-gp/MDR-1 have a large range of overlapping substrate specificities [17]. It has been suggested that they work in conjunction to form a protective barrier, with CYP3A4 metabolising the drug, while P-gp/MDR1 recycles unmetabolised drug from the enterocyte back into the lumen, thus allowing CYP3A4

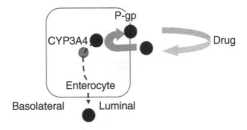

Figure 5.4   *A cartoon to show the combined impact of P-g and CYP3A4 on the potential fate of a drug.*

prolonged access to the drug for metabolism [18]. This is shown in the cartoon in Figure 5.4, where the drug may be effluxed by the P-gp to limit entry into the enterocyte, yet once it enters it is exposed to CYP3A4, which will metabolise the drug. Both P-gp and CYP3A4 are thus limiting the apparent permeability of the drug from the luminal to basolateral side of the cell.

Due to the overlap in structure activity with CYP3A4, it is difficult to define if the low absorption of a compound is due to P-gp/MDR-1 efflux back into the intestinal lumen or due to CYP metabolism in the gut [19]. Location of P-gp/MDR-1 at other sites of the body can also affect systemic exposure and thus present difficulties in assigning a direct link to an intestinal effect. Digoxin is a well-recognised drug substrate for P-gp/MDR-1 and is often used as a probe substrate to identify P-gp/MDR-1 transporter interactions. It is also a substrate that is not susceptible to CYP metabolism and thus simplifies interpreting the mechanism.

5.2.4   Transporters of Greatest Relevance to Oral Biopharmaceutics

Table 5.1 and Figure 5.5 provide details of the major intestinal transporters relevant to oral drug absorption. Two comprehensive review articles provide an overview of intestinal transporters and the impact on drug absorption [20, 21]. The main SLC transporters expressed in the intestine are peptide transporter 1 (PepT1), organic anion polypetide transporters (OATP1A2 and OATP2B1), monocarboxylate transporter 1 (MCT1), sodium multivitamin transporter (SMVT), organic cation and carnitine transporters (OCTN1 and OCTN2), organic cation and carnitine transporter 2 (OCTN2), and the concentrative nucleoside transporters 1 and 2 (CNT1 and CNT2). These SLC intestinal transporters are located at the apical brush-border membrane of the intestinal enterocyte (Figure 5.4), with the exception of MCT1, which is also expressed at the basolateral membrane. The equilibrative nucleoside transporters 1 and 2 (ENT 1 and 2) are also SLC transporters expressed at the basolateral membrane. Plasma membrane monoamine transporter is an SLC influx transporter located on the apical membrane, which is reported to transport metformin, a drug which displays dose-dependent saturable absorption, with absorption by both passive and facilitated transport [22].

The main ABC transporters found in the intestine are multi-drug resistance 1 (MDR-1) often referred to as P-glycoprotein (P-gp), breast cancer resistance protein (BCRP) and multi-resistance proteins 1-5 (MRP1-5). P-gp/MDR-1, MRP2, and BCRP are all located at the apical brush border membrane, whereas MRP1, 3 and 5 are located at the basolateral

Table 5.1 *Human intestinal transporters shown to be involved in the transport of drugs.*

Transporter protein	Gene	Orientation	Drug substrates
P-gp/MDR1	ABCB1	Apical efflux	Actinomycin D, cerivastatin, colchicine, cyclosporine A, daunorubicin, digoxin, docetaxel, doxorubicin, erythromycin, etoposide, fexofenadine, imatinib, indinavir, irinotecan, ivermectin, lapatinib, loperamide, losartan, nelfinavir, oseltamivir, paclitaxel, quinidine, ritonavir, saquinavir, sparfloxacin, tamoxifen, terfenadine, topotecan, verapamil, vinblastine, vincristine
BCRP	ABCG2	Apical efflux	Abacavir, ciprofloxacin, dantrolene, dipyridamole, enrofloxacin, erlotinib, etoposide, furosemide, gefitinib, genistein, glyburide, grepafloxacin, hydrochlorothiazide, imatinib, irinotecan, lamivudine, lapatinib, methotrexate, mitozantrone, prazosin, rosuvastatin, tamoxifen, triamterene, zidovudine
MRP1	ABCC1	Basolateral efflux	Daunorubicin, doxorubicin, epirubicin, grepafloxacin, methotrexate, vincristine
MRP2	ABCC2	Apical efflux	Indinavir, methotrexate, ritonavir, saquinavir, vinblastine
MRP3	ABCC3	Basolateral efflux	Etoposide, methotrexate
MRP4	ABCC4	Efflux	Cefizoxine, topotecan
PepT1	SLC15A1	Apical uptake	Ampicillin, bestatin, captopril, cephalexin, enalapril, fosinopril, oseltamivir, valaciclovir
OATP1A2	SLC01A2	Apical uptake	Fexofenadine, levofloxacin, methotrexate, ouabain, rosuvastatin, saquinavir
OATP2B1	SLC02B1	Apical uptake	Atorvastatin, bosentan, fluvastatin, glyburide, pitavastatin, pravastatin, montelukast, rosuvastatin
MCT1	SLC16A1	Apical uptake	Arbaclofen placarbil, carindacillin, gabapentin enacarbil, ketoprofen, naproxen, phenethicillin, propicillin
SMVT	SCL5A6	Apical uptake	Gabapentin enacarbil
OCTN1	SLC22A4	Apical uptake	Quinidine, verapamil
OCTN2	SLC22A5	Apical uptake	Cephaloridine, imatinib, ipratropium, tiotropium, quinidine, verapamil
CNT1	SCL28A1	Apical uptake	Cytarabine, gemcitabine, zidovudine
CNT2	SLC28A2	Apical uptake	Clofarabine, fluorouridine, ribavirin
ENT1	SLC29A1	Basolateral efflux	Cladribine, clofarabine, cytarabine, gemcitabine, ribavirin
ENT2	SLC29A2	Basolateral efflux	Clofarabine, gemcitabine, zidovudine

Source: Table based on data from references [10, 21, 23–26].

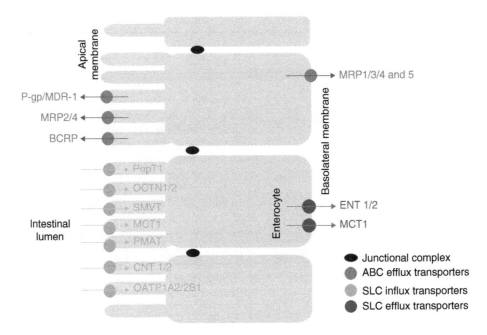

Figure 5.5   *Cellular location of intestinal transporters involved in oral drug absorption.*

membrane, with MRP4 having dual location at both the apical brush border and basolateral intestinal membrane. The apically located ABC transporters (P-gp/MDR-1, MRP2 and BCRP) have been shown to limit absorption of numerous drugs such as the antibiotics, chemotherapy treatments, cardiac drugs, statins and immunosuppressants and HIV proteases [20] (Figure 5.5).

5.2.5   Regulatory Overview of Transporter Effects on Biopharmaceutics

The impact of potential transporter-based DDIs on the overall exposure and safety of that drug should be assessed during development. Both the FDA [27] and EMA [28] have issued guidance providing details of the pre-clinical and clinical studies that must be performed to evaluate the transporter-based, drug interaction risks. The EMA guidance recommends that initially the transporter potential be evaluated *in vitro* using Caco-2 cells and if active transport is observed, then the transporter involved should be identified, if possible. If a clinically relevant transporter DDI is thought to be likely, a clinical study using a strong inhibitor is recommended. If the specific transporter is subject to genetic polymorphisms, *in vivo* studies with specific genotypes are suggested. The FDA states that all investigational drugs should be evaluated *in vitro* for being a potential substrate for P-gp/MDR-1 or BCRP. Whether a drug is an inhibitor of P-gp/MDR-1, BRCP and OATP1B1 should also be investigated *in vitro* based on clinically significant interactions for drugs, which are substrates of these transporters.

5.2.6 Regional Expression and Polymorphism of Intestinal Transporters and Impact of Drug Variability

Expression of transporters can vary between regions of the GI tract and between individuals. A recent review has pooled data on the regional relative and absolute expression of drug transporters in the adult intestine [4, 29]. The solute carrier peptide transporter 1 (PepT1) showed the highest abundance in the jejunum, while multidrug resistance-associated protein (MRP)-2 was the highest abundance ATP-binding cassette transporter [29]. Polymorphisms of specific transporters may also affect expression levels and functionality, thus having the potential to impact drug absorption.

Regional expression and polymorphism of transporters can contribute to variability in the *in vivo* performance of drug products, either directly or in conjunction with other causative factors such as drugs and food that alter transit times within the gastrointestinal tract.

5.3 Applications and Limitations of Characterisation and Predictive Tools for Permeability Assessment

The following section illustrates the wide variety of techniques available for the study/prediction of intestinal permeability and, in particular, highlights the differing levels of complexity, reliability and physiological relevance between the various techniques. Figure 5.6 highlights examples of such techniques organised by aspects such as biological relevance or throughput.

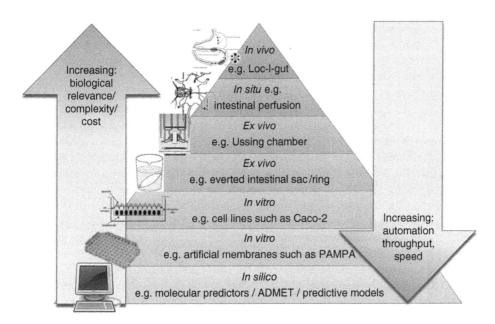

Figure 5.6 *Examples of tools used to predict intestinal permeability.*

5.3.1   *In Silico* Tools: Predictive Models for Permeability

In silico computational tools are based on calculations and predictions from molecular structures and can be used to predict passive permeability in support of the development of molecules with suitable properties [30]. These tools are commercially available and include GastroPlus; SimCYP and ADMET Predictor.

Parameters such as molecular weight, partition coefficients at specific pHs, solubility, lipophilicity and polar surface area are measured and incorporated into the model. This allows for virtual screening of large numbers of potential drug candidates. Screening with *in silico* tools can support the selection of a limited number of potential drug molecules with appropriate physicochemical properties for optimised permeability.

More complex multivariate data analysis models, based on a number of molecular properties or 'descriptors' combined, can improve the applicability of the predictions and enable the creation of quantitative structure–property relationships (QSPR) [31].

5.3.2   *In Vitro* Tools

5.3.2.1   *PAMPA*

Early stage screening tools have been developed to study transcellular passive permeability – the most common mechanism drug molecules use to cross cell membranes. These models measure the permeability of the drug molecule through artificial membranes e.g. a semipermeable plastic or cellulose filter coated with lipids to mimic the lipid bilayer found in the human epithelial wall. These artificial membrane systems offer the advantage of being reproducible, rapid and simple, as well as being both high-throughput and cost-effective to run.

One such *in vitro* tool is the parallel artificial membrane permeability assay (PAMPA), which has been studied widely [32]. The PAMPA system involves a membrane formed from a lipid (such as phospholipids or lecithin) contained within an inert organic solvent. The lipid/solvent solution is stored within a porous hydrophobic filter plate. The filter plate is then filled with the donor solution – containing the test drug, and is suspended over the receptor plate, which contains the acceptor buffer solution. The donor represents the dissolved drug at the membrane surface and the acceptor is on the other side so will enable calculation of the amount of drug that has crossed the membrane. The plates are incubated for a period of time, sometimes with stirring, shaking or rotationally introduced convection to account for the potential impact of an aqueous unstirred boundary layer for rapidly permeating molecules [33]. At specific time points, the amount of drug is measured in all compartments. The removed sample volume is then immediately replaced with fresh, prewarmed physiological buffer. Apparent Permeability (P_{app}) or flux is calculated as the rate of drug accumulation in the acceptor chamber normalised for membrane surface area. The amount of drug present in the lipid membrane can then be calculated through mass balance (the difference between that accounted for in the donor and acceptor phases).

PAMPA is generally used early in development to screen drugs into approximate rankings or categories such as high, intermediate and low permeability and has been demonstrated to correlate with gastrointestinal permeation. Figure 5.7 shows a correlation plot based on a pH 6.5 PAMPA model, utilising a 20% soy lecithin-dodecane membrane with 35 mM sodium lauryl sulphate in the acceptor compartment [34].

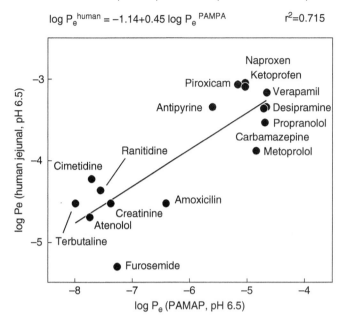

Figure 5.7 *Correlation between human jejunal and PAMPA permeabilities. Source: From Grüneberg and Güssregen [34] / With permission of John Wiley & Sons.*

The main limitations of PAMPA are that it can only approximate the measurement of transcellular passive diffusion-based permeability since the system has no active transport, paracellular pathway (permeation of material through tight junctions between epithelial cells) or metabolising enzyme capability. Another limitation is that studies must be performed at room temperature since the membrane becomes unstable at the more physiologically relevant 37 °C. A further disadvantage of the PAMPA system, along with other artificial membranes, is that the pores of the filter plate where lipid multi-lamellar bilayers form are much longer than an epithelial cell membrane lipid bilayer, thereby slowing the apparent permeation rate compared to the equivalent *in vivo* situation. Calculations and pH shift studies have been employed to address the impact of this issue [35].

Other artificial systems that measure lipophilicity characteristics include immobilised artificial membranes and the related immobilised liposome chromatography (ILC), micellar electrokinetic chromatography and biopartitioning micellar chromatography. All of these techniques result in lipophilicity indices, which may better relate to membrane interactions and passive permeability than simple octanol/water partitioning evaluations [34].

5.3.2.2 Cell Lines

The most routinely used *in vitro* methods for assessing intestinal drug permeability are cell-culture-based monolayer models. Among these, one of the most commonly used is the Caco-2 human colon carcinoma-based cell line. When this cell line is grown on a

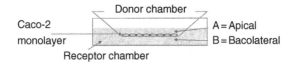

Figure 5.8   *Schematic of drug transport assay in Caco-2 cell monolayers cultured on a culture insert containing a permeable membrane.*

microporous filter (see Figure 5.8) under standard culture conditions for around 16–21 days, monolayers are formed, which have morphological and functional characteristics similar to that of human intestinal epithelial cells [36].

The Caco-2 cell line is the only one to have been found to undergo enterocytic differentiation and become polarised during culturing to resemble the human epithelial cell wall. This results in a morphology with tight cellular junction formation and microvilli at the apical surface, alongside functional expression of many transporters and enzymes found in the small intestine. Transporters expressed include efflux transporters, such as P-gp/MDR1 [37]. It should be noted that the transporters expressed can be different depending upon the culture conditions used. FDA guidance also recommends cell lines such as Caco-2 for *in vitro* mechanistic investigation of drug–drug interactions [27]. Caco-2 cell line investigations can be performed with different pH and/or osmolality in the apical and basolateral chambers to mimic different regions of the gastrointestinal tract *in vivo*. The impact of different formulation components on pH and permeability can also be studied. In a similar manner to PAMPA studies, Caco-2 cell permeability investigations are performed with the monolayer held between an apical and a basolateral chamber into which test compound or blank receiver buffer media can be added. Apical secretion, or efflux, is indicated by a greater amount of drug appearing in the apical chamber following basolateral application of test compound compared to that found in the basolateral chamber following apical application of the same amount of drug. The P_{app} of a test compound is calculated by dividing the quantity of compound reaching the receiver chamber (normalised to the filter surface area per unit time) by the initial drug concentration in the donor chamber. P_{app} can be used to rank compounds in terms of oral absorption potential.

An advantage of the Caco-2 cell line over artificial membranes is the ability to assess paracellular tight-junction permeability, which is thought to be important in the absorption of low-molecular-weight hydrophilic drugs. Tight junctions can be assessed using hydrophilic-marker paracellular compounds such as mannitol and/or through monitoring the transepithelial electrical resistance (TEER) of the monolayers. Caco-2 cells mainly display tight junctions, which match the smaller of the two pore populations found in the human small intestine. This cell line is therefore limited in its ability to identify molecules with significant levels of paracellular permeability *in vivo* in comparison to alternative cell lines, such as the conditionally immortalised rat foetal intestinal cell line 2/4/A1 [38]. However, it should be noted that the 2/4/A1 cell line is more difficult to culture, is poorly differentiated morphologically and does not express many of the enzymes and transporters found in the Caco-2 cell line. The 2/4/A1 cell line should therefore be used to study passive transport only.

Standard culture Caco-2 cell lines show low and unstable expression levels of the major metabolising enzymes of the Cytochrome P450 [39]. Clones of Caco-2 cells such as TC-7 express greater levels of enzymes such as Cytochrome P4501A1 and UDP-glucuronosyltransferase (UGT) and can be used as an alternative intestinal model [40].

A disadvantage the majority of cell lines used for modelling intestinal permeability (including Caco-2) is the lack of a mucus layer, produced *in vivo* by goblet cells. The human intestinal epithelial colon carcinoma cell line, HT29, can be grown to become a polarised monolayer of mucus secreting cells. This can be used to investigate the impact of mucus found *in vivo*, which can be a barrier to intestinal permeability. Attempts made to co-culture Caco-2 cells with HT29-H and HT29-MTX (HT29 clones), to more closely mimic the *in vivo* cell population diversity, were not successful due to insufficient mixing between cell lines [41].

Another popular cell line used to investigate intestinal drug permeability is the Madin-Darby canine kidney cell line (MDCK). Monolayers produced from this cell line have the advantage of being fully differentiated after only around three to seven days of culturing. Despite being derived from a dog kidney epithelium with resulting potential differences in transporter expression, correlation to Caco-2 in terms of passive transport has been reported [40]. It should be noted that there are two sub clones of MDCK with high or low resistance in terms of TEER values, and that this cell line is not suitable for prediction of active transport or efflux due to the potentially different transporters expressed in dog. Transporters, however, are expressed at low levels and so transfected MDCK cells have been used to investigate specific transport mechanisms such as P-gp/MDR1 impact on permeability through MDR1-MDCK cells. An alternative cell line suitable for transfection is the Lewis lung carcinoma-porcine kidney cell (LLC-PK1). Both MDCK and LLC–PK1 cells have shown improved stability and enzymatic activity when transfected with CYP 3A4 compared to Caco-2 transfected cells.

Caco-2 cell permeability is recommended in regulatory guidelines as a suitable *in vitro* cell line to predict permeability and can be used in assigning a BCS classification for certain biowaivers, see Chapter 9. However, it should be noted that there are details on how the cell data should be validated using known drugs as references for high medium and low permeability, as well as the need for bilateral transport and concentration-dependent effects within the guidance [42].

5.3.3   *Ex Vivo* Tools

Although the *in vitro* tools described previously can address specific mechanistic questions and inform decisions regarding intestinal permeability, models utilising living and functioning tissue isolated from an animal or human, also known as *ex vivo* models, reflect the *in vivo* situation in a more comprehensive way. *ex vivo* small intestinal tissue models have beneficial features, which differentiate them from the cell-based *in vitro* models, such as a mucus layer, representative paracellular permeability, expression of transport proteins, and gut metabolising enzymes. They can be readily used to evaluate permeability in different regions of the gastrointestinal tract and compare intestinal permeability between species.

The two main *ex vivo* models used to investigate intestinal permeability are diffusion chamber-based models such as the Ussing chamber [43] and the everted intestinal sac [44].

5.3.3.1   *Ussing Chambers*

Ussing chambers are the most commonly used diffusion chamber model for intestinal permeability studies. Ussing chambers are available in different sizes but are all based on the

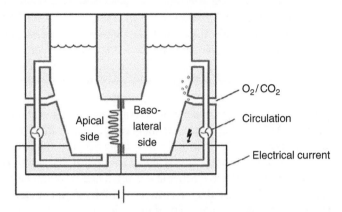

Figure 5.9   *Schematic of a small piece of intestinal epithelial tissue mounted in the Ussing chamber. Source: From Westerhout et al. [45] / With permission of Springer Nature.*

same principle of using an excised section of intact mucosal intestinal tissue. The tissue is cut into small segments of appropriate size and opened to form a flat epithelial sheet. This sheet is mounted, usually vertically, between two chambers containing physiological buffer oxygenated with carbogen (95% O_2, 5% CO_2) and maintained at 37°C (see Figure 5.9).

The study commences when test drug is added to the donor chamber, and test samples are removed at defined time intervals from the receiver chamber for analysis and quantification. Bidirectional transport studies can be conducted with the drug being exposed initially to either the mucosal or the serosal (basolateral) surface of the tissue. Continuous monitoring of the integrity and activity of the excised tissue membrane can be performed through the use of electrodes positioned in each chamber of the apparatus, measuring the TEER, potential difference and short-circuit current. The provision of electrodes and functionality of the tissue membrane used within Ussing chambers additionally allows measurement of changes in ion transport across the membrane in the presence of drug compounds.

Ussing chamber models also allow the study of intestinal permeability and drug–drug interactions in combination with intestinal metabolism. The opportunity to utilise human tissue from a target patient population, which will confer mechanistic features of the disease state, is a distinct advantage models such as the Ussing chamber can offer. As with *in vitro* cell lines, permeability can be studied under different physiological conditions (e.g. pH and osmolality) using Ussing chambers. Since the intestinal tissue segments used in the Ussing chamber model retain the morphological architecture and physiological features of the intestine, the complex interplay of the multicellular environment is also retained. Human tissue studies using the Ussing chamber can be used to troubleshoot unanticipated clinical findings, e.g. where preclinical species did not show the same pharmacokinetic observations as found in human clinical studies.

Many of the limitations and disadvantages of Ussing chamber studies are associated with the practical aspects of excision and preparation of the tissue itself, which can be very difficult to source. One major challenge is the coordination and logistics of sourcing fresh, viable tissue, especially human tissue from both healthy and specific patient groups. Inter-individual variability between tissue segments and membranes can also be an issue, particularly as these models are low throughput and take prolonged periods of time to set up and run. Another limitation of the use of *ex vivo* intestinal segments is the limited viability of the tissue to around 2–2.5 hours during experiments.

Comparison of rat and human small intestinal active and passive permeability using Ussing chambers seems, however, to show good correlation [46].

Attempts to address the disadvantage that the classical Ussing chamber model does not allow simultaneous preparation and analyses of a large number of epithelial tissue segments have been made more recently. Systems that use multiple horizontally orientated membranes have been developed, such as TNO's porcine intestinal tissue-based InTESTine™ medium-throughput multi-well system. This system uses pig tissue, which demonstrates high similarity in anatomy and physiology to human intestinal tissue, and allows potential studies such as parallel regional intestinal permeability assessments to be performed [47].

5.3.3.2   Everted Intestinal Sac/Ring

Many of the advantageous attributes of the Ussing chamber model with animal tissue as the barrier can also be applied to the everted intestinal sac (or everted gut sac) model. In this model, either a 2–6 cm section or the whole of the intestine is rapidly removed from an anaesthetised animal (typically a rat) and flushed with physiological buffer before being everted (turned inside-out) over a glass rod or tube. The everted intestinal tissue is then either tied at both ends to create a sac or cannulated with tubing in an apparatus designed to make subsequent sampling easier, as illustrated in Figure 5.10 [48].

In this model, the mucosal surface becomes the outside layer, while the serosal surface becomes the inside surface. Physiological buffer is filled both inside the sac (or cannulated

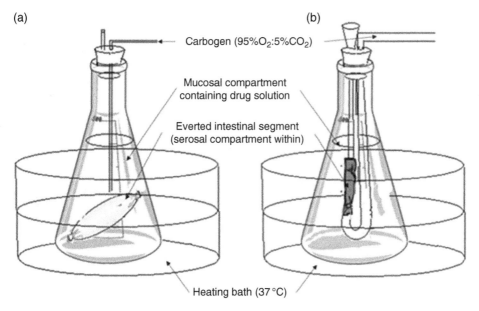

(a)   (b)

Carbogen (95%O_2:5%CO_2)

Mucosal compartment
containing drug solution

Everted intestinal segment
(serosal compartment within)

Heating bath (37 °C)

Figure 5.10   *Schematic of ex vivo everted intestinal sac experiment. (a) Intestinal sac tied at both ends (b) Modified apparatus with cannulation.*

everted tissue 'tube') and outside in the outer container of either sac or cannulation apparatus.

The everted intestinal sac model is rapid, relatively inexpensive, reproducible and can be particularly useful for analysis of low permeability compounds as the volume of fluid within the serosal compartment, i.e. inside the sac or tube, is low. This model provides a relatively large surface area for permeability, has a mucus layer present and is most commonly used to assess the impact of enzyme activity and transporters. The data generated are affected by the donor animal's age, sex, species, disease state, diet and treatments. The major limitations and disadvantages of the technique are maintaining tissue viability, potential morphological damage caused during harvesting and eversion, plus the retention of an intestinal muscle layer (muscularis mucosa), which can affect both the permeation observed and the oxygenation efficiency of the model. Studies are generally run at 37 °C and gentle shaking can be applied to minimise tissue damage. Drug molecules under test can be analysed both in the fluid inside the everted intestinal sac and in the surrounding chamber, and in the tissue itself. The everted sac viability and integrity can be monitored through glucose concentration measurements from the solutions outside and inside of the sac.

5.3.4   *In Situ* Tools

There are a number of intestinal perfusion methods available, which offer the advantages of providing an intact intestinal mucosa, nerve system, lymphatic system, and blood flow, alongside expression of enzymes and transporters. This allows for the assessment of permeability in viable tissue under conditions mimicking the *in vivo* environment closely. In these studies, animals are typically anaesthetised and heated to maintain 37 °C using pads and lamps. An additional advantage of intestinal perfusion models is that regional differences in permeability and metabolism can be investigated without interference from gastric emptying and/or small intestinal transit times, which can impact *in vivo* permeability and absorption studies.

Rat *in situ* perfusion models have been shown to correlate, at least in terms of rank ordering of compounds, with *in vivo* human data for both passive and carrier-mediated drug transport [49]. Tissue viability and integrity are potential limitations, as with other *ex vivo* models. A further limitation of running such a model is that specialist surgical manipulation is required to expose the abdominal cavity and withdraw the region of the intestine to be cannulated at both ends for perfusion.

5.3.4.1   *Closed-Loop Intestinal Perfusion*

In the closed-loop intestinal perfusion model, a selected region of intestine is cannulated and washed with perfusion solution before being filled with perfusion solution containing a drug or formulation. The intestinal lumen content is then sampled at pre-determined time intervals, and the concentration of drug is analysed [50]. Within 10–30 seconds of sampling, the perfusion solution is returned to the isolated intestine lumen. A potential limitation of the closed-loop perfusion model is that the drug is exposed to the entire mucosal surface of the intestinal segment throughout the study. This may not reflect the *in vivo* situation, where a drug product will transit through the small intestine. However, it may allow more precise control of drug concentration and benefit the assessment of low permeability/high-efflux compounds through higher disappearance rates.

5.3.4.2 Single-Pass Intestinal Perfusion

The single-pass (or open-loop) intestinal perfusion model is similar in set up to the closed-loop model except that an infusion pump is used to perfuse the liquid containing drug, at around 0.1–0.3 mL/min continuously down a fixed length intestinal segment. Although permeability estimates generated from both closed and single-pass perfusion methods are similar when normalised for perfused volume and intestinal length, the single-pass method has been shown to be more robust and reproducible [51]. Samples collected from the ileal-end cannulation are assayed for drug and the permeability estimated through calculation of the concentration difference between ingoing and outgoing perfusate, once outgoing concentration has stabilised and steady-state has been achieved. Perfusion studies can be run *in situ* with the closure of the bile duct to prevent entero-hepatic recirculation of drug as part of building a mechanistic understanding of the process of absorption.

5.3.4.3 Intestinal Perfusion with Venous Sampling

An adaptation of the closed-loop and single-pass intestinal perfusion models incorporates plasma sampling through cannulation and drainage of a vein, such as the mesenteric vein of an intestinal segment, with donor blood replacement from the jugular vein (Figure 5.11). This provides a means to quantify drug flux through the intestinal wall rather than only quantifying uptake by the wall and is sometimes referred to as the 'auto-perfused' method. In this technique, intestinal permeability can be estimated from the rate of drug appearing in the blood, and it can be performed with either closed- or open-loop methodologies.

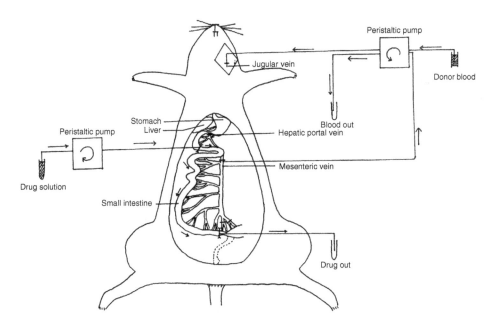

Figure 5.11 *Schematic drawing of in situ intestinal perfusion with venous sampling. Source: From Luo et al. [52] / With permission of Elsevier.*

5.3.4.4   Vascularly Perfused Intestinal Models

A further level of model complexity has been investigated through isolation and cannulation of an intestinal segment, alongside cannulation of a supply artery such as the mesenteric artery plus a drainage vein. Perfusate monitoring, after drug input from the lumen of the intestinal segment and from the artery, can be compared to quantify intestinal-mediated extraction of drug.

5.4   *In Vivo* Tools

A number of open and closed *in vivo* perfusion techniques have been reported for estimating intestinal permeability, drug dissolution, secretion and metabolism in humans. The most commonly used of these techniques is the Loc-I-Gut method, which consists of a six-channel polyvinyl disposable tube [53]. Two of these channels are linked to balloons used to occlude a 10-cm segment of the intestine: one channel is for drug perfusate infusion, another is for luminal content sampling and the final two channels are for viability marker administration of compounds such as phenol red as mentioned earlier [53]. The balloons, when inflated through the multichannel tube, create a closed section of intestine, which has the advantage of minimal contamination from luminal fluid from the rest of the intestine, plus improved control of hydrodynamics within the occluded segment, as shown in Figure 5.12.

Another advantage of this technique is that transit time, and regional pH differences do not influence studies. Potential factors to account for when running this method include correcting for any binding of the drug to the tubing, chemical and/or enzymatic degradation of drug in the luminal contents and possible accumulation of drug in the intestinal wall,

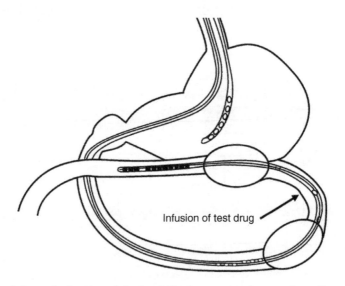

Infusion of test drug

Figure 5.12   *Schematic drawing of the Loc-I-Gut instrument. Source: From Knutson et al. [54] / With permission of American Chemical Society.*

which could affect sink conditions across the epithelium. Drug permeability is estimated from the rate of disappearance of drug from the perfused intestinal segment. The technique also allows the measurement of absolute bioavailability through the use of IV dosing and peripheral blood sampling and analysis for drug in the systemic circulation. Major limitations with the *in vivo* human techniques such as the Loc-I-Gut are the high cost and ethical issues associated with running such a model. As a result, the Loc-I-Gut is not used routinely, e.g. for early permeability screening, but is applied in later development phases, for example to generate permeability values for the *in vitro–in vivo* correlations. It is also used to investigate enzyme and transporter expression levels and functionality measurements or to evaluate food–drug interactions, drug dissolution and intestinal drug secretion [55].

The Loc-I-Gut was used to develop an *in vivo* human permeability database as a basis of BCS for oral IR dosage forms. This exemplifies that this technique is viewed as a gold standard for intestinal permeability and absorption models.

5.5 Conclusion

In conclusion, permeability is a useful biopharmaceutical parameter, which can be used within drug discovery and development to form part of the drug development strategy and assist in formulation optimisation. Permeability can be predicted from relatively simple methods using the structure or assessed in more complex experimental methods, such as *in vitro* assessment and *in vivo* studies.

References

[1] Thelen, K. and Dressman, J.B. (2009). Cytochrome P450-mediated metabolism in the human gut wall. *J. Pharm. Pharmacol.* **61** (5): 541–558.
[2] Paine, M.F., Hart, H.L., Ludington, S.S. et al. (2006). The human intestinal cytochrome P450 "pie". *Drug Metab. Dispos.* **34** (5): 880–886.
[3] Thörn, M., Finnström, N., Lundgren, S. et al. (2005). Cytochromes P450 and MDR1 mRNA expression along the human gastrointestinal tract. *Br. J. Clin. Pharmacol.* **60** (1): 54–60.
[4] Paine, M.F., Khalighi, M., Fisher, J.M. et al. (1997). Characterization of interintestinal and intraintestinal variations in human CYP3A-dependent metabolism. *J. Pharmacol. Exp. Ther.* **283** (3): 1552–1562.
[5] von Richter, O., Burk, O., Fromm, M.F. et al. (2004). Cytochrome P450 3A4 and P-glycoprotein expression in human small intestinal enterocytes and hepatocytes: a comparative analysis in paired tissue specimens. *Clin. Pharmacol. Ther.* **75** (3): 172–183.
[6] Kempf, D.J., Marsh, K.C., Kumar, G. et al. (1997). Pharmacokinetic enhancement of inhibitors of the human immunodeficiency virus protease by coadministration with ritonavir. *Antimicrob. Agents Chemother.* **41** (3): 654–660.
[7] Christiansen, A., Backensfeld, T., Denner, K., and Weitschies, W. (2011). Effects of non-ionic surfactants on cytochrome P450-mediated metabolism *in vitro*. *Eur. J. Pharm. Biopharm.*: official journal of Arbeitsgemeinschaft für Pharmazeutische Verfahrenstechnik e.V **78**: 166–172.
[8] Kohl, C. (2008). The importance of gut wall metabolism in determining drug bioavailability. In: *Drug Bioavailability: Estimation of Solubility, Permeability, Absorption and Bioavailability*, Methods and Principles in Medicinal Chemistry, 2ee, vol. **40** (eds. H. van de Waterbeemd and B. Testa), 333–357. Wiley;VCH Verlag GmbH & Co. KGaA.
[9] Tubic-Grozdanis, M., Hilfinger, J.M., Amidon, G.L. et al. (2008). Pharmacokinetics of the CYP 3A substrate simvastatin following administration of delayed versus immediate release oral dosage forms. *Pharm. Res.* **25** (7): 1591–1600.

[10] Giacomini, K.M., Huang, S.-M., Tweedie, D.J. et al. (2010). Membrane transporters in drug development. *Nat. Rev. Drug Discov.* **9** (3): 215–236.

[11] Hediger, M.A., Clémençon, B., Burrier, R.E., and Bruford, E.A. (2013). The ABCs of membrane transporters in health and disease (SLC series): introduction. *Mol. Asp. Med.* **34** (2): 95–107.

[12] Russel, F.G.M. (2010). Transporters: importance in drug absorption, distribution, and removal. In: *Enzyme- and Transporter-Based Drug-Drug Interactions: Progress and Future Challenges* (eds. K.S. Pang, A.D. Rodrigues and R.M. Peter), 27–49. New York, NY: Springer.

[13] Huls, M., Russel, F.G.M., and Masereeuw, R. (2009). The role of ATP binding cassette transporters in tissue defense and organ regeneration. *J. Pharmacol. Exp. Ther.* **328** (1): 3.

[14] Leslie, E.M., Deeley, R.G., and Cole, S.P. (2005). Multidrug resistance proteins: role of P-glycoprotein, MRP1, MRP2, and BCRP (ABCG2) in tissue defense. *Toxicol. Appl. Pharmacol.* **204** (3): 216–237.

[15] Akamine, Y., Yasui-Furukori, N., Ieiri, I., and Uno, T. (2012). Psychotropic drug-drug interactions involving P-glycoprotein. *CNS Drugs* **26** (11): 959–973.

[16] Aller, S.G., Yu, J., Ward, A. et al. (2009). Structure of P-glycoprotein reveals a molecular basis for poly-specific drug binding. *Science (New York, N.Y.)* **323** (5922): 1718–1722.

[17] Zhou, S.F. (2008). Drugs behave as substrates, inhibitors and inducers of human cytochrome P450 3A4. *Curr. Drug Metab.* **9** (4): 310–322.

[18] Cummins, C.L., Jacobsen, W., and Benet, L.Z. (2002). Unmasking the dynamic interplay between intestinal P-glycoprotein and CYP3A4. *J. Pharmacol. Exp. Ther.* **300** (3): 1036–1045.

[19] Kivistö, K.T., Niemi, M., and Fromm, M.F. (2004). Functional interaction of intestinal CYP3A4 and P-glycoprotein. *Fundam. Clin. Pharmacol.* **18** (6): 621–626.

[20] Estudante, M., Morais, J.G., Soveral, G., and Benet, L.Z. (2013). Intestinal drug transporters: an overview. *Adv. Drug Deliv. Rev.* **65** (10): 1340–1356.

[21] Varma, M.V., Ambler, C.M., Ullah, M. et al. (2010). Targeting intestinal transporters for optimizing oral drug absorption. *Curr. Drug Metab.* **11** (9): 730–742.

[22] Zhou, M., Xia, L., and Wang, J. (2007). Metformin transport by a newly cloned proton-stimulated organic cation transporter (plasma membrane monoamine transporter) expressed in human intestine. *Drug Metab. Dispos.* **35** (10): 1956–1962.

[23] Doan, K.M.M., Humphreys, J.E., Webster, L.O. et al. (2002). Passive permeability and P-Glycoprotein-mediated efflux differentiate Central Nervous System (CNS) and non-CNS marketed drugs. *J. Pharmacol. Exp. Ther.* **303** (3): 1029.

[24] Polli, J.W., Wring, S.A., Humphreys, J.E. et al. (2001). Rational use of *in vitro* P-glycoprotein assays in drug discovery. *J. Pharmacol. Exp. Ther.* **299** (2): 620–628.

[25] Klaassen, C.D. and Aleksunes, L.M. (2010). Xenobiotic, bile acid, and cholesterol transporters: function and regulation. *Pharmacol. Rev.* **62** (1): 1–96.

[26] Murakami, T. and Takano, M. (2008). Intestinal efflux transporters and drug absorption. *Expert Opin. Drug Metab. Toxicol.* **4** (7): 923–939.

[27] FDA (2020). Guidance for industry: *in vitro* drug interaction studies — cytochrome P450 enzyme- and transporter-mediated drug interactions. U.S. Department of Health and Human Services, Food and Drug Administration, Center for Drug Evaluation and Research (CDER). https://www.fda.gov/media/134582/download (accessed 30 November 2020).

[28] EMA (2012). Guideline on the investigation of drug interactions. CPMP/EWP/560/95/Rev. 1 Corr. 2** Committee for Human Medicinal Products (CHMP). https://www.ema.europa.eu/en/documents/scientific-guideline/guideline-investigation-drug-interactions-revision-1_en.pdf (accessed 11 October 2020).

[29] Harwood, M.D., Zhang, M., Pathak, S.M., and Neuhoff, S. (2019). The regional-specific relative and absolute expression of gut transporters in adult caucasians: a meta-analysis. *Drug Metab. Dispos.* **47** (8): 854.

[30] Lipinski, C.A., Lombardo, F., Dominy, B.W., and Feeney, P.J. (2001). Experimental and computational approaches to estimate solubility and permeability in drug discovery and development settings. *Adv. Drug Deliv. Rev.* **46** (1–3): 3–26.

[31] Egan, W.J., Merz, K.M., and Baldwin, J.J. (2000). Prediction of drug absorption using multivariate statistics. *J. Med. Chem.* **43** (21): 3867–3877.

[32] Kansy, M., Senner, F., and Gubernator, K. (1998). Physicochemical high throughput screening: parallel artificial membrane permeation assay in the description of passive absorption processes. *J. Med. Chem.* **41** (7): 1007–1010.

[33] Molloy, B.-J., Tam, K.Y., Matthew Wood, J., and Dryfe, R.A.W. (2008). A hydrodynamic approach to the measurement of the permeability of small molecules across artificial membranes. *Analyst* **133** (5): 655–659.

[34] Grüneberg, S. and Güssregen, S. (2004). Drug bioavailability. Estimation of solubility, permeability, absorption and bioavailability. (Series: Methods and Principles in Medicinal Chemistry, Vol. 18; series editors: R. Mannhold, H. Kubinyi, and G. Folkers). Edited by Han van de Waterbeemd, Hans Lennernäs and Per Artursson. *Angew. Chem. Int. Ed.* **43** (2): 146–147.

[35] Avdeef, A., Bendels, S., Di, L.i. et al. (2007). PAMPA—critical factors for better predictions of absorption. *J. Pharm. Sci.* **96** (11): 2893–2909.

[36] Hidalgo, I.J., Raub, T.J., and Borchardt, R.T. (1989). Characterization of the human colon carcinoma cell line (Caco-2) as a model system for intestinal epithelial permeability. *Gastroenterology* **96** (3): 736–749.

[37] Hunter, J. and Hirst, B.H. (1997). Intestinal secretion of drugs. The role of P-glycoprotein and related drug efflux systems in limiting oral drug absorption. *Adv. Drug Deliv. Rev.* **25** (2): 129–157.

[38] Linnankoski, J., Mäkelä, J., Palmgren, J. et al. (2010). Paracellular porosity and pore size of the human intestinal epithelium in tissue and cell culture models. *J. Pharm. Sci.* **99** (4): 2166–2175.

[39] Brimer, C., Dalton, J.T., Zhu, Z. et al. (2000). Creation of polarized cells coexpressing CYP3A4, NADPH cytochrome P450 reductase and MDR1/P-glycoprotein. *Pharm. Res.* **17** (7): 803–810.

[40] Caro, I., Boulenc, X., Rousset, M. et al. (1995). Characterisation of a newly isolated Caco-2 clone (TC-7), as a model of transport processes and biotransformation of drugs. *Int. J. Pharm.* **116** (2): 147–158.

[41] Hilgendorf, C., Spahn-Langguth, H., Regårdh, C.G. et al. (2000). Caco-2 versus Caco-2/HT29-MTX co-cultured cell lines: permeabilities via diffusion, inside- and outside-directed carrier-mediated transport. *J. Pharm. Sci.* **89** (1): 63–75.

[42] ICH (2020). ICH M9 guideline on biopharmaceutics classification system-based biowaivers. https://www.ema.europa.eu/en/documents/scientific-guideline/ich-m9-biopharmaceutics-classification-system-based-biowaivers-step-5_en.pdf (accessed 5 September 2020).

[43] Ussing, H.H. and Zerahn, K. (1951). Active transport of sodium as the source of electric current in the short-circuited isolated frog skin. *Acta Physiol. Scand.* **23** (2–3): 110–127.

[44] Wilson, T.H. and Wiseman, G. (1954). The use of sacs of everted small intestine for the study of the transference of substances from the mucosal to the serosal surface. *J. Physiol.* **123** (1): 116–125.

[45] Westerhout, J., Wortelboer, H., and Verhoeckx, K. (2015). Ussing chamber. In: *The Impact of Food Bioactives on Health: in vitro and ex vivo Models* (eds. K. Verhoeckx, P. Cotter, I. López-Expósito, et al.), 263–273. Cham: Springer International Publishing.

[46] Lennernäs, H. (2014). Human *in vivo* regional intestinal permeability: importance for pharmaceutical drug development. *Mol. Pharm.* **11** (1): 12–23.

[47] Westerhout, J., van de Steeg, E., Grossouw, D. et al. (2014). A new approach to predict human intestinal absorption using porcine intestinal tissue and biorelevant matrices. *Eur. J. Pharm. Sci.* **63**: 167–177.

[48] Ungell, A.-L. (1997). *in vitro* absorption studies and their relevance to absorption from the GI tract. *Drug Dev. Ind. Pharm.* **23** (9): 879–892.

[49] Salphati, L., Childers, K., Pan, L. et al. (2001). Evaluation of a single-pass intestinal-perfusion method in rat for the prediction of absorption in man. *J. Pharm. Pharmacol.* **53** (7): 1007–1013.

[50] Doluisio, J.T., Billups, N.F., Dittert, L.W. et al. (1969). Drug absorption I: an in situ rat gut technique yielding realistic absorption rates. *J. Pharm. Sci.* **58** (10): 1196–1200.

[51] Schurgers, N., Bijdendijk, J., Tukker, J.J., and Crommelin, D.J.A. (1986). Comparison of four experimental techniques for studying drug absorption kinetics in the anesthetized rat in situ. *J. Pharm. Sci.* **75** (2): 117–119.

[52] Luo, Z., Liu, Y., Zhao, B. et al. (2013). ex vivo and in situ approaches used to study intestinal absorption. *J. Pharmacol. Toxicol. Methods* **68** (2): 208–216.

[53] Lennernäs, H., Fagerholm, U., Raab, Y. et al. (1995). Regional rectal perfusion: a new *in vivo* approach to study rectal drug absorption in man. *Pharm. Res.* **12** (3): 426–432.

[54] Knutson, T., Fridblom, P., Ahlström, H. et al. (2009). Increased understanding of intestinal drug permeability determined by the LOC-I-GUT approach using multislice computed tomography. *Mol. Pharm.* **6** (1): 2–10.

[55] Kesisoglou, F., Chung, J., van Asperen, J., and Heimbach, T. (2016). Physiologically based absorption modeling to impact biopharmaceutics and formulation strategies in drug development-industry case studies. *J. Pharm. Sci.* **105** (9): 2723–2734.

6

Dissolution

Hannah Batchelor[1] and James Butler[2]

[1] *Strathclyde Institute of Pharmacy and Biomedical Sciences, University of Strathclyde, Glasgow, United Kingdom*
[2] *Biopharmaceutics, Product Development and Supply, GlaxoSmithKline R&D, Ware, United Kingdom*

6.1 Introduction

Dissolution is a key method in the oral biopharmaceutics toolkit. This chapter aims to demonstrate the value of dissolution within biopharmaceutics and to highlight the types of apparatus used to better understand the mechanisms associated with drug product dissolution and the interplay with GI physiology. It will also provide some basic guidance on the development of a dissolution method to better understand a product.

6.2 Purpose of Dissolution Testing

Dissolution testing of drug products provides information on the rate and extent of drug release from a formulation under a given set of test conditions. Dissolution is of particular importance for low solubility drugs, for which dissolution is likely to affect pharmacokinetics (PK), and for extended release (ER) products where the rate of drug release from the dosage form by design controls the amount of drug in solution available for absorption. Modifying the release of the drug in these scenarios is likely to alter the PK profile. Through careful design, an ER product can generate a PK profile that provides a more stable and longer lasting plasma profile of the drug for the patient. ER formulations can

Biopharmaceutics: From Fundamentals to Industrial Practice, First Edition. Edited by Hannah Batchelor.
© 2022 John Wiley & Sons Ltd. Published 2022 by John Wiley & Sons Ltd.

reduce side effects and improve adherence to therapy, by enabling once daily dosing for drugs that have a short elimination half-life. Similarly, the PK of a poorly soluble drug can be advantageously improved by careful product design, with dissolution as the key investigational tool.

A well-designed, discriminatory dissolution test can provide insights into the impact of formulation composition and manufacturing process parameters on the release of the active drug. Dissolution is therefore key in ensuring batch-to-batch consistency in the manufacture of a drug product and plays an important role in risk-based assessment of the manufacturing process and its validation.

A **clinically relevant** dissolution test enables correlation of dissolution to *in vivo* absorption of a drug in cases where dissolution is the rate-limiting step. In this case, dissolution can be used to risk assess post-approval formulation changes, predict bioequivalence of alternative formulations, and as a surrogate for *in vivo* drug release.

A **biorelevant** dissolution test is designed to mimic key aspects of the dissolution media to reflect the *in vivo* conditions, thus increasing the probability of the *in vitro* test detecting clinically relevant changes

An ideal dissolution test would be both discriminatory and clinically relevant, with just enough biorelevance to maintain clinical relevance without adding unnecessary complexity. However, often during drug development multiple dissolution test designs, including a range of biorelevant methods are developed to provide mechanistic understanding of key factors.

6.2.1 Dissolution Versus Solubility

Solubility refers to the capacity of solute to dissolve into a solvent and is measured at equilibrium conditions where the maximum solute is dissolved. Dissolution, in contrast, is a kinetic process that measures the rate at which the solute dissolves in a solvent. Equilibrium solubility can be determined at the completion of dissolution in cases where an excess of solute is present such that the system reaches saturation.

The dissolution process typically follows first-order kinetics, beginning with a rapid increase in solubilised material due to the concentration gradient between the solute and the solvent. As the solvent approaches saturation with solute, the rate of dissolution slows until it plateaus at the equilibrium solubility. However, saturation is not always reached in a biopharmaceutics context; thus, if the concentration of solute upon complete dissolution is well below the saturated concentration equilibrium solubility, complete dissolution will be observed.

Dissolution is often performed in a closed system, meaning that as drug dissolves it may reach the saturated solubility for that solute–solvent system. Within the body, the drug can be absorbed, removing dissolved drug, and so is not a closed system. To replicate the *in vivo* scenario, dissolution methods are often designed such that the presence of already dissolved drug does not affect the ability for dissolution of more drug within the solvent. This often means that there is a need for excess solvent or solubilising agents to ensure that the solubility is maintained during the test. This is known as operating at 'sink conditions': i.e. sufficient media to ensure that dissolution of drug is not limited by the solubility of the drug in the solvent.

In practice, sink conditions refers to 3–10 times the amount of solvent required to achieve solubility of a drug [1]. For a tablet containing 20 mg of drug with a solubility of 0.1 mg/mL, saturation solubility would occur in 200 mL; thus, a dissolution test to provide sink conditions at 5× the solubility would require a total volume of 1000 mL. Historically, sink conditions were strongly recommended in dissolution testing to provide reproducible conditions. However, to better replicate *in vivo* conditions and to avoid the risk of under-discrimination, it may be preferable to select conditions which approach saturation solubility, as long as at least 80% of the drug dissolves during the test.

6.3   History of Dissolution Testing

The originators of dissolution theory were Noyes and Whitney whose experiments in the 1890s provided the foundation for the equation that bears their name. Their work used a simple setup where conditions were standardised to record the rate of solubilisation of poorly soluble compounds [2]. They standardised the surface area for dissolution (2 cm diameter circle), the volume of dissolution media (100 mL), the temperature (25 °C) and rotation speed (60 rpm). Samples were taken at 10, 30 and 60 minutes. The legacy of this work was the contextualisation of the relationship between dissolution rate and saturation solubility in the Noyes–Whitney equation.

$$\frac{dC}{dt} = k(C_s - C)$$

where k is a constant, C is the concentration at time t and C_s is the saturated solubility.

The relevance of their work to pharmaceutical products only began to be exploited in the 1950s by scientists at Beecham Laboratories who recognised that dissolution within the gastrointestinal tract was the rate limiting factor for orally administered aspirin [3]. It was noted that the crystals that remained following the disintegration of an aspirin tablet would dissolve at a rate that depended on temperature, pH of solvent and degree of agitation [3]. This work recognised that the disintegration step resulted in particles rather than a cylindrical surface area for dissolution, as well as the complexities in analysing a compound subject to degradation by hydrolysis. The resulting data were presented in a dissolution curve similar to dissolution data presentation today. The experiments conducted on aspirin tablets were done at 37 °C to better represent body temperature. The pH was varied from 1.8 to 9.2, and it was recognised that the gastric pH would be from 1 to 3 whereas the duodenum and jejunum were considered to have pH values from 6 to 8. The volume of fluid was discussed in this paper suggesting that the solubility of aspirin at pH 1.8 would require a volume of 120 mL for complete dissolution; thus, a recommendation to follow each five-grain tablet with half a pint of water. The paper further discussed the absorption of the dissolved drug within the intestine where the volume of fluid within the intestine was linked to the surface area of the intestine and the concentration gradient to drive absorption [3]. This work highlighted key features of dissolution testing including solvent, agitation rate and temperature as critical to understanding the rate of release from a tablet.

The Noyes–Whitney equation was further developed and the constant k was found to relate to the exposed surface area of the dissolving drug (S), the rate of stirring and the temperature [4]. The mass-transfer model of diffusion reveals that the diffusion coefficient (D) of the drug, an inherent drug property, also affects the dissolution rate. At the particle scale, a concentration gradient was described, from the solid drug particle surface, through a thin film or unstirred layer of dissolving drug, to the bulk fluid where the drug is homogeneously dissolved. The thickness of this unstirred layer is given the term h. Thus, the revised Noyes–Whitney equation can be written:

$$\frac{dC}{dt} = \frac{DS}{h}\left(C_s - C\right)$$

The link between dissolution and pharmacokinetics was first reported in 1957 [5]. Comparative dissolution of theophylline salts was conducted to compare the data found by 'clinical workers comparing blood levels'. This divergence between the prototype biopharmaceutical scientists in the 1950s to their clinical colleagues is still in place in many companies today. It was noted that despite having equivalent doses of theophylline, blood levels from the choline salt were higher than from the ethylenediamine salt; thus, dissolution was used to better understand this finding. The description of the testing apparatus reveals that a disk of pure drug was formed and only a circular surface exposed; dissolution was conducted in a 600 mL beaker with a stirrer at 500 rpm. The media was circulated at a rate of 250 mL/min. An alternative dissolution apparatus with a 500 mL volume at 25 °C was also used. Dissolution media included 0.1 N HCl as well as buffers at pH 6.7 and 8.87. The dissolution rate of the salts was found to correlate to the relative blood levels previously reported. This demonstrated the 'predictive' potential of biorelevant dissolution testing.

An explicit link to formulation properties, dissolution and gastrointestinal absorption was reported by Levy in 1964 observing that patients needed to be 'titrated to a brand of tolbutamide and could not simple switch brands without subsequent re-titration [6]. This dissolution method used 200 mL of pH 7.0 buffer at 37 °C. Within this paper, there was a recommendation that pharmaceutical manufacturers use dissolution to assure batch-to-batch reproducibility and that an official dissolution rate standard for tolbutamide tablets should be introduced. This paved the way for dissolution to become a compendial method and a critical tool in the development of orally administered products.

6.4 Compendial (Pharmacopeial) Dissolution Apparatus

6.4.1 USP1 and 2 Apparatus

Dissolution testing was introduced as part of the monograph for six products within the US Pharmacopoeia in 1971. The first method was the basket/stirred vessel test (USP Apparatus 1) with the paddle method (USP Apparatus 2) being introduced in 1978. The design of this apparatus has not changed since its introduction and is still a major part of pharmaceutical product development for oral dosage forms. The design is subject to specifications that are detailed in the US, European and Japanese pharmacopoeias; methodology was harmonised via the ICH Q4B guidelines in 2006 [7].

In brief, the apparatus consists of a glass (or other inert transparent material) cylindrical vessel with a hemispherical bottom of usually 1 L capacity; this vessel is heated to 37 °C

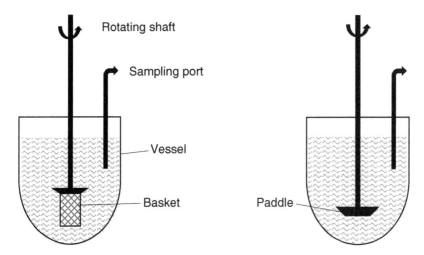

Figure 6.1   *Cartoon of USP dissolution apparatus; type 1 on the left with a basket and type 2 on the right with a paddle.*

with a cover to limit evaporation. A shaft with a basket is inserted centrally into the vessel that is stirred at a fixed rate. A sampling port enables samples to be collected at predetermined time points. A cartoon of the apparatus is shown in Figure 6.1.

The rotation speed and temperature of liquid are precisely controlled within this apparatus and calibration is essential for regulatory use of USP1 and 2 apparatus.

USP1, the basket apparatus is used for testing capsules or tablets, particularly in cases where these may float. The basket ensures that the product is submerged in the dissolution media to maintain good contact. The mesh in the basket is required to have 0.22–0.31 mm wire diameter with wire openings of 0.36–0.44 mm; however, in certain cases finer or coarser mesh is used.

In USP2, the paddle apparatus, the tablet is dropped into the vessel and will settle at the low point in the hemispherical shape. However, due to the poor hydrodynamics within the vessel, coning of the undissolved tablet material as it disintegrates can be an issue. Coning refers to the insoluble material forming a cone at the base of the dissolution vessel under the paddle (at the centre base of the hemispherical vessel); this zone is poorly mixed and can lead to incomplete dissolution, particularly for products with high amounts of insoluble excipients. Although coning is more common in the paddle apparatus, it can also be present where fragments fall through the mesh of the basket apparatus. A peaked vessel can be used to overcome this limitation; in this case, the vessel has an inverted peak at the centre base that prevents cone formation [8].

Sinkers, which are inert materials, can be attached to dosage forms that may otherwise float in the USP2 apparatus. Typically, these are fine wire helices that wrap around the dosage form.

Although the volume of the vessel in USP1 and 2 apparatus is 1 L, the volume of media used is often 900 mL or in some cases 500 mL. There are adaptations that enable larger vessels to be used (up to 4 L) and smaller vessels (down to 100 mL) with geometry-matched baskets and paddles.

6.4.2   USP3 Apparatus

USP3 reciprocating cylinder dissolution apparatus was introduced in the USP in 1991 [9]; it is mechanically similar to a disintegration tester with motion in a vertical plane, shown as a cartoon in Figure 6.2. This apparatus consists of a reciprocating cylinder that contains the dosage form under test; this cylinder has a mesh base such that as it is dipped into the dissolution fluid the dosage form is agitated within the media at a fixed rate of dips per minute. The up-down motion occurs over a stroke distance of 10 cm and the dosage form remains completely immersed within the dissolution media. The entire apparatus is maintained at 37 °C to replicate body temperature. The apparatus is set up such that there are sequential vessels that can each be filled with a unique dissolution media, useful to mimic gastrointestinal transit. Typically, vessels contain ~250 mL of dissolution media. This use of sequential vessels can overcome solubility limitations for certain products and can also replicate the change in composition of fluids as the product moves through the gastrointestinal tract.

   This method was established as an advance to the pre-existing rotating bottle method where the dosage form moves through the dissolution media as the bottle is rotated [10]. It was designed to address issues with the hydrodynamics associated with USP1 and 2 apparatus where it is known that the bulk media is poorly mixed. It was thought that movement

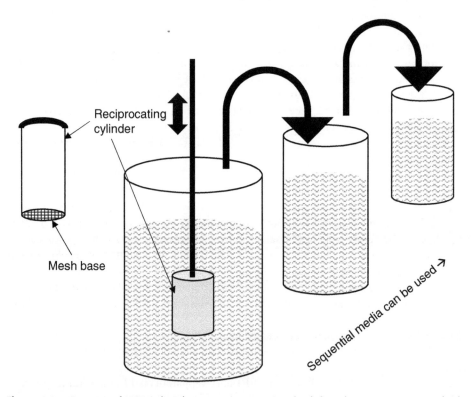

Figure 6.2   *Cartoon of USP3 dissolution apparatus. On the left is the reciprocating cylinder unit with a mesh at the base; the right-hand side shows the sequential nature of the unit where multiple media can be used with this equipment.*

Table 6.1   *Example experimental setup within USP3 apparatus to mimic fasted state.*

Row	GI region	pH	Speed (dips/min)	Duration (min)
1	Stomach	1.6	8	60
2	Small intestine (proximal)	6.5	4	40
3	Small intestine (medial)	6.8	4	60
4	Small intestine (distal)	7.4	4	80
5	Ascending colon	7.8	4	>120

Source: Modified from Andreas et al. [12].

of the dosage form rather than the media would provide more reliable conditions for testing. It also eliminated the coning phenomena observed in USP2 apparatus. More recently, it has been used together with biorelevant media to simulate gastrointestinal transit [11].

Typical conditions for testing within USP3 apparatus to mimic gastrointestinal transit are listed in Table 6.1.

6.4.3 USP4 Apparatus

The USP4 flow-through method was introduced in the USP in 1995. This apparatus consists of a flow-through cell that houses the dosage form; glass beads are used at the entry to control the fluid flow and a filter is used to prevent drug loss at the exit, as shown in Figure 6.3. Fluid flows through the cell at a carefully controlled flow rate, typically in a range of 240–960 mL/h and can be operated as an open or closed system. In the closed system, the fluid is recirculated throughout the test whereas in the open system fresh media

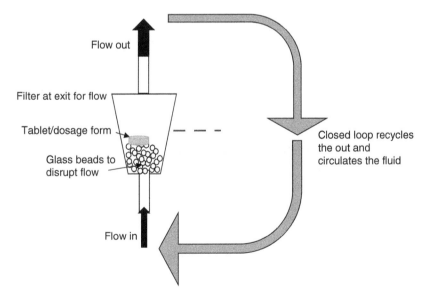

Figure 6.3   *Cartoon of USP4 apparatus. This can be run in an open system or as a closed loop.*

is used throughout the experimental process. This apparatus offers advantages that include the ability to assess the dissolution of low solubility compounds without reaching sink conditions (in open loop mode), the ability to use very low dissolution volumes (in closed loop mode), and increased discrimination by providing relatively gentle agitation in comparison to the other commonly used methods. There is also potential to change the composition of the media under flow to allow for a pH (or other) change during the same experimental process; in this case, the system has to be run in the open rather than the closed loop mode. The apparatus is temperature controlled so that it can be run at 37 °C.

6.4.4   USP5 Apparatus

USP5 apparatus describes a paddle over disk apparatus that was introduced to evaluate transdermal products. It has a similar design to similar to USP2. In USP5, the transdermal patch is secured to a disk at the base of the vessel to minimise the effect of the dead zone beneath the plate and the base of the vessel. The temperature is 32 °C rather than 37 °C to better reflect skin temperature. The media volume can be up to 900 mL, although smaller volumes are usually used.

6.4.5   USP6 Apparatus

USP6 apparatus is a variant of USP1. In this case, the basket is replaced with a stainless-steel cylinder. It is typically used to test topical or transdermal products. The temperature is 32 °C rather than 37 °C to better reflect skin temperature.

6.4.6   USP7 Apparatus

USP Apparatus 7 is similar in appearance to USP3. It evolved to test transdermal and other modified release products including osmotic pumps, beads and arterial stents. The volume within the vessels is smaller than that found in USP3, typical volumes are 50 mL (compared to 250 mL in USP3). The stroke distance is shorter at 2 cm compared to 10 cm for USP3 and the dip rate is typically in the range 5–40 dpm. If used for dermal/topical products, the temperature is 32 °C rather than 37 °C to better reflect skin temperature. Even smaller volume apparatus for USP7 are also available.

An overview of USP dissolution apparatus and the products they are used to evaluate is shown in Table 6.2.

6.4.7   Intrinsic Dissolution Rate (IDR) Apparatus

IDR determination can be used in the selection of optimal form or version (e.g. salts) of the drug substance to be used within a formulation. In this test, it is pure drug substance that is evaluated rather than the formulated drug product. Intrinsic dissolution can be valuable in comparing the relative dissolution rates of polymorph or salts of a given drug [13, 14].

The intrinsic dissolution rate (IDR) is defined as the amount of drug dissolved per unit time per unit area and is expressed as a rate, in contrast to the thermodynamic equilibrium reported by solubility data of the pure drug substance.

Apparatus for IDR is described by pharmacopoeias and includes the rotating disk system (or Wood's apparatus); the rate of release into an aqueous solution at a given pH value can then be measured [14]. This is a modified USP1/2 setup whereby a disc of drug with a

Table 6.2 *Overview of USP dissolution apparatus (ER = extended release; IR = immediate release).*

USP apparatus	Description	Rotation speed/ dip speed	Typical dosage form
Type 1	Basket apparatus	50–200 rpm	Immediate release (IR) tablets and capsules, chewable tablets, ER tablets and multiparticulates
Type 2	Paddle apparatus	25–100 rpm	IR capsules and tablets, chewable tablets and ER tablets
Type 3	Reciprocating cylinder	4–35 dpm	ER tablets/granules
Type 4	Flow through	N/A	ER tablets, poorly soluble drug containing tablets, granules and multiparticulates, implants
Type 5	Paddle over disk	25–100 rpm	Transdermal
Type 6	Cylinder	N/A	Transdermal
Type 7	Reciprocating holder	5–40 rpm	ER (non-disintegrating) and transdermal

known surface area is rotated in a dissolution vessel and the concentration at set time points measured. The shaft within the apparatus is replaced with one that holds the disk of drug such that the surface area of exposed drug is kept constant.

6.4.8 Micro-dissolution Apparatus

Solubility and IDR data are often required early in the development process where there may be limitations on the amount of drug available. Miniaturised systems are available where data can be generated using milligram quantities of drug (e.g. ~5 mg instead of ~500 mg) [15]. These systems may use UV fibre optic dip probes to read concentration over time rather than taking samples as the volume loss would be significant in these small volume systems [16].

Small-scale adaptations are available for the USP1 and 2 apparatus to reduce the volume to 100 mL in cases (i) where drug may be in short supply; (ii) where analytical methods are insensitive or (iii) to replicate paediatric populations.

6.5 Dissolution Media Selection

Before initiating the development of a dissolution test for quality control purposes, the solubility of the drug should be considered to understand whether the drug solubility will limit dissolution.

Surfactants may be added to the dissolution media to improve the solubility and enable dissolution of >80% of the drug.

Surfactants are amphiphilic molecules that possess both polar (hydrophilic) and non-polar (hydrophobic) regions, which when dissolved in aqueous solutions at concentrations above their critical micelle concentration (CMC) will self-assemble into aggregates. These aggregates minimize the contact between water and the hydrophobic regions, and can aid in the solubilisation of hydrophobic drugs to increase the dissolution rate and improve the solubilisation of the drug within the media. Sodium dodecyl sulphate (SDS) is commonly used as a surfactant in dissolution media; however, SDS can lead to interactions with some drugs. Polysorbate 80 is another popular surfactant and is the surfactant of choice in Japan.

Large-scale vessels are an alternative to surfactant addition where drug solubility is low. Two- and four-litre dissolution vessels are available for USP1 and 2 apparatus. Flow-through systems (USP4) and the reciprocating cylinder apparatus (USP3) can also be used to change volumes more flexibly if needed, and may be considered for extended release formulations.

Co-solvents have also been used historically to generate improve drug solubility, although these are increasingly an uncommon approach as it is unlikely to be acceptable to regulatory bodies, with the notable exception of alcohol, which has been recommended as a co-solvent specifically to study formulation robustness when taken with alcoholic drinks.

6.5.1   Biphasic Dissolution Media

Biphasic dissolution media uses a layer of an organic solvent, such as octanol, that is immiscible with the aqueous layer effectively creating a two-phase dissolution test. Drug dissolution occurs within the lower aqueous layer and then the solubilised drug diffuses into the upper organic layer, preventing saturation of solubility within the aqueous phase. However, despite the appeal of a biphasic system in providing a mimic of drug absorption in the intestine, these test methods also add complexity, and are typically incompatible with surfactant containing media. It is also important to consider how the setup affects the agitation and other aspects of the dissolution test [17]. In addition, there are hazards associated with the use of octanol as it is an irritant with a penetrating aromatic odour.

6.6   Dissolution Agitation Rates

The rate of agitation (e.g. stirrer speed, dip rate) is known to impact dissolution rate; the more efficient the mixing, the quicker the rate of dissolution, as the unstirred layer that surrounds the dissolving particle will be reduced. Hydrodynamic conditions are dictated by the apparatus and the experimental procedure selected. The type of formulation can also vary in sensitivity to agitation/hydrodynamics. For instance, erodible dosage forms are typically much more sensitive to agitation compared to a rapidly disintegrating formulation; thus, it is important to consider the mechanistic aspects of a formulation within the choice of dissolution test.

Dissolution apparatus has closely controlled agitation rates either via rotation of the paddle or basket or by controlling the dip rate or fluid flow. The poor mixing within USP1 and 2 apparatus has been reported previously, specifically relating to the formation of a cone at the base of the round-bottomed vessel (see Section 6.4.1) [18].

USP type 1 and 2 apparatus does not provide a 'well stirred tank' model and there are known differences in the mixing rates within and across the vessel [19]. The mixing is more intense close to the paddle/basket. Therefore, it is critical that the sampling points are consistently and uniformly positioned. The impact of this non-uniform mixing is exacerbated when the dissolution media is more viscous [20].

Typical ranges of agitation rates used for each USP apparatus are listed in Table 6.2. The hydrodynamics in Type 3 apparatus is controlled by both the dip rate and the dimensions of the mesh used within the apparatus [21]. Within USP4 apparatus, the flow rate, dimensions of the flow cell and configuration of glass beads can all affect the hydrodynamics [22]. In addition, the orientation of the tablet within the cell has been demonstrated to have an effect on the dissolution observed [23].

Note that none of the pharmacopeial apparatus are specifically designed to mimic *in vivo* dynamics, which are more complex, intermittent, variable in intensity and driven by peristaltic movement.

6.7   Reporting Dissolution Data

Dissolution data are typically reported as the percentage of drug from within the dosage form that has been released cumulatively with time. For immediate release (IR) products, this typically results in a non-linear (first-order) relationship. Comparison of dissolution data, especially when determining if a dissolution profile is inadequate, needs careful consideration, as this can be used to imply non-equivalence, and may inform dissolution product specifications. Studies in the 1950s reported the time taken for 50% of the drug to be released as a comparative measure for dissolution data [3]. Nowadays, the so-called Q value is typically used for a single-point dissolution test, representing the quantity or amount of dissolved drug, expressed as a percentage of the label content of the dose unit. Within monographs for IR products, this is typically set at 85% released within 15–45 minutes.

Sometimes, there is a need to compare dissolution more accurately, for example, for BCS biowaiver determinations (Chapter 9) [24], for extended release products or those with a narrow therapeutic index. Often, there is a need to know two profiles are similar over the entire time course rather than at a single time point. There is also a need to consider the impact of a difference, what should be considered statistically significant, when is the difference clinically significant and how to verify correlation to *in vivo* data.

The f_1/f_2 comparison was introduced by Moore and Flanner [25]. These parameters compare two dissolution profiles nominally termed the reference (R) and test (T) compared over time (t).

$$f_1 = \left(\frac{\sum_{t=1}^{n} Rt - Tt}{\sum_{t=1}^{n} Rt} \right) \times 100$$

$$f_2 = 50 \times \log \left(\sqrt{\left[1 + \frac{1}{n} \sum_{r=1}^{n} wt \left(Rt - Tt \right) \right]} \times 100 \right)$$

f_1 is the difference factor and f_2 the similarity factor; these are derived from average absolute differences (for f_1) and mean squared differences (for f_2).

f_1 provides information on the cumulative differences between two dissolution curves at all timepoints; it relates to the average absolute difference. The greater the value of f_1, the more different the profiles are; values close to 0 indicate similarity, and for dissolution comparison, values less than 15 are typically considered to be similar.

f_2 is based on the reciprocal of mean square root transformed differences in data points at all time points; it relates to the overall similarity where a value of 100 would indicate identical dissolution profiles. However, due to the nature of dissolution testing apparatus, there is some instrumental variability making identical profile unlikely even for tablets or capsules from the same batch, thus f_2 values greater than 50 are regarded as similar. An f_2 value of 50 is equivalent to an average difference of 10% between two dissolution profiles.

As f_1 and f_2 test differences between the curves, it is important that sufficient data points are collected prior to the plateau observed as dissolution approaches complete release. It is recommended only one data point is included after 85% release and that there are at least three data points in total. Due to the inherent variability in dissolution, the mean data from 12 unit should be used and specific regulatory guidelines are detailed [26].

There are some excellent papers on statistical testing and dissolution if further information is required [27, 28].

Another mathematical tool, the dissolution number (D_n) can be a useful parameter for describing drug dissolution [29]. It is defined as the ratio of mean residence time in the gastrointestinal tract to mean dissolution time. Rapidly dissolving products will have a high dissolution number whereas those with a low dissolution number show slow dissolution. D_n can be used to optimise a formulation's release rate as the dissolution number relates to the drug's solubility (C_s), diffusivity (D), density (ρ) and particle size radius (r_0). The intestinal transit time T_{si}, which is usually assumed to be 199 minutes, is approximately the time taken for unabsorbed material in the GI tract to transit the small intestine. This approach is especially useful to show the impact of drug particle size on dissolution and why control of particle size can be critical for some low solubility drugs.

$$D_n = \left(\frac{3D}{r_0^2} \right) \left(\frac{C_s}{\rho} \right) (T_{si})$$

6.8   *In Vitro In Vivo* **Relationships and Correlations (IVIVR/IVIVC)**

Demonstration of a relationship/ correlation between *in vitro* dissolution data and *in vivo* clinical data (typically C_{max} and/or AUC) is the ultimate goal of much dissolution experimentation. A proven IVIVC can negate the need for costly *in vivo* studies and thus speed up the development timeline for a new product or avoid the need for human studies to demonstrate bioequivalence.

The theoretical basis for an IVIVC to link *in vitro* dissolution to *in vivo* pharmacokinetic data may be used in cases where dissolution is the rate-limiting step to absorption. This is most likely to be the case for ER dosage forms. The underlying assumption is that once a drug is dissolved, it will be absorbed; meaning that the percentage dissolved at any given time can predict the percentage absorbed. To generate these data require a pharmacokinetic study with intensive sampling during the absorption phase to compare to the dissolution data set. For drugs where the relationship between the fraction dissolved and the fraction absorbed is more complex, e.g. where absorption rate varies with time, or a drug undergoes degradation in the GI tract, more sophisticated tools (e.g. PBPK modelling) may be required to describe the link between the dissolution and the pharmacokinetics.

6.8.1 Convolution and Deconvolution of Dissolution Data

Convolution is the process of obtaining/predicting a pharmacokinetic profile for a drug from the dissolution data, whilst deconvolution is the opposite i.e. extracting the dissolution profile from the pharmacokinetic data.

Deconvolution of *in vivo* pharmacokinetic data to provide theoretical dissolution rates was introduced in the 1960s [30]. These models assumed a single compartment pharmacokinetic model where dissolution was the rate-limiting step. Mathematically, this is complex, although there are several software packages that can help. The complexity arises as drug dissolution is a function of the solubility of the drug and the product properties whereas the pharmacokinetic elimination is a factor associated with the drug alone. Often it is assumed that dissolution and absorption are linearly related, yet there is inevitably elimination during the absorption phase of a pharmacokinetic profile and accounting for this within the deconvolution is the complicated aspect.

Convolution techniques can be more simply applied where the pharmacokinetic parameters of volume of distribution and oral bioavailability are known; these methods can be used to design in input rate via dissolution to provide the desired pharmacokinetic profile. When designing controlled/modified release dosage forms, convolution methods are commonly applied to dissolution data. Examples of reports of convolution are available [31, 32].

Categories of IVIVC were introduced by the FDA for application to extended release oral dosage forms in 1997 [33]. IVIVCs require that human PK data to be generated on formulation variants with significantly different release rates.

For level A correlation, the fraction of drug absorbed is plotted against the fraction of the dose dissolved to establish a point to point relationship. This is the most preferred IVIVC but is also the most challenging to establish. A level A IVIVC demonstrates that the dissolution rate *in vivo* can be reliably predicted from the *in vitro* dissolution profile. This type of correlation is more likely to be demonstrated for a modified release product. A simple representation is provided in Figure 6.4.

Level B IVIVC uses the mean dissolution time to compare to the mean *in vivo* GI residence time or mean *in vivo* dissolution time. Rather than showing a time-matched linear relationship as in level A, the level B uses the entire data set but as a single comparator.

Level C uses a single-point comparison, for example, the percent dissolved at time *t* is compared to a PK parameter, e.g. C_{max}, or AUC (see Chapter 2). In some cases, several timepoints are used to correlate to different PK parameters, but unlike the level A IVIVC, correlation between the entire *in vitro* and *in vivo* profile is not established.

In all cases, the choice of dissolution testing method is critical to ensure that it provides discriminatory data that show differences in the formulation variants, to develop the correlation to *in vivo* PK data.

Given the complexity of the drug absorption process, robust IVIVCs are challenging to achieve in practice. Increasingly, so-called IVIVE (*in vitro–in vivo* extrapolation) using PBPK modelling are being used as these models can capture the multiple factors that influence the relationship between dissolution and pharmacokinetics [34].

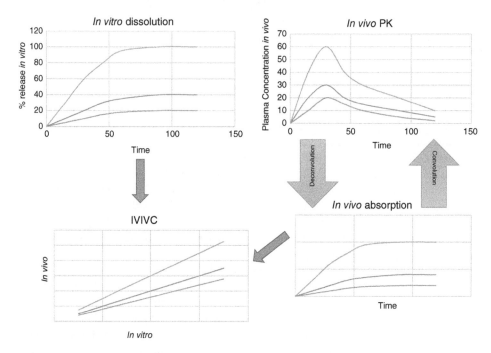

Figure 6.4   *A schematic overview of a level A IVIVC.*

6.9   Evolution of Biorelevant Dissolution Testing

Biorelevant dissolution is used to deliberately mimic the conditions available for dissolution to prospectively predict likely *in vivo* performance. In contrast, biopredictive, or clinically relevant dissolution describes the scenario where *in vivo* performance is demonstrated to be deducible from *in vitro* dissolution, irrespective of the biorelevance of the media used. However, the use of an appropriate level of biorelevance will significantly increase the probability of bioprediction [35]. Therefore, the starting point for dissolution to predict the *in vivo* performance of products is to ensure that the testing conditions appropriately match those found *in vivo*. Even in the first dissolution tests described in the 1950s–1960s, efforts were made to replicate the different pH values (and to some extent, volumes) found in the stomach and small intestine. However, additional factors can be critical for *in vivo* dissolution: the media in the gastrointestinal tract; the dynamic nature of gastrointestinal tract and the absorption process; i.e. the impact of drug removal via permeability though the intestinal mucosa. These factors are discussed separately below.

Biorelevant dissolution is continually adapting to incorporate the latest data on gastrointestinal conditions to best reflect the human/patient.

6.9.1   Biorelevant Dissolution Media

As stated in Chapter 4, it has been demonstrated that replication of the composition of the gastrointestinal media improves the ability to predict *in vivo* performance of drug products.

Table 6.3 Summary of USP compendial dissolution media.

Dissolution media	Composition	Comments
Simulated gastric fluid (SGF)	2 g NaCl 3.2 g purified pepsin 7 mL HCl Water to 1 L pH = 1.2 Osmolality = 180 ± 3.6 mOsm/kg Surface tension 33.7 mN/m	Higher than measured pepsin concentration Higher than measured surface tension [36]
SGF-blank	As above but without the pepsin	
Simulated intestinal fluid (SIF)	6.81 g KH_2PO_4 68.05 g 77 mL 0.2 N NaOH 10 g Pancreatin Water to 1 L pH = 6.8 Osmolality = 113 mOsm/kg	The original medium had a pH of 7.5 yet this was changed to pH 6.8 in 1996 [37]
SIF-blank	As above but without pancreatin	

Source: From USP [38].

The development of simulated intestinal fluids has been the subject of much research due to the majority of medicines being formulated as oral solid dosage forms, where there is a need to replicate the GI conditions to better understand product performance.

Compendial dissolution media that replicate GI fluids may be used as a simplest form of biorelevant media; these are summarised in Table 6.3.

These compendial media reflect the pH of the stomach and small intestine but do not incorporate surfactants that aid in the wetting and/or solubilisation of drugs, nor do they mimic the fed state.

Several attempts have been made to better replicate gastric fluid by characterisation then replication of aspirated human gastric fluid. A recipe for fasted state simulated gastric fluid (FaSSGF) was published in 2005 [39]; this shows good correlation to solubility data from human gastric aspirates. The composition of FaSSGF is provided in Table 6.4.

Similarly, as compendial SIF represents a simplification of intestinal fluid and does not fully reflect physiological conditions, a simulated fasted state intestinal fluid was developed based on characterisation of human small intestinal aspirates [40]. Since the development of the first Fasted State Simulated Intestinal Media (FaSSIF V1) in 1998 [41], there have been at least two further versions reported: FaSSIF V2 [42] and FaSSIF V3 [43]. These updated fluids better reflect the complexity within intestinal fluids by incorporation of bile salts, phospholipids and fatty acids that form colloidal structures (mixed micelles) that can solubilise lipophilic drugs. The composition of each of these fluids is presented in Table 6.5.

In the first two iterations of FaSSIF, sodium taurocholate was used as a representative bile salt present due to ease of availability and cost. The second version was introduced to improve the stability of the fluid and better reflect the composition within the intestine [42].

Table 6.4 *Composition and physicochemical characteristics of FaSSGF.*

Component	Concentration
Sodium taurocholate	80 μM
Lecithin	20 μM
Pepsin	0.1 mg/mL
NaCl	34.2 mM
HCl conc.	qs ad 1.6
Deionized water	Ad 1.0 L
pH	1.6
Osmolality	120.7 ± 2.5 mOsmol/kg
Surface tension	42.6 mN/m

Source: Modified from Vertzoni et al. [39].

Table 6.5 *Composition and physicochemical characteristics of three versions of FaSSIF.*

	FaSSIF V1	FaSSIF V2	FaSSIF V3
Cholesterol (mM)	—	—	0.2
Lecithin (mM)	0.75	0.2	0.035
Lysolecithin (mM)	—	—	0.315
Sodium glycocholate (mM)	—	—	1.4
Sodium oleate (mM)	—	—	0.315
Sodium taurocholate (mM)	3	3	1.4
Hydrochloric acid			
Maleic acid (mM)	—	19.12	10.26
Potassium dihydrogen phosphate (mM)	28.65	—	—
Sodium chloride (mM)	105.85	68.62	93.3
Sodium hydroxide (mM)	10.5	34.8	16.56
pH	6.5	6.5	6.7

Source: Based on Dressman et al. [41], Jantratid et al. [42], and Fuchs et al. [43].

In FaSSIF v2, the osmolality is somewhat lower to reflect *in vivo* data and the buffer system was switched from phosphate to maleate [42, 44]. The updates in FaSSIF v3 account more accurately for the amount and type of micelle-forming components, as these have a significant influence on the solubility of highly lipophilic drugs. The recipe reported was better able to match the solubility of drugs in aspirated human intestinal fluid. FaSSIF v3 also has a lower surface tension compared to previous versions which is considered to better aid in the wetting of poorly soluble drugs [43]. Recent work has compared all three variants and suggests that the FaSSIF v1 giving the best correlations to aspirated human intestinal fluid [45]. It is thus important to understand how biorelevant media affects the solubility and to understand, particularly for poorly soluble drugs, how well the simulated media reflects the micellar solubilisation to obtain the best prediction.

Phosphate buffers are widely used in dissolution testing as buffers due to their robust buffer capacity at the pH of intestinal fluids [46]. However, the common ion effect can lead to precipitation of certain drugs, so alternative buffer systems can be employed; in FaSSIF versions 2 and 3, the phosphate buffer is replaced by the maleic acid. The use of a bicarbonate buffer has been proposed to better match the relatively low buffer capacity seen in intestinal aspirates; however, this requires special conditions to prevent oxidation that make testing complex [47].

There is a great deal of literature evidence that demonstrates that lipophilic drug solubility is increased in biorelevant fluids that contain bile salt micelle components compared to in buffers alone, an overview is provided by Bou-Chacra et al. [46].

The fed state introduces further changes to the gastrointestinal environment. For instance, there is an increase in pH in the stomach, a slight reduction in pH in the duodenum, and intestinal buffer capacity and osmolality show a sharp increase. The secretion of bile is stimulated leading to increased concentrations of solubilising components [40]. Preliminary work used milk as a prototype media to represent the gastric fed state [48]. Subsequently, more complex gastric media were developed to represent gastric media at different time points after high-calorie meal intake, see Table 6.6.

Fed-state-simulated intestinal fluid (FeSSIF) was introduced in 1998 by Galia et al [40]. The major differences between FaSSIF and FeSSIF are the buffer system which is a phosphate buffer in FASSIF versus an acetate buffer in FeSSIF, the pH and the higher concentration of taurocholate and lecithin in FeSSIF. Due to the changing nature of the fed state, three snapshot media were created to better replicate the timeframe post ingestion. An update, FeSSIF v2 was introduced subsequently as a composite of the three 'snapshot' media where the second-generation FeSSIF contained two digestion products: glyceryl monooleate and sodium oleate compared to the original FeSSIF [42]. The composition of the FeSSIF media is shown in Table 6.7.

For poorly soluble drugs, the increased bile and lecithin present in FeSSIF compared to FaSSIF can lead to faster dissolution and higher solubility [44], and is the reason why many insoluble, lipophilic drugs show improved absorption in the fed state.

Although most drugs are absorbed within the small intestine, a number are designed to release drug for absorption in the colon; therefore, there is a need for a simulated fluid to represent conditions within the colon. Simulated colonic fluid based on relevant pH values and short chain fatty acid concentrations was introduced by Fotaki et al. [49] and is shown in Table 6.8.

Table 6.6 *Composition and physicochemical characteristics of FeSSGF.*

Composition	FeSSGF = middle	Early FeSSGF	Late FeSSGF
NaCl (mM)	237.02	148	122.6
Acetic acid (mM)	17.12	—	—
Sodium acetate (mM)	29.75	—	—
Ortho-phosphoric acid (mM)	—	—	5.5
Sodium dihydrogen phosphate (mM)			32
Milk/acetate buffer	1 : 1	1 : 0	1 : 3
HCl/NaOH (qs ad).	pH 5.0	pH 6.4	pH 3
pH	5.0	6.4	3.0
Osmolality (mOsmol/kg)	400	559	300
Buffer capacity (mEq/pH/L)	25	21.33	25

Source: Modified from Jantratid et al. [42].

Table 6.7 *Composition and physicochemical characteristics of FeSSIF media published in the literature.*

Composition	FeSSIF	FeSSIF Early	FeSSIF Middle	FeSSIF Late	FeSSIF-V2
Bile salt (taurocholate) (mM)	15.00	10.00	7.50	4.50	10.00
Phospholipid (lecithin) (mM)	3.75	3.00	2.00	0.50	2.00
Acetic acid (mM)	144.00	—	—	—	—
Sodium chloride (mM)	173.00	145.20	122.80	51.00	125.50
Maleic acid (mM)	—	28.60	44 50	8.09	55.02
Sodium hydroxide (mM)	101.00	52.50	65.30	72.00	81.65
Glyceryl mono-oleate (mM)	—	6.50	5.00	1.00	5.00
Sodium oleate (mM)	—	40.00	30.00	0.80	0.80
pH	5	6.5	5.8	5.4	5.8
Osmolality (mOsmol/kg)	635.00	400.00	390.00	240.00	390.00 ± 10
Buffer capacity (mEq/pH/L)	76.00	25.00	25.00	15.00	25.00

Source: Modified from Jantratid et al. [42].

Table 6.8 *Composition and physicochemical characteristics of simulated colonic fluid (SCoF).*

Component	Value
Acetic acid (mmol)	170
NaOH (mmol)	~157
Distilled water (ml)	qs 1000
pH	5.8
Osmolality (mOsmol/kg)	295
Buffer capacity (mEq/L/pH)	29.1
Ionic strength	0.16

Source: From Fotaki et al. [49] / With permission of Elsevier.

Overall however, for targeted release formulations, the choice of media depends upon the major sites for drug release and should be selected based on the formulation design.

6.9.2   Dissolution Testing to Mimic GI Transit

Following ingestion of an orally administered product, the stomach is the first significant location for dissolution. After gastric emptying, the drug/dosage form enters the small intestine, where most of the absorption of dissolved drug typically occurs. Finally, any undissolved/unabsorbed material transits to the colon, which although has much reduced fluid volumes, can be an important region for drug absorption for ER formulations prior to elimination. Even within the transit through each of these GI compartments, there are significant changes over time and location which alter drug dissolution.

Efforts have been made to replicate GI transit within dissolution testing to better mimic the passage of an orally administered product, the dynamic element of dissolution, using multi-compartmental models. Transit from the stomach to the small intestine is the most obvious environmental change a dissolving drug/dosage form is exposed to. The significance of this transition for the behaviour of formulations containing pH-dependent excipients, such as delayed release dosage forms has long been recognised, and methods to mimic this transition are recommended in pharmacopoeia [38].

This change from gastric to intestinal environment is also a key step for many IR drug formulations containing poor solubility weak acids and weak bases. Dissolution in the intestine can be significantly altered by initial behaviour (e.g. form changes) in the stomach. A dual media, two-step media addition method for the evaluation of low solubility drugs, using biorelevant micellar components, using a paddle method has been published to harmonise a biorelevant test method for use in industry [50].

USP3 reciprocating cylinder apparatus was introduced to the USP in 1991 to evaluate the dissolution of modified release products [9]. Some of these products were designed to release specifically in one region of the GI tract, for example, using a pH-dependent coat to target drug release to a specific region of the intestine. The USP3 apparatus consists of a series of vessels which can contain different fluids that each represent a unique environment meaning that the dosage form is exposed to multiple simulated environments over time. The ability to sequentially alter the pH, molarity, viscosity and surfactant concentration provides improved replication of the *in vivo* conditions [51]. Both fasted and fed state transit can be considered [52]. However, this method is mostly only suitable for non-disintegrating dosage forms as transfer relies on maintaining the dosage form in a dip tube above a mesh, which is used to transfer the dosage form into the series of fluids.

USP4 flow through has also been used with a range of media to replicate transit within the GI tract [49]. This apparatus design enables changes in media by switching of the input fluid and is therefore useful for the evaluation of modified release dosage forms. The flow-through apparatus has somewhat greater flexibility to add smaller and disintegrating dosage forms than the reciprocating cylinder apparatus (USP3).

A further level of sophistication in mimicking GI transit can be introduced by considering transit, not as a series of instantaneous media changes, but continuous transfer over time. This is important for stomach to intestinal transfer of some immediate release drugs, such as poorly soluble weak bases, which may supersaturate under intestinal conditions following dissolution in gastric conditions. A simple dissolution-transfer model based on USP type 2 apparatus was published in 2004 [53]. This model simulated drug transfer from the stomach into the small intestine where representative gastric and small intestinal media was used to assess dissolution. More sophisticated transfer models have since been developed, such as the BioGIT apparatus, which mimics transfer from the gastric to the small intestine whilst additionally using media from an intestinal media reservoir to maintain *in vivo* relevant fluid volume and composition [54].

Other transfer models available include the artificial stomach duodenal (ASD) model which is a dynamic model where the fluid composition changes to reflect typical *in vivo* conditions in the upper GI tract [55]. It has been used to demonstrate prediction of *in vivo* drug product performance [56].

In more advanced dissolution systems, dynamic transit to match that of the GI tract can be mimicked to better predict dosage form behaviour. The TIM-1 system replicates the dynamic fluid composition in the stomach as well as three sections of the intestine (duodenum, jejunum and ileum) [57]. The motility is also somewhat replicated via the use of peristalsis to promote mixing. Drug can transit through this system and samples are taken to measure dissolution as it progresses over physiological time frames. The TinyTIM apparatus is similar in design, but only has a single intestinal compartment [57].

The SimuGIT® model combines a gastric and small intestinal component [58]. The gastric component is a stirred fermenter where the fluid composition matches that in humans. There is transfer of this fluid to an intestinal compartment where fluid composition is controlled yet motility is not replicated.

6.9.3 Dissolution Testing to Mimic Motility/Hydrodynamic Conditions

GI motility is not homogenous there are periods of high and low agitation within all regions from the stomach to the colon. This is very difficult to replicate *in vitro*. Dissolution in USP1 and 2 apparatus uses a single agitation rate throughout the duration of the test. Within the USP3 and 4 apparatus, there is scope to change the agitation during the test period; however, *in vivo* motility is more complex and varies considerably with the motility cycle of the GI tract as well as with transit. In addition, most simpler dissolution methods do not replicate peristalsis, the motility of most importance to mixing and agitation *in vivo*.

The Reynolds number (Re) is a dimensionless quantity that relates the relative flow of liquid at the surface of a solid. A higher Re number indicates greater flow of fluid. In drug dissolution methods, a higher agitation rate will increase the value of Re within the test system. In the stomach, Reynolds numbers in the range 0.01–30 have been reported based on a 1 cm diameter tablet with a typical gastric viscosity and propagation speed of contraction waves [59]. Reported Reynolds values in USP1 and 2 apparatus are dependent on location in the vessel, but are up to several orders higher, indicating hydrodynamic conditions *in vivo* are not well replicated in these simple apparatus. It is possible to use USP1 or 2 apparatus yet suspend the tablet towards the exterior of the vessel to provide a more representative Re [59], although the significant variations in motility will still not be replicated.

Another advanced dissolution system, the dynamic gastric model (DGM) was designed to replicate physical forces within the stomach, initially to mimic digestion, but subsequently applied to dosage form behaviour. Similar to TIM systems, it also mimics the dynamic nature of the gastric media and transit time [60]. It has been used to demonstrate the impact of forces on the release from oral tablets [61]. A similar biorelevant model that replicates the stomach is the artificial gastric digestive system that was primarily designed to replicate the composition of fluid during digestion yet also replicates mechanical forces within the stomach [62]. An alternative gastric digestion simulator replicates peristalsis and offers a transparent exterior such that the contents can be visualised during motion [63]. The human gastric simulator was designed to simulate the amplitude and frequency of peristaltic movement of the stomach wall and also replicates the gastric fluid composition [64]. A modification to TIM models, the TIMagc, simulates the dynamic conditions of phase dependent physiological motility, whilst retaining the secretion and emptying properties designed for the TIM-1 stomach [65].

Another relatively complex *in vitro* model, the stress dissolution tester, replicates motility with a pattern of irregular movement and forces identified *in vivo* and applies these to oral dosage forms [66]. A simpler method has been patented that can be incorporated into USP2 apparatus where the dosage form is held within the dissolution media in a bespoke device where forces of varying duration and intensity can be applied throughout the test [67]. The stress testing apparatus has been used to improve the prediction of drug release of ER erodible matrices, dosage forms for which it is challenging to mimic the impact of motility, especially food-induced motility [68].

A dynamic colon model was developed to replicate the motility and hydrodynamics within the colon and this has been shown to correlate to *in vivo* motility patterns [69]. This apparatus is also able to incorporate relevant media to represent fluids within the colon.

6.9.4   Dissolution Testing to Incorporate Permeability

Methods that combine dissolution with permeability are of value in biopharmaceutics as these are key interlinked parameters. Biphasic dissolution media can also provide a relatively simplistic mimic of permeability for lipophilic drugs, as the dissolved drug partitions into the organic layer which can mimic the absorption of the drug [70].

In preliminary studies of a more complex combination, samples taken from a conventional dissolution apparatus have been directly administered to Caco-2 cells to assess the permeability [71]. A continuous, in-line system was introduced where samples were pumped directly from the dissolution vessel into a side-by-side diffusion cell where permeation across Caco-2 cells has also been evaluated [72]. This required some adaptation to ensure that the dissolution media was appropriate for direct contact with the Caco-2 cells to support the permeability study. Similar studies have used rat intestinal tissue [73] or artificial hexadecane membrane [74] in place of Caco-2 cells to assess drug permeability. The use of flow-through (USP4) dissolution apparatus can simplify the incorporation of in-line permeability testing [75].

Two miniaturised apparatus, µFlux and µDiss have been reported [76, 77] that can simultaneously measure dissolution and passive permeability. These are different to standard dissolution testing apparatus and are of greatest value for early stage drug evaluation rather than to evaluate dissolution of a drug product. In addition, there are ongoing effort to develop combined systems including at the microscale with lab-on-a-chip and organ-on-a-chip technology [78, 79].

A major limitation in most systems where the incorporation of permeability has been attempted is the very small surface area available for permeability compared to that in the intestine. Hollow fibre membranes have been used to overcome this surface area limitation and may better reflect the ratio of surface area to dissolved media within the GI tract [80]. A system of hollow fibres is used within the TIM systems to incorporate the permeability aspect of this system. However, fibres, like any of the non-biological methods for permeation assessment, will only reflect the passive permeability and do not account for transporter or efflux aspects, which may be critical for some drugs. Other attempts to improve the prediction of permeability through making the relative surface area for permeation greater include the Artificial Membrane Insert (AMI) approach and the Permeapad® technology [81, 82].

In-silico models and PBPK approaches can be used to predict permeability. By combining an *in-silico* prediction of permeability and dissolution data, it may be possible to adequately mimic the interplay between the two; this is discussed in further detail in Chapter 12.

6.10   Conclusions

The historic link of dissolution being the rate-limiting step in the absorption of poorly soluble drugs was established in the 1950s [5]. Compendial dissolution apparatus provides reproducible testing systems that will provide detailed information on batch-to-batch variability yet may not always replicate the *in vivo* system adequately for some drugs and formulations. Advances based on compendial apparatus and using novel systems have provided biorelevant testing systems to better replicate conditions within the GI tract. These biorelevant systems and media have helped in the understanding of oral drug behaviour, and have aided in generating *in vitro in vivo* correlations with clinical data [83].

References

[1] Rohrs, B. (2001). Dissolution method development for poorly soluble compounds. *Dissolution Technologies* **8**: 6–12.

[2] Noyes, A.A. and Whitney, W.R. (1897). The rate of solution of solid substances in their own solutions. *Journal of the American Chemical Society* **19** (12): 930–934.

[3] Edwards, L.J. (1951). The dissolution and diffusion of aspirin in aqueous media. *Transactions of the Faraday Society* **47**: 1191–1210.

[4] Bruner, L. and St. Tolloczko (1900). Über die auflösungsgeschwindigkeit fester körper. *De Gruyter Oldenbourg* **35** (1): 283–290.

[5] Nelson, E. (1957). Solution rate of theophylline salts and effects from oral administration. *Journal of the American Pharmaceutical Association* **46** (10): 607–614.

[6] Levy, G. (1964). Effect of dosage form properties on therapeutic efficacy of tolbutamide tablets. *Canadian Medical Association Journal* **90** (16): 978–979.

[7] ICH (2010). ICH guideline Q4B annex 7 (R2) to note for evaluation and recommendation of pharmacopoeial texts for use in the ICH regions on dissolution test – general chapter. EMA/CHMP/ICH/645469/2008. https://www.ema.europa.eu/en/documents/scientific-guideline/ich-guideline-q4b-annex-7-r2-note-evaluation-recommendation-pharmacopoeial-texts-use-ich-regions_en.pdf (accessed March 2021).

[8] Baxter, J.L., Kukura, J., and Muzzio, F.J. (2006). Shear-induced variability in the United States Pharmacopeia Apparatus 2: modifications to the existing system. *The AAPS Journal* **7** (4): E857–E864.

[9] Esbelin, B., Beyssac, E., Alache, J.M. et al. (1991). A new method of dissolution *in vitro*, the "bio-dis" apparatus: comparison with the rotating bottle method and *in vitro*: *in vivo* correlations. *Journal of Pharmaceutical Sciences* **80** (10): 991–994.

10] Borst, I., Ugwu, S., and Beckett, A.H. (1997). New and extended application for USP drug release apparatus. *Dissolution Technology* **4**: 1–6.

[11] Reppas, C., Vrettos, N.-N., Dressman, J. et al. (2020). Dissolution testing of modified release products with biorelevant media: an OrBiTo ring study using the USP apparatus III and IV. *European Journal of Pharmaceutics and Biopharmaceutics* **156**: 40–49.

[12] Andreas, C.J., Tomaszewska, I., Muenster, U. et al. (2016). Can dosage form-dependent food effects be predicted using biorelevant dissolution tests? Case example extended release nifedipine. *European Journal of Pharmaceutics and Biopharmaceutics* **105**: 193–202.

[13] Etherson, K., Dunn, C., Matthews, W. et al. (2020). An interlaboratory investigation of intrinsic dissolution rate determination using surface dissolution. *European Journal of Pharmaceutics and Biopharmaceutics* **150**: 24–32.

[14] Issa, M.G. and Ferraz, H.G. (2011). Intrinsic dissolution as a tool for evaluating drug solubility in accordance with the biopharmaceutics classification system. *Dissolution Technologies* **18** (3): 6–13.

[15] Avdeef, A. and Tsinman, O. (2008). Miniaturized rotating disk intrinsic dissolution rate measurement: effects of buffer capacity in comparisons to traditional Wood's apparatus. *Pharmaceutical Research* **25** (11): 2613–2627.

[16] Tsinman, K., Avdeef, A., Tsinman, O., and Voloboy, D. (2009). Powder dissolution method for estimating rotating disk intrinsic dissolution rates of low solubility drugs. *Pharmaceutical Research* **26** (9): 2093–2100.

[17] Phillips, D.J., Pygall, S.R., Cooper, V.B., and Mann, J.C. (2012). Overcoming sink limitations in dissolution testing: a review of traditional methods and the potential utility of biphasic systems. *Journal of Pharmacy and Pharmacology* **64** (11): 1549–1559.

[18] Mirza, T., Joshi, Y., Liu, Q., and Vivilecchia, R. (2005). Evaluation of dissolution hydrodynamics in the USP, Peak™ and flat-bottom vessels using different solubility drugs. *Dissolution Technologies* **12** (1): 11–16.

[19] McCarthy, L.G., Bradley, G., Sexton, J.C. et al. (2004). Computational fluid dynamics modeling of the paddle dissolution apparatus: agitation rate, mixing patterns, and fluid velocities. *AAPS PharmSciTech* **5** (2): e31–e31.

[20] Stamatopoulos, K., Batchelor, H.K., Alberini, F. et al. (2015). Understanding the impact of media viscosity on dissolution of a highly water soluble drug within a USP2 mini vessel dissolution apparatus using an optical planar induced fluorescence (PLIF) method. *International Journal of Pharmaceutics* **495** (1): 362–373.

[21] Yu, L.X., Wang, J.T., and Hussain, A.S. (2002). Evaluation of USP apparatus 3 for dissolution testing of immediate-release products. *AAPS PharmSci* **4** (1): E1–E1.

[22] Shiko, G., Gladden, L.F., Sederman, A.J. et al. (2011). MRI studies of the hydrodynamics in a USP4 dissolution testing cell. *Journal of Pharmaceutical Sciences* **100** (3): 976–991.

[23] Shiko, G., Gladden, L.F., Sederman, A.J. et al. (2011). MRI studies of the hydrodynamics in a USP4 dissolution testing cell. *Journal of Pharmaceutical Sciences* **100** (3): 976–991.

[24] ICH (2019). ICH harmonised guideline M9: biopharmaceutics classification system-based biowaivers, questions and answers. https://database.ich.org/sites/default/files/M9_QAsAnnex_Step4_2019_1116_0.pdf (accessed March 2021).

[25] Moore, J. and Flanner, H. (1996). Mathematical comparison of curves with an emphasis on *in vitro* release profiles. *Pharmaceutical Technology* **20** (6): 64–74.

[26] FDA (1997). *Guidance for Industry: Dissolution Testing of Immediate Release Solid Oral Dosage Forms*. U.S. Department of Health and Human Services Food and Drug Administration Center for Drug Evaluation and Research (CDER) https://www.fda.gov/media/70936/download.

[27] Shah, V.P., Tsong, Y., Sathe, P., and Liu, J.-P. (1998). *in vitro* dissolution profile comparison—statistics and analysis of the similarity factor, f2. *Pharmaceutical Research* **15** (6): 889–896.

[28] Duan, J.Z., Riviere, K., and Marroum, P. (2011). *in vivo* bioequivalence and *in vitro* similarity factor (f2) for dissolution profile comparisons of extended release formulations: how and when do they match? *Pharmaceutical Research* **28** (5): 1144–1156.

[29] Oh, D.M., Curl, R.L., and Amidon, G.L. (1993). Estimating the fraction dose absorbed from suspensions of poorly soluble compounds in humans: a mathematical model. *Pharmaceutical Research* **10** (2): 264–270.

[30] Loo, J.C. and Riegelman, S. (1968). New method for calculating the intrinsic absorption rate of drugs. *Journal of Pharmaceutical Sciences* **57** (6): 918–928.

[31] Bose, A. and Wui, W.T. (2013). Convolution and validation of *in vitro–in vivo* correlation of water-insoluble sustained-release drug (domperidone) by first-order pharmacokinetic one-compartmental model fitting equation. *European Journal of Drug Metabolism and Pharmacokinetics* **38** (3): 191–200.

[32] Rastogi, V., Yadav, P., Lal, N. et al. (2018). Mathematical prediction of pharmacokinetic parameters-an *in-vitro* approach for investigating pharmaceutical products for IVIVC. *Future Journal of Pharmaceutical Sciences* **4** (2): 175–184.

[33] FDA (1997). Guidance for industry extended release oral dosage forms: development, evaluation, and application of *in vitro/in vivo* correlations. https://www.fda.gov/media/70939/download (accessed March 2021).

[34] Chen, Y., Jin, J.Y., Mukadam, S. et al. (2012). Application of IVIVE and PBPK modeling in prospective prediction of clinical pharmacokinetics: strategy and approach during the drug discovery phase with four case studies. *Biopharmaceutics Drug Disposal* **33** (2): 85–98.

[35] Markopoulos, C., Andreas, C.J., Vertzoni, M. et al. (2015). *In-vitro* simulation of luminal conditions for evaluation of performance of oral drug products: choosing the appropriate test media. *European Journal of Pharmaceutics and Biopharmaceutics* **93**: 173–182.

[36] Efentakis, M. and Dressman, J.B. (1998). Gastric juice as a dissolution medium: surface tension and pH. *European Journal of Drug Metabolism Pharmacokinetics* **23** (2): 97–102.

[37] Gray, V. and Dressman, J. (1996). Change of pH requirements for simulated intestinal fluid TS. In: *Pharmacopeial Forum*. Rockville, MD: US Pharmacopeial Convention 12601 Twinbrook PKWY.

[38] USP (2016). *United States Pharmacopeia and National Formulary (USP 41-NF 36)*. United States Pharmacopeial Convention.

[39] Vertzoni, M., Dressman, J., Butler, J. et al. (2005). Simulation of fasting gastric conditions and its importance for the *in vivo* dissolution of lipophilic compounds. *European Journal of Pharmaceutics and Biopharmaceutics* **60** (3): 413–417.

[40] Galia, E., Nicolaides, E., Hörter, D. et al. (1998). Evaluation of various dissolution media for predicting *in vivo* performance of class I and II drugs. *Pharmaceutical Research* **15** (5): 698–705.

[41] Dressman, J.B., Amidon, G.L., Reppas, C., and Shah, V.P.J. (1998). Dissolution testing as a prognostic tool for oral drug absorption: immediate release dosage forms. *Pharmaceutical Research* **15** (1): 11–22.

[42] Jantratid, E., Janssen, N., Reppas, C., and Dressman, J.B. (2008). Dissolution media simulating conditions in the proximal human gastrointestinal tract: an update. *Pharmaceutical Research* **25** (7): 1663.

[43] Fuchs, A., Leigh, M., Kloefer, B. et al. (2015). Advances in the design of fasted state simulating intestinal fluids: FaSSIF-V3. *European Journal of Pharmaceutics and Biopharmaceutics* **94**: 229–240.

[44] Klein, S. (2010). The use of biorelevant dissolution media to forecast the *in vivo* performance of a drug. *The AAPS Journal* **12** (3): 397–406.

[45] Klumpp, L., Leigh, M., and Dressman, J. (2020). Dissolution behavior of various drugs in different FaSSIF versions. *European Journal of Pharmaceutical Sciences* **142**: 105138.

[46] Bou-Chacra, N., Melo, K.J.C., Morales, I.A.C. et al. (2017). Evolution of choice of solubility and dissolution media after two decades of biopharmaceutical classification system. *The AAPS Journal* **19** (4): 989–1001.

[47] Fadda, H.M., Merchant, H.A., Arafat, B.T., and Basit, A.W. (2009). Physiological bicarbonate buffers: stabilisation and use as dissolution media for modified release systems. *International Journal of Pharmaceutics* **382** (1-2): 56–60.

[48] Macheras, P., Koupparis, M., and Apostolelli, E. (1987). Dissolution of 4 controlled-release theophylline formulations in milk. *International Journal of Pharmaceutics* **36** (1): 73–79.

[49] Fotaki, N., Symillides, M., and Reppas, C. (2005). *in vitro* versus canine data for predicting input profiles of isosorbide-5-mononitrate from oral extended release products on a confidence interval basis. *European Journal of Pharmaceutical Sciences* **24** (1): 115–122.

[50] Mann, J., Dressman, J., Rosenblatt, K. et al. (2017). Validation of dissolution testing with biorelevant media: an orbito study. *Molecular Pharmaceutics* **14** (12): 4192–4201.

[51] Klein, S., Rudolph, M., and Dressman, J. (2002). Drug release characteristics of different mesalazine products using USP apparatus 3 to simulate passage through the GI tract. *Dissolution Technology* **9** (4): 6–12.

[52] Andreas, C.J., Chen, Y.-C., Markopoulos, C. et al. (2015). *in vitro* biorelevant models for evaluating modified release mesalamine products to forecast the effect of formulation and meal intake on drug release. *European Journal of Pharmaceutics and Biopharmaceutics* **97**: 39–50.

[53] Kostewicz, E.S., Wunderlich, M., Brauns, U. et al. (2004). Predicting the precipitation of poorly soluble weak bases upon entry in the small intestine. *Journal of Pharmacy and Pharmacology* **56** (1): 43–51.

[54] Kourentas, A., Vertzoni, M., Khadra, I., et al. (2016). Evaluation of the impact of excipients and an albendazole salt on albendazole concentrations in upper small intestine using an *in vitro* biorelevant gastrointestinal transfer (BioGIT) system. *Journal of Pharmaceutical Sciences*. **105**(9):2896–2903. doi: https://doi.org/10.1016/j.xphs.2016.04.037

[55] Polster, C.S., Wu, S.-J., Gueorguieva, I., and Sperry, D.C. (2015). Mechanism for enhanced absorption of a solid dispersion formulation of LY2300559 using the artificial stomach duodenum model. *Molecular Pharmaceutics* **12** (4): 1131–1140.

[56] Carino, S.R., Sperry, D.C., and Hawley, M. (2010). Relative bioavailability of three different solid forms of PNU-141659 as determined with the artificial stomach-duodenum model. *Journal of Pharmaceutical Sciences* **99** (9): 3923–3930.

[57] Verwei, M., Minekus, M., Zeijdner, E. et al. (2016). Evaluation of two dynamic *in vitro* models simulating fasted and fed state conditions in the upper gastrointestinal tract (TIM-1 and tiny-TIM) for investigating the bioaccessibility of pharmaceutical compounds from oral dosage forms. *International Journal of Pharmaceutics* **498** (1): 178–186.

[58] Rivas-Montoya, E., Miguel Ochando-Pulido, J., Manuel López-Romero, J., and Martinez-Ferez, A. (2016). Application of a novel gastrointestinal tract simulator system based on a membrane bioreactor (SimuGIT) to study the stomach tolerance and effective delivery enhancement of nanoencapsulated macelignan. *Chemical Engineering Science* **140**: 104–113.

[59] Abrahamsson, B., Pal, A., Sjöberg, M. et al. (2005). A novel *in vitro* and numerical analysis of shear-induced drug release from extended-release tablets in the fed stomach. *Pharmaceutical Research* **22** (8): 1215–1226.

[60] Wickham, M., Faulks, R., and Mills, C. (2009). *in vitro* digestion methods for assessing the effect of food structure on allergen breakdown. *Molecular Nutrition and Food Research* **53** (8): 952–958.

[61] Mason, L.M., Chessa, S., Huatan, H. et al. (2016). Use of the dynamic gastric model as a tool for investigating fed and fasted sensitivities of low polymer content hydrophilic matrix formulations. *International Journal of Pharmaceutics* **510** (1): 210–220.

[62] Liu, W., Fu, D., Zhang, X. et al. (2019). Development and validation of a new artificial gastric digestive system. *Food Research International* **122**: 183–190.

[63] Kozu, H., Nakata, Y., Nakajima, M. et al. (2015). Analysis of disintegration of agar gel particles with different textures using gastric digestion simulator. *Japan Journal of Food Engineering* **16** (2): 161–166.

[64] Kong, F. and Singh, R.P. (2010). A human gastric simulator (HGS) to study food digestion in human stomach. *Journal of Food Science* **75** (9): E627–E635.

[65] Bellmann, S., Lelieveld, J., Gorissen, T. et al. (2016). Development of an advanced *in vitro* model of the stomach and its evaluation versus human gastric physiology. *Food Research International* **88**: 191–198.

[66] Garbacz, G., Wedemeyer, R.-S., Nagel, S. et al. (2008). Irregular absorption profiles observed from diclofenac extended release tablets can be predicted using a dissolution test apparatus that mimics *in vivo* physical stresses. *European Journal of Pharmaceutics and Biopharmaceutics* **70** (2): 421–428.

[67] Burke, M.D., Maheshware, C.R., and Zimmerman, B.O. (2006). *Pharmaceutical Analysis Apparatus and Method in AU2005304889A1*. Australia: SmithKline Beecham.

[68] Danielak, D., Milanowski, B., Wentowski, K. et al. (2020). Physiologically based dissolution testing in a drug development process—a case study of a successful application in a bioequivalence study of trazodone ER formulations under fed conditions. *AAPS PharmSciTech* **21** (5): 161.

[69] Stamatopoulos, K., Batchelor, H.K., and Simmons, M.J.H. (2016). Dissolution profile of theophylline modified release tablets, using a biorelevant dynamic colon model (DCM). *European Journal of Pharmaceutics and Biopharmaceutics* **108**: 9–17.

[70] Shi, Y., Gao, P., Gong, Y., and Ping, H. (2010). Application of a biphasic test for characterization of *in vitro* drug release of immediate release formulations of celecoxib and its relevance to *in vivo* absorption. *Molecular Pharmaceutics* **7** (5): 1458–1465.

[71] Ginski, M.J. and Polli, J.E. (1999). Prediction of dissolution–absorption relationships from a dissolution/Caco-2 system. *International Journal of Pharmaceutics* **177** (1): 117–125.

[72] Ginski, M.J., Taneja R, Polli JE. Prediction of dissolution-absorption relationships from a continuous dissolution/caco-2 system. *AAPS PharmSci* 1999;1(2):E3. doi: https://doi.org/10.1208/ps010203

[73] He, X., Sugawara, M., Kobayashi, M. et al. (2003). An *in vitro* system for prediction of oral absorption of relatively water-soluble drugs and ester prodrugs. *International Journal of Pharmaceutics* **263** (1–2): 35–44.

[74] Fliszar, K.A. and Foster, N. (2008). Examination of metformin hydrochloride in a continuous dissolution/HDM system. *International Journal of Pharmaceutics* **351** (1): 127–132.

[75] Motz, S.A., Schaefer, U.F., Balbach, S. et al. (2007). Permeability assessment for solid oral drug formulations based on Caco-2 monolayer in combination with a flow through dissolution cell. *European Journal of Pharmaceutics and Biopharmaceutics* **66** (2): 286–295.

[76] Berthelsen, R., Byrialsen, J.P., Holm, R. et al. (2016). Development of a μ dissolution-permeation model with in situ drug concentration monitoring. *Journal of Drug Delivery Science and Technology* **35**: 223–233.

[77] Borbás, E., Balogh, A., Bocz, K. et al. (2015). *in vitro* dissolution-permeation evaluation of an electrospun cyclodextrin-based formulation of aripiprazole using μFlux™. *International Journal of Pharmaceutics* **491** (1-2): 180–189.

[78] Pocock, K., Delon, L., Bala, V. et al. (2017). Intestine-on-a-chip microfluidic model for efficient *in vitro* screening of oral chemotherapeutic uptake. *ACS Biomaterials Science & Engineering* **3** (6): 951–959.

[79] Bein, A., Shin, W., Jalili-Firoozinezhad, S. et al. (2018). Microfluidic organ-on-a-chip models of human intestine. *Cellular and Molecular Gastroenterology and Hepatology* **5** (4): 659–668.

[80] Hate, S.S., Reutzel-Edens, S.M., and Taylor, L.S. (2017). Absorptive dissolution testing of supersaturating systems: impact of absorptive sink conditions on solution phase behavior and mass transport. *Molecular Pharmaceutics* **14** (11): 4052–4063.

[81] Berben, P., Brouwers, J., and Augustijns, P. (2018). Assessment of passive intestinal permeability using an artificial membrane insert system. *Journal of Pharmaceutical Sciences* **107** (1): 250–256.

[82] Ilie, A.-R., Griffin, B.T., Brandl, M. et al. (2020). Exploring impact of supersaturated lipid-based drug delivery systems of celecoxib on *in vitro* permeation across Permeapad⊠ membrane and *in vivo* absorption. *European Journal of Pharmaceutical Sciences* **152**: 105452.

[83] Lue, B.M., Nielsen, F.S., Magnussen, T. et al. (2008). Using biorelevant dissolution to obtain IVIVC of solid dosage forms containing a poorly-soluble model compound. *European Journal of Pharmaceutics and Biopharmaceutics* **69** (2): 648–657.

7

Biopharmaceutics to Inform Candidate Drug Selection and Optimisation

Linette Ruston

Advanced Drug Delivery, Pharmaceutical Sciences, R&D,
AstraZeneca, Macclesfield, United Kingdom

7.1 Introduction

This chapter will focus on the application of biopharmaceutics principles to inform drug discovery and candidate drug selection. Whilst candidate selection decisions are primarily driven by the pharmacological profile of the molecule, a balance must be achieved between pharmacology and pharmaceutical/ADME properties to ensure the selected molecule is suitable for clinical development. This can sometimes be referred to as the 'drugability' of a potent compound [1].

Biopharmaceutics in the early phases of drug discovery is focused on understanding what the drug product ultimately needs to deliver to treat the disease, defining the molecular properties required to achieve the desired exposure profile and influencing compound design to achieve these properties. This is shown in Figure 7.1.

Biopharmaceutics: From Fundamentals to Industrial Practice, First Edition. Edited by Hannah Batchelor.
© 2022 John Wiley & Sons Ltd. Published 2022 by John Wiley & Sons Ltd.

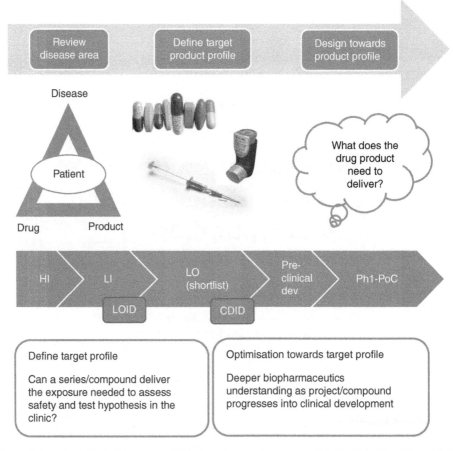

Figure 7.1   *Biopharmaceutics considerations in drug discovery (HI, hit identification; LI, lead identification; LO, lead optimisation; Ph1, Phase 1 clinical trial; PoC, proof of concept; LOID, lead optimisation investment decision; CDID, candidate drug investment decision).*

7.2   Oral Product Design Considerations During Early Development

Route of administration is obviously a primary consideration during early drug development, if an oral product is the appropriate choice for the patient and commercial market then the physicochemical and ADME properties of the drug must be suitable to deliver oral bioavailability, whilst injectable or inhaled molecules would require different attributes. It is important to remember that although required molecular properties may differ, application of biopharmaceutic principles and thinking is independent of administration route (inhaled and parenteral biopharmaceutics are covered in Chapters 14 and 15). This chapter will predominantly focus on oral delivery of small molecule drugs as these are still most common during development. A summary of the key considerations for the product design during early development is presented in Table 7.1.

Table 7.1 *Summary of oral drug product design considerations during early development pathways.*

Product design considerations		
Target access and engagement (organ, tissue, cell, protein)	Duration of treatment	Competitors and differentiation strategy
Therapeutic index/potential side effects	Dose frequency and patient compliance	Line and generation of therapy
PK–PD and efficacy relationship	Stage of disease	Patient affordability
Synergy with other treatments/ combination schedules	Hospital or home setting	Payers/real-world evidence
Drug–drug interactions	Anatomy/physiology impairment	Storage requirements

Recent publications from the pharmaceutical industry describe various measures of compound quality and define selection criteria for candidate compounds based on holistic analysis of company data and commercial drugs [2–4]. One example is AstraZeneca's decision-making framework, termed the 5Rs, which is based on thorough technical evaluation of five key areas: right target, right tissue, right safety, right patient and right commercial potential [5]. In a nutshell, understanding the relationship between drug exposure at the target site and downstream pharmacological response underpins human dose prediction, regardless of target class, chemotype, administration route or delivery system.

7.3   Biopharmaceutics in Drug Discovery

Early biopharmaceutics assessment is generally based on minimal testing to provide a fit for purpose data set. Typically, many molecules are manufactured on a small scale only providing sufficient material for initial testing. Medium to high throughput *in vitro* screening tools are used to evaluate pharmacology, safety, ADME and physical properties which enables hypothesis-led molecular design and rapid data-driven decisions, often termed the design, make, test, analyse (DMTA) cycle.

Predicted MAD (maximum absorbable dose) (see Section 4.3) in the context of estimated dose to man (eD2M) [6] provides a quick and simple way to highlight potential exposure risks. If the margin between MAD and eD2M is large this represents a low risk whereas if the MAD is close to eD2M then this represents a higher risk. Calculations can be set up to run automatically in company databases giving project teams an early view of biopharmaceutics risk without detailed biopharmaceutics experiments or extensive modelling. Note that when using this system a poorly soluble compound may be acceptable in cases where the anticipated dose is very low. In the absence of an estimated dose, perhaps due to insufficient understanding of the relationship between pharmacokinetics and pharmacodynamics (PK/PD) in the early stages, biopharmaceutic property space and risk can be evaluated based on nominal dose values and used to guide and optimise molecular design. Typically this will result in improving solubility where possible.

More detailed pharmacological and physicochemical profiling is undertaken later in drug discovery as compounds are optimised towards candidate drug selection (Figure 7.2). Knowledge deepens as lead molecules progress through test cascades, from early pharmacological and ADME screening towards detailed mechanistic profiling of promising drug candidates.

Fundamental understanding of the relationship between drug concentration and pharmacological effect allows definition of a target exposure profile. For example, in some cases a rapid onset is required to treat an acute issue whereas continuous exposure drug therapy is required to treat a chronic illness. The biopharmaceutics risk assessment roadmap, BioRAM [6], described this in more detail and is discussed in Chapter 8. Once defined for a given disease target the feasibility of achieving the target exposure profile can be assessed through molecular and/or formulation design (Figure 7.3).

7.3.1   Pre-Clinical Studies

Model compounds are routinely used as probes to gain understanding of disease pharmacology in pre-clinical studies. Pharmacological activity in response to the extent and duration of target engagement is monitored in enzyme, cell or animal models to establish

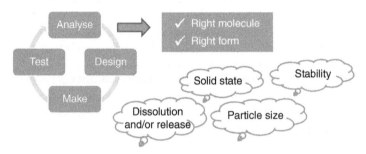

Figure 7.2   *Schematic of biopharmaceutics activity demonstrating that deeper knowledge is obtained as the compound progresses through the development process (HI, hit identification; LI, lead identification; PoC, proof of concept).*

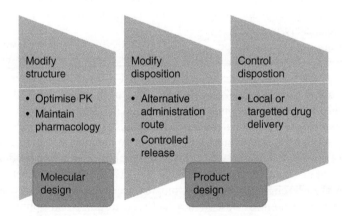

Figure 7.3   *Simple overview of how molecular or product design can provide the required target exposure profile.*

dose-response relationships and define exposure profile requirements. Enabling formulation strategies or alternative administration routes may be employed for compounds that have sub-optimal ADME properties for pre-clinical *in vivo* studies. One approach is to use sub-cutaneous (SC) continuous infusions to prolong exposure of short half-life probe molecules thus allowing exploration of biological activity in response to extended target cover, or to dose via an alternative administration route (SC or intraperitoneal) to mitigate intestinal or metabolic instability. Sustained release oral formulations for pre-clinical studies are less well established but gaining interest. Further reading on this topic is available [7].

Note that animal disease models can show sensitivity to excipients commonly used in human formulations, and care should be taken to ensure that the formulation components do not induce physiological or pharmacological responses that would confound interpretation of PK/PD results.

It is important to gain broad understanding of the factors that drive and limit exposure such that appropriate mitigation strategies may be identified during candidate identification and optimisation. A solubility limited compound may require enabling formulation techniques to enhance solubility/dissolution and promote supersaturation in the gastro-intestinal environment, whereas formulations that enhance permeation across the intestinal barrier may be employed to deliver oral bioavailability of peptides.

Early assessment of feasibility for controlled release is becoming more common – challenging PK means that formulation options to mitigate a short half-life may need to be explored early in development. Key factors are solubility, permeability and stability, which must be considered in context of the dose required to achieve target therapeutic concentrations. Standard tests such as caco-2 permeability and pH-dependent solubility can be used, in combination with *in silico* models based on simple release profiles, predicted human PK parameters and target potency.

The Developability Classification System (DCS), described in Chapter 8 is an extension of the BCS to aid selection of appropriate formulation approaches, classifying poor solubility compounds as solubility or dissolution rate limited [8]. Poor solubility compounds require formulation approaches to enhance solubility (examples include salts/cocrystals, pH-adjusted solutions, surfactant/complexation, stabilised amorphous form), whereas particle size reduction (micronisation, nanosuspension) may be used to mitigate dissolution limitations [9].

7.4   Biopharmaceutics Assessment

Biopharmaceutics assessment of potential drug candidates is underpinned by understanding of fundamental properties and their influence on exposure: solubility, permeability, lipophilicity, pK_a, molecular size and solid form.

7.4.1   Solubility

Solubility is described in detail in Chapter 4. Solubility in simple aqueous buffer is one of the first properties measured for all new compounds. Early solubility screening methods are designed to spare drug substance, often starting from stock solutions rather than solid samples. Solubility in biorelevant media (simulated intestinal fluids such as FaSSIF) and pH-solubility profiles including intrinsic solubility of the neutral form of ionisable molecules is determined around shortlisting stage (see Figure 7.1), or for key probe compounds during lead identification or optimisation. This provides mechanistic information to facilitate prediction of *in vivo* behaviour. It is important to identify changes in pH and or physical form during measurement: atypical pH-solubility profiles that don't follow theory can indicate complex solution behaviour such as aggregation or micellisation that may influence absorption and complicate mechanistic understanding of rate limiting factors. The required solubility for a given drug depends on other characteristics, notably dose and permeability. Further reading on biorelevant solubility screening in drug discovery is available [10–12].

7.4.2 Permeability

Permeability is discussed in detail in Chapter 5. Briefly, permeability measurement during lead optimisation is usually conducted *in vitro* using cell based models such as Caco-2 or MDCK cell lines to assess inherent membrane permeability and/or availability of active transport mechanisms. *In silico* property based predictions and/or *in vitro* physicochemical membrane systems such as PAMPA can also be used. Complex *in vivo* intestinal perfusion models are less common in discovery, although are routinely used to screen permeation enhancing formulations. Ussing chamber data may be generated to assess behaviour in intact intestinal tissue, including concentration-dependent effects of transporters or metabolising enzymes.

7.4.3 Dissolution

Dissolution is discussed in more detail in Chapter 6. Dissolution is measured for each API across the physiological pH range. Small scale dissolution apparatus (for example, PION's uDiss) may be applied due to limited compound availability to study API dissolution and precipitation behaviour [13]. This can include the risks of precipitation on pH-shift between the acidic stomach and more neutral intestinal pH.

7.4.4 Biopharmaceutics Classification System

The biopharmaceutics classification system (BCS) uses the two factors that govern the rate and extent of drug absorption: solubility and intestinal permeability [14]. This is described in much more detail in Chapter 9. In brief, it classifies drugs into high and low based on both their solubility and permeability where class 1 compound are both highly soluble and highly permeable and those in class 4 show low permeability and solubility. However, it is possible for drugs to be close to the cut off borders for the high classification yet still fall into BCS class 4. This is shown in Figure 7.4 where the relative risk associated with a BCS class 4 drug increases the further from BCS1 the compound sits on this figure.

7.4.5 Lipophilicity

$\text{Log} D_{7.4}$ measured at fixed pH rather than $\log P$ of neutral species, is routinely determined for all newly synthesised molecules. It is useful to guide molecular design, is a key descriptor in structure activity/property relationships and informs ADME understanding.

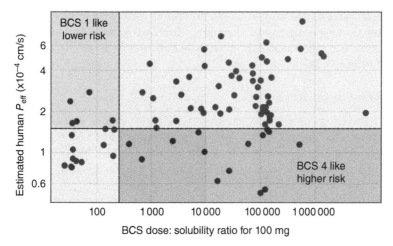

Figure 7.4 *Example BCS visualisation for an early discovery project. Human P_{eff} is estimated from measured in vitro or predicted in silico permeability. Dose : solubility ratio is calculated from early solubility measurement assuming nominal 100 mg dose in 250 mL. Markers show the spread of biopharmaceutics property space for the project compounds, indicate likely biopharmaceutics risk and can guide appropriate formulation strategies.*

Lipophilicity is an important consideration in formulation design and selection for pre-clinical studies and clinical development. Highly lipophilic molecules typically have poor aqueous solubility, often termed 'grease-balls', and are likely to require enabling formulations.

7.4.6 p*K*$_a$

p*K*$_a$, may be calculated or measured. Determination of the p*K*$_a$ underpins understanding of ionisation behaviour in physiological fluids across a relevant pH range. Understanding of the ionisation behaviour is fundamental to solubility, dissolution and permeability behaviour. It is particularly relevant to GI lumen for oral dosing due to the pH changes along the gastro-intestinal tract.

7.4.7 Molecular Size

API MW > 500 is associated with poor absorption (Lipinski's rule of 5) [15]. A higher MW is also linked to molar volume which is used as a guide for diffusivity, that can be important in permeability/dissolution and *in silico* modelling.

7.4.8 Crystallinity

In an ideal world, the most thermodynamically stable crystalline form (polymorph) would be progressed from discovery into clinical development to minimising risk of poor physical/chemical stability. However, crystalline form change can arise through changes in synthetic route, purification and isolation procedures, during manufacture or under stressed storage conditions and may be associated with changes in formulation performance.

In some instances the dissolution rate of crystalline material may be too low to deliver the desired *in vivo* exposure and an alternative form may be sought to improve solubility. Approaches include salts, cocrystals or stabilised amorphous forms. This is discussed in more detail in Chapter 8.

7.4.9 *In Vivo* Pre-Clinical Studies

Lead molecules with promising pharmacology profiles and good *in vitro* ADME characteristics (solubility, dissolution and permeability) will be progressed to *in vivo* PK screening, allowing evaluation of exposure profiles in the context of target blood levels for *in vivo* safety, pharmacodynamic or efficacy studies in animal models (typically rodent, dog, pig or monkey).

In the early stages, solutions or simple suspension formulations provide straight-forward options for *in vivo* studies in pre-clinical species and allow crude rank-ordering of compounds based on pharmacokinetic properties including overall exposure. Solutions or simple suspensions are convenient to handle for *in vivo* pharmacology and safety studies. In addition, solutions remove the need to consider polymorphs as the drug is already solubilised. Due to the early phase it is important to consider the batch to batch variability associated with the drug product as the synthetic pathway and manufacturing conditions will not yet have been optimised.

7.4.10 *In Silico* Modelling

In silico modelling is used to inform pre-clinical formulation strategy, for example, solubilisation, supersaturation or particle size reduction. *In silico* mechanistic physiologically based pharmacokinetic (PBPK) absorption modelling based on robust *in vitro*/predicted physicochemical property inputs is recommended to assess oral absorption limitations. Commercial software such as GastroPlus or SimCYP may be used, and several companies/ groups have developed their own versions of these modelling tools [16].

Key input parameters for prediction of oral absorption are MW, solubility in aqueous and biorelevant media, pK_a, and permeability. Absorption sensitivity to dose, dosage form (solution, suspension), solubility, permeability and particle size can be assessed *in silico* using physiological models that reflect fed or fasted prandial states, influence of disease, GI surgery, comedication and excipients.

Biomodels and simulation via *in silico* modelling is discussed in detail in Chapter 12. Note that additional reading on this topic can be found [17–19].

7.4.11 Human Absorption/Dose Prediction

It is important to understand the basis for human dose prediction and be aware of assumptions about extent and duration of target engagement required for efficacy. It is important to consider whether pharmacological response is driven by sustained concentrations above a minimum value (C_{trough}), overall exposure (AUC), transient high exposure (C_{max}) or time averaged coverage, and are there any complex biological relationships?

Human PK parameters, predicted from *in vitro* systems using measured or calculated properties and appropriate scaling factors are combined with potency data to give early estimates of human efficacious dose [20]. PK and dose estimates are refined as potential drug candidates are explored and profiled in more detail. Early estimates may assume complete absorption, or apply a more conservative fraction absorbed (Fabs) estimate if solubility and/or permeability suggest limitations are likely.

7.5 Output of Biopharmaceutics Assessment

The ultimate aim of biopharmaceutics input in early development of pharmaceutical formulations is identification of risk factors that may limit the bioavailability of a drug. Thus an integrated assessment, combining *in vitro* measurements, *in silico* modelling and evaluation of *in vivo* data from pre-clinical species is produced as part of the risk assessment data package.

The following properties would indicate a moderate biopharmaceutics risk:

- dose dependent absorption
- solubility limited absorption,
- sensitive to solubility and GI conditions such as variable gastric pH, food intake and bile salt concentrations, as well as pharmaceutical properties including primary particle size.

An example of the predicted fraction absorbed for a model compound is shown in Figure 7.5. This data shows how the sensitivity to a particular parameter can be modelled; in this case to the dose in part (a) and to the solubility in part (b). The details of model compound A are provided: weak base, pK_a 4, solubility at pH 6.5 = 20 µg/mL, FaSSIF solubility = 30 µg/mL, high permeability: predicted efficacious human dose 200 mg.

This information suggests solubility limited absorption, yet the higher solubility in FaSSIF is promising and there is likely to be even higher solubility in FeSSIF.

In silico modelling can show at what dose the solubility will limit the fraction absorbed; in this case with compound A, at doses over 50 mg and solubility less than 80 µg/mL the fraction absorbed decreases.

This type of modelling helps to predict non-linear pharmacokinetics and can provide insights into formulation strategies. For example, an increase in the apparent solubility would lead to a greater fraction absorbed; thus using an enabled formulation would achieve the required exposure. Achieving exposure during pre-clinical testing is important as often the tests are designed to better understand the maximum tolerated dose whereby the maximum exposure is required. In cases where oral absorption may limit exposure a drug may be given by an alternative route (e.g. intravenously). The use of *in silico* models at this stage provides a rapid insight into the likely testing required and provides comprehensive information for a risk assessment for each compound.

Similar principles can be applied to a range of scenarios that may limit the fraction absorbed. This may include permeability limitations including those associated with saturation of intestinal transporters or intestinal metabolism.

This detailed absorption modelling can be used to inform further *in vitro* or *in vivo* studies, and to guide form selection and formulation strategy for safety assessment and clinical development.

7.5.1 New Modalities/Complex Delivery Systems Within Early Development

In order to address unmet medical needs, pharmaceutical company portfolios are expanding from traditional small molecules towards increasingly challenging chemical space [21].

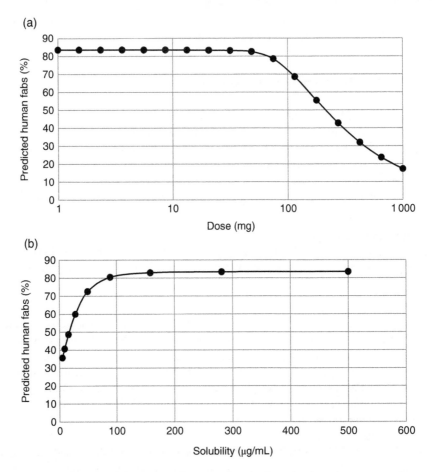

Figure 7.5 *Predictions of the human fraction absorbed (in the fasted state) as a function of (a) dose and (b) solubility.*

New modalities (biologics including peptides, monoclonal antibodies and mRNA) require drug delivery systems to overcome barriers to absorption. In contrast to small molecule drug projects where their formulation is likely to change over the course of discovery and development, the selected delivery system must be in place very early, typically before lead optimisation. The project team then develop the delivery system and active molecule as one, with the intended clinical formulation in place prior to safety and efficacy assessment.

7.6   Influence/Optimise/Design Properties to Inform Formulation Development

The formulation strategy of an API depends on physicochemical properties and factors that limit the rate or extent of absorption *in vivo*. Formulation is primarily used to overcome poor aqueous solubility, although permeation enhancers may be employed to overcome intrinsically poor membrane permeation, or recently co-dosing with efflux pump inhibitors [22, 23]. Solubility enhancement is a common strategy in early development, which can be used without detailed mechanistic understanding of absorption in early drug projects [24]. However, understanding must be developed if the compound progresses as there can be a complex interplay of factors including: solubility; dissolution; precipitation; permeability; supersaturation; micelle and excipient interactions as well as GI physiology. A simple guide for rational formulation strategy has been reported previously [25]. This is presented in Figure 7.6.

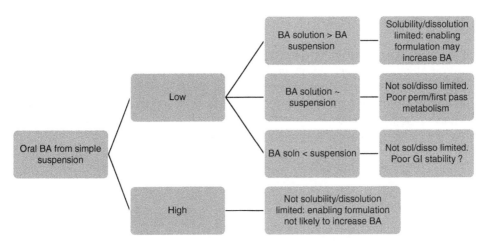

Figure 7.6   *Rational formulation selection decision tree for early development.*
Source: Adapted from Ayad [25].

Table 7.2   *Summary of the enabling technologies that can be applied to a compound based on the rate limiting step for its absorption.*

Rate limiting factor for absorption	BCS classification	Food effect	Enabling strategy
Dissolution	N/A	+ve	Particle size reduction
Unstirred water layer permeability limited	1	None	None required – good absorption
Epithelial membrane permeability limited	3	−ve	Pro-drugs
Solubility - unstirred water layer permeability limited	2	+ve	Solubility enhancement (SEDDS, nano, cyclodextrins)
Solubility - epithelial permeability limited	4	~+ve	Solubility enhancement: supersaturatable API and formulations

Source: Adapted from Sugano and Terada [26].

7.6.1   Fraction Absorbed Classification System

The fraction absorbed classification system (FaCS) introduced by Sugano in 2015, built on the BCS and developability classification (DCS) to evaluate the relative importance of permeability and solubility limitations, highlight risks and define appropriate enabling strategies [26]. The implication of the FaCS on enabling strategy to optimise formulation is shown in Table 7.2. The FaCS is particularly useful in early stages of product development. (Further details on the DCS, which is typically used in later development can be found in Chapter 8.)

7.7   Conclusion

Collaboration in multidisciplinary project teams is key to achieving shared understanding of exposure requirements and developing human PK and dose prediction. Biopharmaceutics scientists can provide mechanistic absorption predictions that take account of pharmaceutical properties of the molecule, administration route, formulation and physiology. These absorption predictions integrate with models describing systemic PK, pharmacology and safety to ultimately provide quantitative human dose estimates.

References

[1] Egner, U. and Hillig, R.C. (2008). A structural biology view of target drugability. *Expert Opin. Drug Discov.* **3** (4): 391–401.

[2] Wang, J. and Skolnik, S. (2009). Recent advances in physicochemical and ADMET profiling in drug discovery. *Chem. Biodivers.* **6** (11): 1887–1899.

[3] Campbell, J.B. (2010). Improving lead generation success through integrated methods: transcending 'drug discovery by numbers'. *IDrugs* **13** (12): 874–879.

[4] Empfield, J.R. and Leeson, P.D. (2010). Lessons learned from candidate drug attrition. *IDrugs* **13** (12): 869–873.

[5] Cook, D., Brown, D., Alexander, R. et al. (2014). Lessons learned from the fate of AstraZeneca's drug pipeline: a five-dimensional framework. *Nat. Rev. Drug Discov.* **13** (6): 419–431.

[6] Selen, A., Dickinson, P.A., Müllertz, A. et al. (2014). The biopharmaceutics risk assessment roadmap for optimizing clinical drug product performance. *J. Pharm. Sci.* **103** (11): 3377–3397.

[7] Shah, S.M., Jain, A.S., Kaushik, R. et al. (2014). Preclinical formulations: insight, strategies, and practical considerations. *AAPS PharmSciTech* **15** (5): 1307–1323.

[8] Butler, J.M. and Dressman, J.B. (2010). The developability classification system: application of biopharmaceutics concepts to formulation development. *J. Pharm. Sci.* **99** (12): 4940–4954.

[9] Bayliss, M.K., Butler, J., Feldman, P.L. et al. (2016). Quality guidelines for oral drug candidates: dose, solubility and lipophilicity. *Drug Discov. Today* **21** (10): 1719–1727.

[10] Bergström, C.A. et al. (2014). Early pharmaceutical profiling to predict oral drug absorption: current status and unmet needs. *Eur. J. Pharm. Sci.* **57**: 173–199.

[11] Augustijns, P., Wuyts, B., Hens, B. et al. (2014). A review of drug solubility in human intestinal fluids: implications for the prediction of oral absorption. *Eur. J. Pharm. Sci.* **57**: 322–332.

[12] Skolnik, S.M., Geraci, G.M., and Dodd, S. (2018). Automated supersaturation stability assay to differentiate poorly soluble compounds in drug discovery. *J. Pharm. Sci.* **107** (1): 84–93.

[13] Scheubel, E., Lindenberg, M., Beyssac, E., and Cardot, J.-M. (2010). Small volume dissolution testing as a powerful method during pharmaceutical development. *Pharmaceutics* **2** (4): 351–363.

[14] Amidon, G.L., Lennernas, H., Shah, V.P., and Crison, J.R. (1995). A theoretical basis for a biopharmaceutic drug classification: the correlation of *in vitro* drug product dissolution and *in vivo* bioavailability. *Pharm. Res.* **12** (3): 413–420.

[15] Lipinski, C.A. (2000). Drug-like properties and the causes of poor solubility and poor permeability. *J. Pharmacol. Toxicol. Methods* **44** (1): 235–249.

[16] Bouzom, F., Ball, K., Perdaems, N., and Walther, B. (2012). Physiologically based pharmacokinetic (PBPK) modelling tools: how to fit with our needs? *Biopharm. Drug Dispos.* **33** (2): 55–71.

[17] Jones, H., Chen, Y., Gibson, C. et al. (2015). Physiologically based pharmacokinetic modeling in drug discovery and development: a pharmaceutical industry perspective. *Clin. Pharmacol. Ther.* **97** (3): 247–262.

[18] Jones, H.M., Dickins, M., Youdim, K. et al. (2012). Application of PBPK modelling in drug discovery and development at Pfizer. *Xenobiotica* **42** (1): 94–106.

[19] Jones, H.M., Gardner, I.B., and Watson, K.J. (2009). Modelling and PBPK simulation in drug discovery. *AAPS J.* **11** (1): 155–166.

[20] Page, K.M. (2016). Validation of early human dose prediction: a key metric for compound progression in drug discovery. *Mol. Pharm.* **13** (2): 609–620.

[21] Ashford, M. (2020). Drug delivery—the increasing momentum. *Drug Deliv. Transl. Res.* **10**: 1888–1894.

[22] Maher, S., Brayden, D.J., Casettari, L., and Illum, L. (2019). Application of permeation enhancers in oral delivery of macromolecules: an update. *Pharmaceutics* **11** (1): 41.

[23] Bohr, A., Nascimento, T.L., Harmankaya, N. et al. (2019). Efflux inhibitor bicalutamide increases oral bioavailability of the poorly soluble efflux substrate docetaxel in co-amorphous anti-cancer combination therapy. *Molecules (Basel, Switzerland)* **24** (2): 266.

[24] Zhang, L., Luan, H., Lu, W., and Wang, H. (2020). Preformulation studies and enabling formulation selection for an insoluble compound at preclinical stage—from *in vitro*, *in silico* to *in vivo*. *J. Pharm. Sci.* **109** (2): 950–958.

[25] Ayad, M.H. (2015). Rational formulation strategy from drug discovery profiling to human proof of concept. *Drug Deliv.* **22** (6): 877–884.

[26] Sugano, K. and Terada, K. (2015). Rate- and extent-limiting factors of oral drug absorption: theory and applications. *J. Pharm. Sci.* **104** (9): 2777–2788.

8

Biopharmaceutics Tools for Rational Formulation Design

Panagiota Zarmpi[1], Mark McAllister[2], James Butler[3] and Nikoletta Fotaki[1]

[1]Department of Pharmacy and Pharmacology, University of Bath, Bath, United Kingdom
[2]Pfizer Drug Product Design, Sandwich, United Kingdom
[3]Biopharmaceutics, Product Development and Supply, GlaxoSmithKline R&D, Ware, United Kingdom

8.1 Introduction

Further to drug candidate optimisation, a product needs to be designed so that the API is properly delivered and achieves the desired clinical outcome without compromising patient health. Product quality is achieved through formulation design and manufacturing process selection and development with the ultimate guiding principle of ensuring consistent performance for the patient [1]. The clinical and pharmacological target product profile is a key component in the design strategy for a formulation as it determines the dose, drug loading and release rate required. The formulation also needs to be designed with the end-user, the patient, in mind. A patient-centric design approach may include delivering the desired target profile to the patient in a way that simplifies the dosing regimen, avoids the administration of unnecessary multiple dosage units and wherever possible restrictions such as the need to adhere to a specific prandial state.

Biopharmaceutics: From Fundamentals to Industrial Practice, First Edition. Edited by Hannah Batchelor.
© 2022 John Wiley & Sons Ltd. Published 2022 by John Wiley & Sons Ltd.

During the lead candidate selection/optimisation (described in Chapter 7), molecular properties are optimised for selectivity and potency so that the desired pharmacological outcome can be achieved at doses that limit toxicity and ideally provide a wide therapeutic window. The molecular design necessary to achieve such a profile may not however be ideal in terms of the physicochemical properties for drug delivery (see Chapter 7 for additional details). Consideration of biopharmaceutics properties or the developability of a compound during the candidate optimisation phase may allow some absorption-limiting properties to be designed out before the compound progresses in the preclinical phase or if this is not possible, then an appropriate enabling formulation strategy can be assessed at an early stage.

Formulation selection and optimisation can then play a key role in circumventing these delivery limitations and assist in successful disease treatment. Appropriate formulation selection and optimisation is not always easy due to the complex interplay of drug physicochemical properties and physiological parameters/processes. The design of more advanced formulations may be required to improve drug delivery and bioavailability. Formulation strategies, therefore, need to account for the biopharmaceutical considerations to design the right product for a given compound. Quality by design principles highlights the need for quality to be built in during product development. In this chapter, the design principles for oral drug products are discussed. Conventional or novel therapy-driven approaches to development are presented in Figure 8.1. Current tools guiding formulation strategies based on drug, product and/or disease considerations are reviewed within this chapter. The goal is to inform readers on biopharmaceutics considerations when promising drug candidates are formulated into drug products.

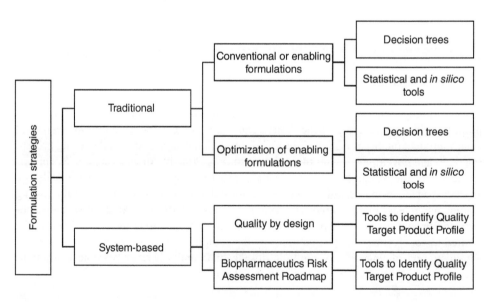

Figure 8.1 *Overview of the formulation strategies and associated biopharmaceutics tools presented in the current chapter.*

8.2   Formulation Development to Optimise Drug Bioavailability

From the vast number of pharmacologically active compounds, only a few molecules reach the market. It is widely recognised that the most potent drug might not always be the most successful [2], as the optimal molecule is the one that achieves the best balance of potency, selectivity, metabolic stability and pharmacokinetic (PK) profile. Biopharmaceutically, poor drug solubility, permeability and metabolic instability in the gastrointestinal tract may contribute to development failures [3]. These challenges often delay product development.

The lipophilic nature of endogenous targets [4] dictates the molecular properties of drugs and currently highly lipophilic drugs are identified as successful drug candidates. As drug physicochemical properties control oral drug delivery, the interplay of drug solubility/lipophilicity is critical for oral drug absorption; dose, solubility, dose to solubility ratio, maximum absorbable dose and lipophilicity have been identified as potential candidate quality parameters [5]. When manipulation of drug properties (such as salt or solid form modification) is not an option to improve oral drug bioavailability, a formulation (conventional or enabling) may be needed. When designing a formulation to address a specific limitation for oral absorption, such as dissolution rate and/or solubility, the design space will be mainly defined by the drug's physicochemical properties. Other factors such as access or availability of technology platforms and historical experience in industrial development labs will also guide the selection of the technology or approach for bio-enhancement. Tools and approaches to guide the choice of the appropriate formulation have been developed to facilitate the decision-making process. In this chapter, this decision-making process is discussed, for example, what would be the best formulation to circumvent the limitations for oral drug bioavailability, what relevant information exists to guide development towards one against another formulation and how the formulation scientist ascertains that the selected formulation will at the end achieve the desired outcome.

8.3   Traditional Formulation Strategies

8.3.1   Decision Making for Conventional or Enabling Formulations

When the oral route has been selected as the preferred route of administration, a number of conventional (e.g. simple capsules, tablets) or enabling (formulations that 'make' a drug bioavailable [6]) approaches are available to the formulator. The development of a conventional formulation is usually preferred, due to ease of manufacturing and speed to the clinic, provided that such a formulation has adequate bioperformance.

8.4   Decision Trees to Guide Formulation Development

8.4.1   Decision Trees Based on Biopharmaceutics Classification System (BCS)

At the early stages of product development, the decision to progress a conventional approach or to develop an enabling formulation can be guided by two key biopharmaceutical drug properties, namely drug solubility and drug permeability. The BCS classification system (described in more detail in Chapter 9) provides a conceptually easy to understand framework to guide formulation strategy. Typically, BCS I drugs (highly soluble/highly permeable) offer the widest range of formulation options and conventional formulations can be used [7, 8], as in general the physicochemical properties of the drug are not considered problematic. BCS I drugs can still pose challenges for manufacturability with factors such as poor flow (which is typically addressed through crystallisation control or selection of a granulation process for the drug product) [9] (Figure 8.2). Wettability and aqueous stability are additional factors to consider when proposing a formulation for a BCS I drug. A decision tree to guide formulation based on the BCS is presented in Figure 8.2.

More strategic considerations are required for the remaining BCS drug classes. Formulation development based on BCS is guided on whether poor bioavailability relates to solubility-limited or permeability-limited issues (or both). When BCS II drugs (poorly soluble/highly permeable) are discussed, drug bioavailability is compromised by the poor drug solubility. Advanced technologies (i.e. coating technologies) or enabling formulations (solid dispersions, lipid formulations, cocrystals, etc.) can be used to improve drug solubility [7, 8]. For such drugs, the impact of drug ionisation on drug solubility and bioavailability can simplify development, as the increased drug solubility of the ionised molecules in different gastrointestinal regions can favour drug bioavailability. The decision tree presented in Figure 8.2 indicates that in cases where the solubility of the ionised species permits the complete drug dissolution in 250 mL of an appropriate media (dose to volume [DV] < 250 mL), formulation development can follow the BCS I drug considerations. Advanced technologies

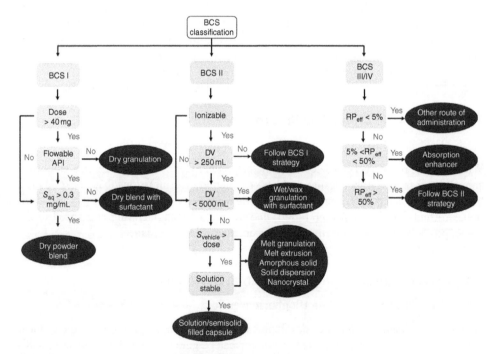

Figure 8.2 *Formulation selection strategies based on BCS. DV, dose to volume, RP_{eff}, relative effective permeability. Source: Modified from Kuentz et al. [9].*

or enabling formulations will be recommended for drugs not meeting the above criterion (Figure 8.2). For BCS III (highly soluble/poorly permeable) and IV (poorly soluble/poorly permeable) drugs, inadequate bioavailability is mainly caused by poor drug permeability and a combination of poor solubility and low permeability, respectively. Usually, strategies for BCS I or II drugs are followed for drugs belonging to these BCS classes (BCS III and IV) and the use of absorption enhancers is suggested [7, 8]. For drugs demonstrating extremely low permeability (relative effective permeability [RP_{eff}]<5% when compared to a model compound that is known to exhibit high permeability), other routes of administration may be needed [7, 9]. In practice, such decisions are based on in-house industrial rules and experience. The dose filters, solubility limits or technology platforms/preferences, as presented in Figure 8.2, may vary (e.g. for a weak base with very high solubility in gastric fluids but very low solubility in intestinal fluids, this guidance would suggest that a simple formulation approach would be supported which is not always the case due to the anticipated drug precipitation in the intestine). Moreover, most BCS IV drugs are formulated to address mainly the solubility limitation, as the permeability range associated with this class of drugs is not always problematic for oral drug absorption.

8.4.2   Decision Trees Based on Developability Classification System (DCS)

The BCS is a good starting point for formulation development and it provides a simplistic view of the quite complex biopharmaceutical mechanisms affecting drug bioavailability. The introduction of the developability classification system (DCS), shown in Figure 8.3, and the refined DCS accounted for additional considerations when oral drug development is of concern [10, 11].

As drug solubility in the presence of solubilising components (e.g. bile salts) may greatly differ from that in simple buffers, the DV is calculated using solubility data generated in fasted state human intestinal fluids (FaHIF) (in cases where this is feasible as FaHIF

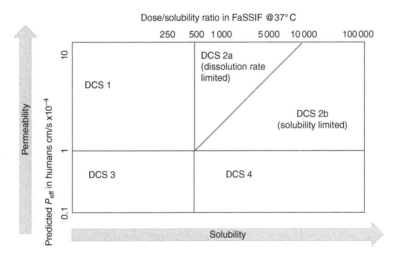

Figure 8.3 *An overview of the Developability Classification System (DCS). Source: Modified from [10]*

samples are difficult to handle and quite expensive) or in media representing the gastrointestinal conditions, for example, fasted state simulated intestinal fluid (FaSSIF) [11, 12]. Furthermore, the DCS recognises the need to classify a drug across a range of dose strengths. During early development, where the therapeutic clinical dose is yet to be determined and the range of doses considered in phase I single- and multiple-ascending dose studies may result in a drug spanning multiple classification categories due to the broad DV values that depend on the dose. To accommodate this, the DCS proposes that three different doses (5, 50 and 500 mg) are used and a separate drug classification is given for each dose. The highly soluble criterion is set to 500 mL, against the more conservative 250 mL of the BCS when the drug is administered with a glass of water and as additional fluid is available once any undissolved drug is transferred to the small intestine [11]. Assessment of drug permeability is made with the use of *in silico* or cell-based *in vitro* models to estimate human jejunal permeability (P_{eff}) with a cut-off value of 1.0×10^{-4} cm/s to define highly permeable drugs [11]. This is of particular relevance in early product development where high throughput methods are more commonly used to estimate permeability. A new subdivision (differing for subdivisions based on the drug's chemical class, for example, DCS IIa/IVa, IIb/IVb and IIc/IVc representing weak acids, weak bases and neutral compounds, respectively [13]) of drug classes is introduced for class II drugs: DCS IIa, for which the high drug permeability compensates their poor solubility and DCS IIb drugs where drug solubility is truly the rate-limiting step for drug absorption. As a first step, a standard investigation is performed with drugs being categorised into the five DCS classes using the aforementioned criteria for each of the three doses. Additional customised investigations are triggered for:

1. DCS IIa (irrespective of their aqueous solubility) or DCS I/III drugs with drug aqueous solubility lower than 100 µg/mL where dissolution rate experiments are recommended. The aim, in this case, is to determine an optimum drug particle size below which drug dissolution rate (k_{diss}) will be higher than drug permeability rate (k_{perm}) and therefore oral drug bioavailability will not be compromised by the slow drug dissolution
2. DCS IIb/weak bases or DCS IV/weak bases where supersaturation/precipitation experiments are recommended to determine whether *in situ* supersaturation may enhance oral drug absorption or whether precipitation inhibitors are required
3. DCS IIb/salt of a weak acid or DCS IV/salt of a weak acid with a $pK_a > pH$ of the small intestine where precipitation experiments are recommended to account for similar processes as per point (2).

These customised investigations aid in the identification of key parameters that, when controlled, would maximise drug delivery [11].

Examples of formulation strategy consideration with the use of the refined DCS are provided in Figure 8.4. Consider the example compound A, for which low and medium doses (5 and 50 mg) are classified as DCS I, dissolution is not considered a rate-limiting step for drug absorption due to the adequate drug solubility ($S > 100$ µg/mL) and a conventional formulation can be manufactured. For higher doses with a compound classification of DCS IIa$_{500mg}$, small-scale dissolution experiments are required which will determine the optimum drug particle size so that $k_{diss} \gg k_{perm}$. Compound B is classified as DCS IIb for all studied doses (the 5 mg dose is on the borderline between IIa and IIb), already indicating that significant efforts are likely to be required to formulate this drug. Small scale

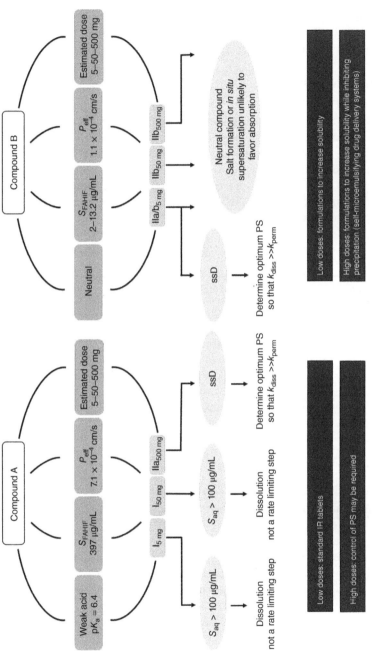

Figure 8.4 Formulation selection strategies based on refined DCS. S_{FAHIF}, drug solubility in fasted state human intestinal fluids, P_{eff}, Drug effective permeability, S_{aq}, drug aqueous solubility, ssD, Small Scale Dissolution experiments, PS, particle size, k_{diss}, drug dissolution rate, k_{perm}, drug permeability rat. Source: Modified from Rosenberger et al. [11].

dissolution needs to be conducted, due to the DCS $\text{IIa}_{5\,\text{mg}}$ classification, which will determine the optimum particle size of the drug under which absorption is not dissolution rate-limited, as per compound A. Strategies to improve drug solubility such as salt formation or *in situ* supersaturation are not an option however, as compound B is neutral. Formulations that increase drug solubility may be required and an additional step of inhibiting potential drug precipitation for the higher drug doses need to be considered [11].

8.4.3 Expanded Decision Trees

Expansions to the BCS or DCS classification have been adopted in the industry based on in-house criteria [14]. Additional considerations for the 'conventional or enabling' decision can evaluate the chemical class of the compound (neutral, weak acid, weak base), pH of the medium, the volume needed to dissolve the dose in humans (VDAD), the dose number (DN: dose/250 mL *solubility) or the relative bioavailability (F_{rel}) of drug administered orally in animals as a suspension vs a solution (Table 8.1). Although not harmonised across the pharmaceutical industry, these approaches are designed to reflect more closely the biorelevant *in vivo* conditions.

Table 8.1 *Biopharmaceutics parameters and their cut-off values guiding the development of a conventional formulation.*

Parameters evaluated	Cut-off values for developing a conventional formulation
VDAD	Neutral: $\text{VDAD}_{\text{pH}\,1-7} < 3\,\text{L}$
$F_{rel,susp/sol}$	Weak acids: $\text{VDAD}_{\text{pH}\,7} < 6\,\text{mL}$
Drug chemical class	Weak bases: $\text{VDAD}_{\text{pH}\,1} < 175\,\text{mL}$
	or
	$F_{rel,susp/sol} > 50\%$
BCS	BCS I/III with DN < 1
DN	BCS II/IV with DN < 10
BCS	BCS I/II/III/IV with $F_{rel,susp/sol} > 50\%$
$F_{rel,susp/sol}$	

*VDAD, volume needed to dissolve the dose in humans; DN, the dose number (DN, dose/250 mL *solubility).*
Source: Based on Zane et al. [14].

8.5 Computational Tools to Guide Formulation Strategies

Given the limited availability of drug-related physicochemical or pharmacokinetic data at the early preclinical stages of drug development, computational methods have been promising in providing initial information to guide formulation strategies [15, 16]. Based on prior knowledge and available drug physicochemical properties and/or pharmacokinetic data, decisions on formulations to be considered can be obtained.

8.5.1 Statistical Tools

Statistical tools aid in correlating drug properties to formulation success. Such an approach has been described in which several physicochemical properties (e.g. pK_a, hydrogen donor acceptors/donors, polar surface area, drug solubility, etc.) of various drugs were correlated to formulation suitability (based on formulation performance and commercial viability [17]). Discriminant analysis and partial least squares were used to identify the relation between drug physicochemical properties and formulation classes. The use of conventional or non-conventional formulations was linked to drug properties; highly soluble drugs with the high polar surface area would benefit from the former while drugs of high dose number, molecular size and Lipinski score from the latter [17]. Decision trees guiding formulation selection can be obtained by such approaches. Mathematical models and statistical tools can also be used to determine the suitable components/parameters when a particular formulation has already been selected [8].

8.5.2 Physiologically Based Pharmacokinetic/Biopharmaceutics Models

In silico tools assist in simulating the expected *in vivo* performance of a formulation. These have been proved beneficial for understanding drug pharmacokinetics and formulation performance [18]. Physiologically based pharmacokinetic (PBPK) models or physiologically based biopharmaceutics models (PBBM) are biopredictive and mechanistic biopharmaceutics models taking into account critical formulation/manufacturing aspects to product performance [19, 20]. These models require a set of input parameters related to the compound and formulation to predict pharmacokinetics in humans. Input parameters at an early stage are obtained from the biopharmaceutical profiling of drugs, cell cultures or animal studies and are updated once *in vivo* data from humans are obtained to ensure adequate predictability [21]. Integration of key drug (pK_a, $\log P$, P_{eff}, etc.), gastrointestinal (transit times, volumes, fasted/fed conditions, etc.) and pharmacokinetic (first-pass extraction, blood/plasma ratio, volume of distribution, etc.) parameters related to oral drug absorption allows the simulation of different absorption/bioavailability scenarios giving insights on potential successful strategies [22]. These models are not only informative of early absorption predictions or formulation optimisation but provide insights into oral drug absorption under atypical conditions (fed state, region-dependent absorption, etc), and patient's needs (e.g. dosing frequency) which are key aspects in holistic formulation strategies [21, 23–26]. For example, for a BCS II drug, drug solubility and particle size were *a priori* expected to affect oral drug bioavailability. Sophisticated strategies, therefore, may have been initiated in an attempt to develop a formulation of superior performance. The use of PBPK simulations proved otherwise; increasing drug solubility or decreasing drug particle size improved the rate but not the extent of oral drug absorption, as the high drug permeability and typical gastrointestinal transit times resulted insufficient drug bioavailability even in the worst-case scenarios [27]. This example illustrates the advantages of these computational tools as not only the selection of a particular formulation is justified but also time and cost-consuming approaches without providing any benefit to oral drug delivery can be avoided. In addition, formulation optimisation based on a patient-centric perspective can be achieved. For example, administration of a 150 mg immediate-release (IR) BCS I drug product revealed that a three times daily (TID) dosing is required to maintain plasma concentrations for improved efficacy. The use of PBPK modelling evaluated the feasibility of developing a 450 mg controlled-release (CR) formulation that would achieve similar plasma concentrations as the IR-TID when administered once daily (Figure 8.5). Tuning the rate for complete drug dissolution between 8 and 16 hours was promising, as comparable plasma concentrations to the IR-TID were predicted within the model [24].

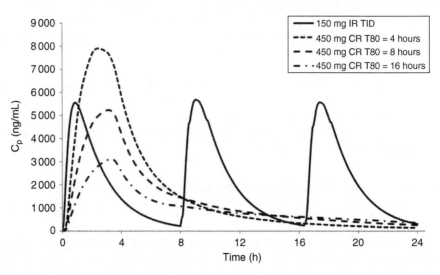

Figure 8.5 *Simulated human PK profiles for three controlled-release (CR) formulations (once daily) as compared to the immediate-release (IR) formulation (three times daily; TID). Source: From Kesisoglou et al. [24] / With permission of Elsevier.*

8.6   Decision-Making for Optimising Enabling Formulations

Once it has been recognised that a conventional formulation cannot adequately address the issues of poor drug bioavailability, the development of enabling formulations is considered. Based on the issue of poor oral drug absorption, a number of strategies are available to ensure that the API is properly delivered; these are listed in Table 8.2.

When oral drug absorption is solubility and/or dissolution rate-limited, a number of possibilities are available to the formulation scientist. The use of surfactants in simple formulations enhances drug solubility, as based on excipient concentrations, an improvement in the dissolution process can be achieved as micelles able to solubilise lipophilic compounds are formed [28, 29]. Lipidic simple (solution, dispersions) or

Table 8.2 *Formulation strategies to improve oral drug absorption.*

Limiting factor for oral drug absorption	Potential approach
Solubility/dissolution	Solubilisation in surfactant micelles or cyclodextrins
	Use of cosolvents
	Lipid-based formulations
	Use of precipitation inhibitors
	Amorphous solid dispersions
Permeability	Use of permeability enhancers

Source: Based on Boyd et al. [4].

enabling (self-emulsifying drug delivery systems) formulations are advantageous as the active is pre-dissolved in lipidic excipients (drug dissolution not the rate-limiting step) and upon digestion, the formation of micelles allows drug solubilisation and delivery to the enterocytes [30]. Achieving drug concentration higher than equilibrium solubility is another possibility to improve oral drug bioavailability. The creation of a so-called supersaturation state (e.g. through amorphous solid dispersions) allows for a temporarily increase in the amount of drug in the gastrointestinal tract and a subsequent increase in the drug flux through the epithelial cells [31]. *In situ* formation of a supersaturated state is also feasible for weakly basic drugs; the higher drug solubility in the stomach leads to higher than equilibrium drug concentrations in the small intestine upon gastric emptying and an increase in drug flux as discussed above. It should be noted that the aforementioned strategy can be compromised by physiological variability in the gastric pH [32]. The use of precipitation inhibitors (e.g. polymeric excipients that delay drug precipitation) might be necessary for such thermodynamically unstable systems as they will eventually precipitate [31]. Efforts to improve drug permeability are more restrictive and focus on the use of permeability enhancers through the paracellular route [33–38].

8.7   Decision Trees for Enabled Formulations

All the recent advances in drug development allow the improvement of oral drug bioavailability through appropriate formulations. Drug physicochemical properties can be used as a surrogate to guide the selection of the appropriate enabling formulation. When poor drug solubility is expected to affect oral drug absorption, formulations that solubilise the drug may be required. Once conventional formulations or salt formation have been deemed unfeasible [8, 39], drug size reduction or use of solubility enhancing vehicle are evaluated. Drug solubility in the vehicle is the driving parameter for formulation selection in the latter. Decision trees for choosing the optimal formulation can evaluate in a step-by-step process drug solubility in a liquid/wax vehicle, drug miscibility with a polymer or drug : polymer solubility in a common volatile solvent [8]. Self-microemulsifying drug delivery systems are selected when the drug is soluble in a liquid/wax vehicle. If not, the miscibility of a drug with a polymer is considered and formulations using polymers are developed. In the case of inadequate miscibility with a polymer, solid dispersions or solid solutions can be used. The stability of the formulation is another confounding factor to select or reject a formulation approach. Other formulation considerations may include, the gradual assessment of drug solubility in (i) surfactants, (ii) co-solvents, (iii) cyclodextrins and (iv) lipids. Once adequate drug solubility in the vehicle is demonstrated, and the formulation to be used is selected [(i) formulation containing surfactant, (ii) co-solvent formulation, (iii) cyclodextrin-based formulation, (iv) lipidic formulation]. Suspension-based formulations will be selected if insufficient drug solubility is demonstrated in all the aforementioned vehicles [39].

The lipophilicity ($\log P$) and melting point (T_m) of a drug are additional critical parameters guiding such selections. These parameters permit the identification of a suitable vehicle that can (or cannot) adequately solubilise a poorly soluble drug. Drugs of high lipophilicity are favourably solubilised in lipids (e.g. lipid-based formulations, use of surfactants). High melting points denote drugs insoluble in both aqueous or lipidic media for which other formulation strategies (e.g. nanoparticles, solid dispersions) may be required [40, 41]. For highly lipophilic drugs ($\log P > 2$–3) of low melting point ($T_m < 160\,°C$), lipid-based formulations, cosolvents or use of surfactants are suggested. When highly lipophilic drugs exhibit high melting points, their high crystal energy may compromise drug solubilisation in any vehicle; therefore, nanosuspensions, inclusion complexes or solid dispersion may be a suitable choice [41]. Drugs of low lipophilicity can be formulated with the use of co-solvents; however, nanoparticle technologies or solid dispersions may be required if a drug's melting point is high [40, 41].

Strategies combining different parameters (e.g. solubility and lipophilicity) have been as well developed. Maps of dose to solubility ratio as a function of $\log P$ can give insights into whether a given formulation will be successful in terms of oral drug absorption or if additional considerations (e.g. crystal modification, alternative formulation) are required. Potential adjustments to dose can also be considered in order to improve formulation performance [8].

8.7.1 Statistical Tools

Selection of the appropriate enabling formulation is usually based on prior experience or requires experimental studies that are often time and resource consuming. The need for advanced computational methods to forecast the appropriate enabling formulation to be used is recognised [26]. Multivariate data analysis (e.g. linear or non-linear regression) or molecular dynamics methods have been developed to assess the 'formulate ability' for an enabling formulation. The multivariate methods adequately predict critical formulation factors (e.g. glass stability in amorphous formulations, compound solubility in lipidic excipients) based on drug's molecular descriptors [42–44]. A more in-depth understanding of the molecular interactions in or post oral administration of an advanced formulation is obtained with molecular dynamics [45, 46]. Such approaches are beneficial at the early stages of development as they rationalise the feasibility of using an enabling formulation for drug delivery. An overview of current developed computational methods for strategising the use of an enabling formulation has been described in the literature in detail [26].

8.7.2 Physiologically Based Pharmacokinetic/Biopharmaceutics Models

The principles of physiologically based pharmacokinetic/biopharmaceutics models can be expanded in determining and justifying the selection of an enabling formulation to improve oral drug bioavailability. The current status of these models for enabling formulations focuses on defining/designing adequate models able to capture a formulation's performance, as a number of additional considerations need to be explored (e.g. drug precipitation in amorphous solid dispersions, lipid digestion in lipidic formulations, etc). Successful examples in literature have been reported [47, 48] and can provide the surrogate to further expand these physiologically based models for the selection of enabling formulations.

8.8   System-Based Formulation Strategies

8.8.1   Quality by Design

Product quality not only relates to the control of materials, manufacturing process and stability during shelf-life but also encompasses consistent and reliable performance in the patient. This relates to the necessary pharmacodynamic and pharmacokinetic outcome to be achieved for disease treatment. The desired onset and duration of a treatment is disease-dependent and the developed formulation needs to be designed to fulfil certain pharmaco-logical/clinical requirements (e.g. receptor binding). The above, along with the absence of harm to the patient, will determine product quality. Innovative approaches to product development, such as Quality by Design (QbD), have recognised the need for designing and building, rather than testing, quality throughout the process of formulation development and assess/control *a priori* the available options and/or risks of designing a successful product [49]. The focus of such system-thinking approaches is to start development with the final goal in mind and design each development step so that this final goal is achieved. In terms of formulation strategy, this goal relates to the desired pharmacodynamic outcome for disease treatment and is summarised in the Quality Target Product Profile (QTPP). QTTP constitutes the basis for setting the strategy for drug and formulation development and facilitating decision-making and regulatory approvals [50]. The QTPP may include, but is not limited to, information regarding [51]:

- Intended use in a clinical setting, route of administration, dosage form, delivery systems
- Dosage strength(s)
- Container closure system
- Therapeutic moiety release or delivery and attributes affecting pharmacokinetic characteristics (e.g. dissolution, aerodynamic performance) appropriate to the drug product dosage form being developed
- Drug product quality criteria (e.g. sterility, purity, stability and drug release) appropriate for the intended marketed product

8.8.2   Tools to Identify Quality Target Product Profile

All the aforementioned parameters are based on prior knowledge from scientific literature and technical experience [52] and provide a design target for an initial formulation. An example of QTPP guiding the development of an enabling formulation for a challenging compound in terms of biopharmaceutics (poorly soluble weak acid, unstable in acidic pH, high hepatic first-pass effect) is presented in Figure 8.6. The selection of each QTPP element is appropriately justified and aims at the development of a robust and high-quality formulation throughout the product's life-cycle [53].

The QTPP can serve as a surrogate to determine additional implications for more challenging situations, such as a formulation to be administered to elderly patients (Figure 8.7). Due to the changes in several gastrointestinal factors and the presence of comorbidities, the classical approaches to development for adult products cannot be entirely expanded to other patient populations [54]. Outlining the critical parameters of such a population will assist in developing the appropriate formulation accounting for all patient-centric factors

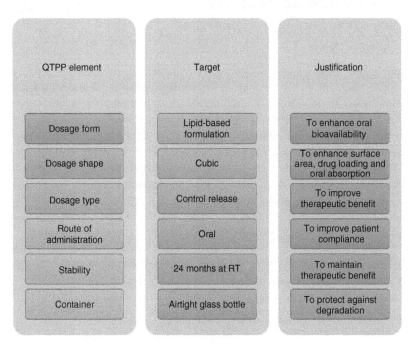

Figure 8.6   *QTPP guiding formulation strategy based on appropriate justifications. RT, room temperature. Source: Modified from Javed et al. [53].*

that may complicate drug administration (e.g. self-dosed or presence of a caregiver, changes in gastrointestinal factors that may affect compliance or absorption) [54]. These considerations are risk assessed later towards the development of a superior formulation.

8.9   Biopharmaceutics Risk Assessment Roadmap (BioRAM)

The classical QbD is mostly manufacturing-oriented in terms of controlling the variability of input materials and processes to ensure the development of products within certain acceptance range criteria. In subsequent steps the critical quality attributes (CQAs) of the product, critical material attributes (CMAs) of raw material and critical process parameters (CPPs) are identified and controlled in order to ensure that the desired product will be consistently manufactured. A more patient-centric approach for the identification of the QTPP is presented in the Biopharmaceutics Risk Assessment Roadmap (BioRAM) which includes patient, caregiver and stakeholder inputs.

8.9.1   Tools to Identify Quality Target Product Profile

BioRAM is a more structured and harmonised QbD paradigm to development based on therapy (not manufacturing) driven considerations [55]. The BioRAM not only benefits from prior knowledge but also from biopharmaceutics and all related sciences to establish the rationale for the development of a particular formulation followed by feasibility, risk assessment and confirmatory studies to confirm or further optimise the suitability of a

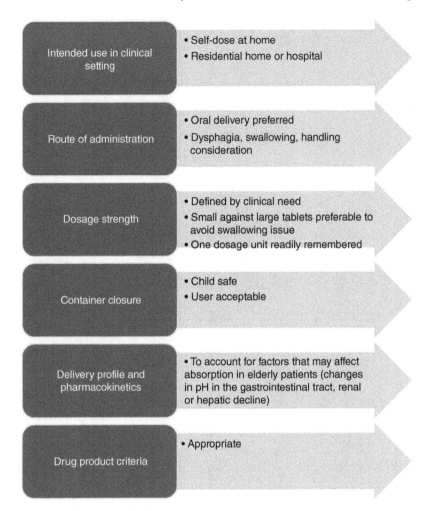

Figure 8.7 *Factors considered in the QTPP of a formulation to be administered to elderly patients. Source: Modified from Page et al. [54].*

product and its QTPP. The desired pharmacokinetic outcome is categorised into four delivery scenarios into which all products are assumed to fit: rapid onset, multiphasic delivery, targeted time-window and targeted prolonged exposure. The delivery scenario and subsequently the formulation achieving the intended scenario are selected based on patient needs and disease treatment. The BioRAM approach to development is summarised in Figure 8.8. An initial QTPP is drawn (Box 1) based on prior knowledge. Key aspects at this stage include [55]:

- First in a class molecule or not (i.e. how much is already understood about how molecules from that pharmacological class can modulate disease).
- Translatability of preclinical PK (such as absorption, clearance), PD and/or disease models.
- Confidence in the estimation and prediction of human PK based on preclinical data.

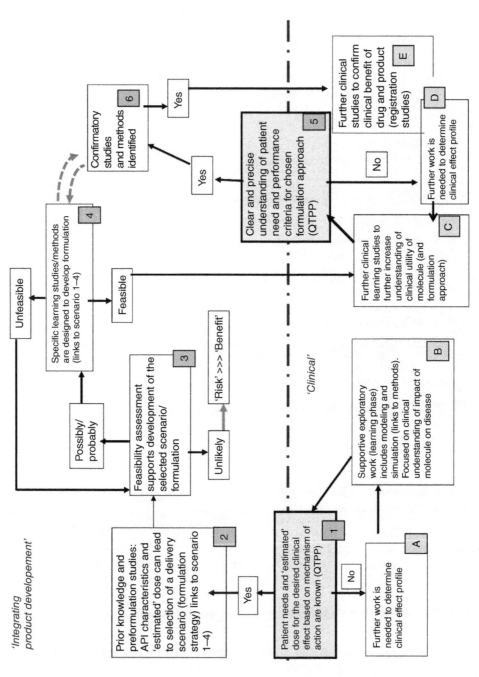

Figure 8.8 The Biopharmaceutics Risk Assessment Roadmap (BioRAM). Boxes 1–6 outline the steps of BioRAM. Boxes A–E represent multidisciplinary input such as clinical efficacy and safety trials, nonclinical studies and/or other studies that provide product performance data. Source: From Selen et al. [55] / With permission of Elsevier.

- Availability of modelling and simulation approaches for the disease and prospective product.
- Variability of input data and understanding of the importance of variability for clinical outcome.

Once the early QTPP has been defined, the formulation strategy (Box 2) is assessed based on prior knowledge and preformulation data. Potential risks leading to product failure are also evaluated. Subsequently, feasibility/risk assessment *in vitro* and *in vivo* studies (Boxes 3–4) are conducted that will confirm (or not) that the selected strategy (based on Box 2) is likely to be successful. The generated knowledge serves to further refine the QTPP (Box 5) to adequately include all patients' needs. Finally, confirmatory studies (Box 6) are conducting to identify the CQAs, CMAs and CPPs ensuring final product quality. The state of the development process is assessed with the introduction of the BioRAM scoring grid. Key questions to be addressed, that subsequently aid the decision-making process, are considered in each BioRAM box. The presence of sufficient data to support a decision and further development translates into positive integer scores. The reasons for 0 scorings are further assessed as i. the inability to score a positive integer due to missing information or ii. the existence of information strongly suggesting that the score is 0 [56]. This can trigger additional considerations (use of prior knowledge, further experimentation, *in silico* models) to draw safer conclusions on the feasibility of a formulation approach. Specific examples of the BioRAM approach to formulation development are presented in subsequent sections.

An example of a BioRAM approach to development is shown in Figure 8.9. For an analgesic compound A, a rapidly dissolving conventional tablet is initially selected based on the drug's therapeutic similarity to ibuprofen and on animal studies showing modest bioavailability when formulating capsules. For successful and fast pain relief, the desirable input delivery rates are identified in subsequent studies to achieve a 70% receptor occupancy for 6 hours. Fast tablet disintegration and dissolution are initially considered optimum. Further feasibility/risk assessment studies revealed a hemodynamic effect suggesting that relatively slower dissolution rates may be required and leading to the development of an enabling prototype formulation. *In vivo* studies in humans revealed a significant and sudden drop in the blood pressure of patients. The risks are considered higher than the benefits post administration of the identified product; therefore, formulation strategies for compound A are discontinued [55].

8.10 Conclusions

Successful treatment not only depends upon the selection of a suitably potent molecule against the therapeutic target, but also the development of an appropriate formulation. Novel drug candidates may be biopharmaceutical challenging, increasing the need for designing sophisticated formulations that will adequately deliver the API into the systemic circulation. Strategising and rationalising the formulation decision-making process is necessary to maximise the probability of success. Traditionally, decisions on the use of a conventional vs enabling or the appropriate enabling formulation are obtained based on rules of thumb deriving from the key drug physicochemical properties (e.g. drug solubility, drug permeability, drug lipophilicity, etc.). More recently, decision trees and computational tools have been

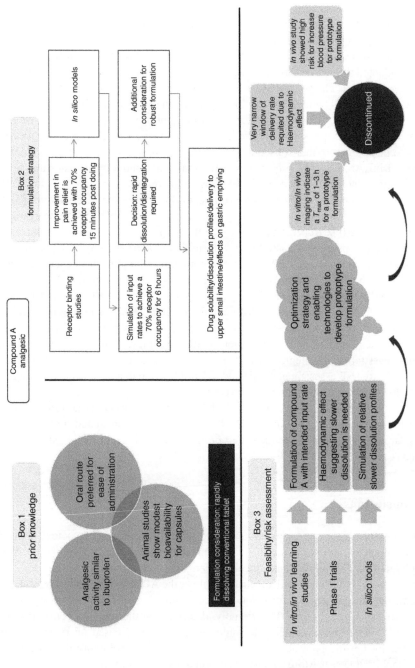

Figure 8.9 *Therapy-driven formulation strategy based on the Biopharmaceutics Risk Assessment Roadmap (BioRAM).*

developed to guide formulation development towards successful formulation at the early stages of development. A desire for *a priori* product quality means a formulation's performance must be linked to pharmacokinetic and pharmacodynamic performance and the patient's needs. These approaches take advantage of prior knowledge, and drug properties to define the ideal characteristics of a product, achieving the desired clinical outcome, and meeting the quality target product profile. Biopharmaceutical tools and/or *in silico/in vivo* studies can then be used to determine the optimum formulation.

References

[1] FDA (2006). Guidance for industry: Q8 pharmaceutical development. https://www.fda.gov/media/71524/download (accessed 2 June 2020).

[2] Darvas, F., Keseru, G., Papp, A. et al. (2002). *In silico* and ex silico ADME approaches for drug discovery. *Curr. Top. Med. Chem.* **2** (12): 1287–1304.

[3] Gribbon, P. and Sewing, A. (2005). High-throughput drug discovery: what can we expect from HTS? *Drug Discov. Today* **10** (1): 17–22.

[4] Boyd, B.J., Bergström, C.A.S., Vinarov, Z. et al. (2019). Successful oral delivery of poorly water-soluble drugs both depends on the intraluminal behavior of drugs and of appropriate advanced drug delivery systems. *Eur. J. Pharm. Sci.* **137**: 104967.

[5] Bayliss, M.K., Butler, J., Feldman, P.L. et al. (2016). Quality guidelines for oral drug candidates: dose, solubility and lipophilicity. *Drug Discov. Today* **21** (10): 1719–1727.

[6] Buckley, S.T., Frank, K.J., Fricker, G. et al. (2013). Biopharmaceutical classification of poorly soluble drugs with respect to "enabling formulations". *Eur. J. Pharm. Sci.* **50** (1): 8–16.

[7] Ku, M.S. (2008). Use of the biopharmaceutical classification system in early drug development. *AAPS J.* **10** (1): 208–212.

[8] Fridgeirsdottir, G.A., Harris, R., Fischer, P.M. et al. (2016). Support tools in formulation development for poorly soluble drugs. *J. Pharm. Sci.* **105** (8): 2260–2269.

[9] Kuentz, M., Holm, R., and Elder, D.P. (2016). Methodology of oral formulation selection in the pharmaceutical industry. *Eur. J. Pharm. Sci.* **87**: 136–163.

[10] Butler, J.M. and Dressman, J.B. (2010). The developability classification system: application of biopharmaceutics concepts to formulation development. *J. Pharm. Sci.* **99** (12): 4940–4954.

[11] Rosenberger, J., Butler, J., and Dressman, J. (2018). A refined developability classification system. *J. Pharm. Sci.* **107** (8): 2020–2032.

[12] Fotaki, N. and Vertzoni, M. (2010). Biorelevant dissolution methods and their applications in *in vitro in vivo* correlations for oral formulations. *Oral Drug Deliv. J.* **4** (1): 2–13.

[13] Tsume, Y., Mudie, D.M., Langguth, P. et al. (2014). The biopharmaceutics classification system: subclasses for *in vivo* predictive dissolution (IPD) methodology and IVIVC. *Eur. J. Pharm. Sci.* **57**: 152–163.

[14] Zane, P., Gieschen, H., Kersten, E. et al. (2019). *In vivo* models and decision trees for formulation development in early drug development: a review of current practices and recommendations for biopharmaceutical development. *Eur. J. Pharm. Biopharm.* **142**: 222–231.

[15] Lai, T., Pencheva, K., Chow, E. et al. (2019). De-risking early-stage drug development with a bespoke lattice energy predictive model: a materials science informatics approach to address challenges associated with a diverse chemical space. *J. Pharm. Sci.* **108** (10): 3176–3186.

[16] Docherty, R., Pencheva, K., and Abramov, Y.A. (2015). Low solubility in drug development: de-convoluting the relative importance of solvation and crystal packing. *J. Pharm. Pharmacol.* **67** (6): 847–856.

[17] Branchu, S., Rogueda, P.G., Plumb, A.P. et al. (2007). A decision-support tool for the formulation of orally active, poorly soluble compounds. *Eur. J. Pharm. Sci.* **32** (2): 128–139.

[18] Agoram, B., Woltosz, W.S., and Bolger, M.B. (2001). Predicting the impact of physiological and biochemical processes on oral drug bioavailability. *Adv. Drug Deliv. Rev.* **50** (Suppl 1): S41–S67.

[19] FDA (2019). WORKSHOP: current state and future expectations of translational modeling strategies to support drug product development, manufacturing changes and controls. https://www.fda.gov/media/128809/download (accessed 2 June 2020).

[20] Bermejo, M., Hens, B., and Dickens, J. (2020). A mechanistic physiologically-based biopharmaceutics modeling (PBBM) approach to assess the *in vivo* performance of an orally administered drug product: from IVIVC to IVIVP. *Pharmaceutics* **12** (1): 1–28.

[21] Kuentz, M. (2009). Drug absorption modeling as a tool to define the strategy in clinical formulation development. *AAPS J.* **11** (1): 32–32.

[22] Kostewicz, E.S., Aarons, L., Bergstrand, M. et al. (2014). PBPK models for the prediction of *in vivo* performance of oral dosage forms. *Eur. J. Pharm. Sci.* **57**: 300–321.

[23] Kesisoglou, F., Balakrishnan, A., and Manser, K. (2016). Utility of PBPK absorption modeling to guide modified release formulation development of gaboxadol, a highly soluble compound with region-dependent absorption. *J. Pharm. Sci.* **105** (2): 722–728.

[24] Kesisoglou, F. et al. (2016). Physiologically based absorption modeling to impact biopharmaceutics and formulation strategies in drug development-industry case studies. *J. Pharm. Sci.* **105** (9): 2723–2734.

[25] Stillhart, C., Pepin, X., Tistaert, C. et al. (2019). PBPK absorption modeling: establishing the *in vitro-in vivo* link-industry perspective. *AAPS J.* **21** (2): 19.

[26] Bergström, C.A.S., Charman, W.N., and Porter, C.J.H. (2016). Computational prediction of formulation strategies for beyond-rule-of-5 compounds. *Adv. Drug Deliv. Rev.* **101**: 6–21.

[27] Kuentz, M., Nick, S., Parrott, N. et al. (2006). A strategy for preclinical formulation development using GastroPlus as pharmacokinetic simulation tool and a statistical screening design applied to a dog study. *Eur. J. Pharm. Sci.* **27** (1): 91–99.

[28] Vinarov, Z., Katev, V., Radeva, D. et al. (2018). Micellar solubilization of poorly water-soluble drugs: effect of surfactant and solubilizate molecular structure. *Drug Dev. Ind. Pharm.* **44** (4): 677–686.

[29] Park, S.-H. and Choi, H.-K. (2006). The effects of surfactants on the dissolution profiles of poorly water-soluble acidic drugs. *Int. J. Pharm.* **321** (1): 35–41.

[30] Mu, H., Holm, R., and Müllertz, A. (2013). Lipid-based formulations for oral administration of poorly water-soluble drugs. *Int. J. Pharm.* **453** (1): 215–224.

[31] Brouwers, J., Brewster, M.E., and Augustijns, P. (2009). Supersaturating drug delivery systems: the answer to solubility-limited oral bioavailability? *J. Pharm. Sci.* **98** (8): 2549–2572.

[32] Kalantzi, L., Goumas, K., Kalioras, V. et al. (2006). Characterization of the human upper gastrointestinal contents under conditions simulating bioavailability/bioequivalence studies. *Pharm. Res.* **23** (1): 165–176.

[33] Bucheit, J.D., Pamulapati, L.G., Carter, N. et al. (2020). Oral semaglutide: a review of the first oral glucagon-like peptide 1 receptor agonist. *Diabetes Technol. Ther.* **22** (1): 10–18.

[34] Buckley, S.T., Bækdal, T.A., Vegge, A. et al. (2018). Transcellular stomach absorption of a derivatized glucagon-like peptide-1 receptor agonist. *Sci. Transl. Med.* **10** (467): eaar7047.

[35] Gradauer, K., Iida, M., Watari, A. et al. (2017). Dodecylmaltoside modulates bicellular tight junction contacts to promote enhanced permeability. *Mol. Pharm.* **14** (12): 4734–4740.

[36] Babadi, D., Dadashzadeh, S., Osouli, M. et al. (2020). Nanoformulation strategies for improving intestinal permeability of drugs: a more precise look at permeability assessment methods and pharmacokinetic properties changes. *J. Control. Release* **321**: 669–709.

[37] Brayden, D.J., Hill, T.A., Fairlie, D.P. et al. (2020). Systemic delivery of peptides by the oral route: formulation and medicinal chemistry approaches. *Adv. Drug Deliv. Rev.* **157**: 2–36.

[38] Dahlgren, D., Cano-Cebrián, M.J., Olander, T. et al. (2020). Regional intestinal drug permeability and effects of permeation enhancers in rat. *Pharmaceutics* **12** (3): 242.

[39] Gopinathan, S., Nouraldeen, A., and Wilson, A.G. (2010). Development and application of a high-throughput formulation screening strategy for oral administration in drug discovery. *Future Med. Chem.* **2** (9): 1391–1398.

[40] Ayad, M.H. (2015). Rational formulation strategy from drug discovery profiling to human proof of concept. *Drug Deliv.* **22** (6): 877–884.

[41] Rabinow, B.E. (2004). Nanosuspensions in drug delivery. *Nat. Rev. Drug Discov.* **3** (9): 785–796.

[42] Alzghoul, A., Alhalaweh, A., Mahlin, D. et al. (2014). Experimental and computational prediction of glass transition temperature of drugs. *J. Chem. Inf. Model.* **54** (12): 3396–3403.

[43] Persson, L.C., Porter, C.J.H., Charman, W.N. et al. (2013). Computational prediction of drug solubility in lipid based formulation excipients. *Pharm. Res.* **30** (12): 3225–3237.

[44] Alskär, L.C., Porter, C.J., and Bergström, C.A. (2016). Tools for early prediction of drug loading in lipid-based formulations. *Mol. Pharm.* **13** (1): 251–261.

[45] Xiang, T.-X. and Anderson, B.D. (2014). Molecular dynamics simulation of amorphous hydroxypropyl-methylcellulose acetate succinate (HPMCAS): polymer model development, water distribution, and plasticization. *Mol. Pharm.* **11** (7): 2400–2411.

[46] Warren, D.B., King, D., Benameur, H. et al. (2013). Glyceride lipid formulations: molecular dynamics modeling of phase behavior during dispersion and molecular interactions between drugs and excipients. *Pharm. Res.* **30** (12): 3238–3253.

[47] Mitra, A., Zhu, W., and Kesisoglou, F. (2016). Physiologically based absorption modeling for amorphous solid dispersion formulations. *Mol. Pharm.* **13** (9): 3206–3215.

[48] Gao, Y., Carr, R.A., Spence, J.K. et al. (2010). A pH-dilution method for estimation of biorelevant drug solubility along the gastrointestinal tract: application to physiologically based pharmacokinetic modeling. *Mol. Pharm.* **7** (5): 1516–1526.

[49] Woodcock, J. (2004). The concept of pharmaceutical quality. *Am. Pharm. Rev.* **7** (6): 10–15.

[50] Yu, L.X. (2008). Pharmaceutical quality by design: product and process development, understanding, and control. *Pharm. Res.* **25** (4): 781–791.

[51] FDA (2009). Guidance for industry: Q8(R2) pharmaceutical development. https://www.fda.gov/media/71535/download (accessed 2 June 2020).

[52] Politis, S.N., Colombo, P., Colombo, G. et al. (2017). Design of experiments (DoE) in pharmaceutical development. *Drug Dev. Ind. Pharm.* **43** (6): 889–901.

[53] Javed, M.N., Kohli, K., and Amin, S. (2018). Risk assessment integrated QbD approach for development of optimized bicontinuous mucoadhesive limicubes for oral delivery of rosuvastatin. *AAPS PharmSciTech* **19** (3): 1377–1391.

[54] Page, S., Coupe, A., and Barrett, A. (2016). An industrial perspective on the design and development of medicines for older patients. *Int. J. Pharm.* **512** (2): 352–354.

[55] Selen, A., Dickinson, P.A., Müllertz, A. et al. (2014). The biopharmaceutics risk assessment roadmap for optimizing clinical drug product performance. *J. Pharm. Sci.* **103** (11): 3377–3397.

[56] Dickinson, P.A., Kesisoglou, F., Flanagan, T. et al. (2016). Optimizing clinical drug product performance: applying biopharmaceutics risk assessment roadmap (BioRAM) and the BioRAM scoring grid. *J. Pharm. Sci.* **105** (11): 3243–3255.

9

Biopharmaceutic Classification System

Hannah Batchelor[1] and Talia Flanagan[2]

[1] Strathclyde Institute of Pharmacy and Biomedical Sciences, University of Strathclyde, Glasgow, United Kingdom
[2] UCB Pharma S.A., Avenue de l'industrie, 1420Braine l'Alleud, Belgium

9.1 Description and History of the BCS

The biopharmaceutic classification system was introduced in the 1990s and is described in a seminal paper by Amidon et al. [1]. The biopharmaceutic classification system (BCS) uses the two factors that govern the rate and extent of drug absorption: solubility and intestinal permeability. The main objective of the BCS was to predict *in vivo* pharmacokinetic behaviour of drug products using measurements of permeability and solubility. The rationale behind its use was to provide a standardised approach to assessing products rather than a product-by-product approach. The principles of the BCS have been widely applied to preclinical, clinical drug development, new drug application (NDA), abbreviated new drug application (ANDA) submissions and post-approval changes for marketed products.

9.2 BCS-Based Criteria for Solubility, Dissolution and Permeability

The definition of solubility within the BCS has been based on the assumption that the entire dose is soluble when ingested with a standard glass of water, as used in clinical testing. In principle, this suggests that solubility is not a rate-limiting factor for subsequent absorption

Biopharmaceutics: From Fundamentals to Industrial Practice, First Edition. Edited by Hannah Batchelor.
© 2022 John Wiley & Sons Ltd. Published 2022 by John Wiley & Sons Ltd.

thus a drug that is soluble is classified as highly soluble whereas one that is not is poorly soluble. There have been many debates about the media that can be used to assess solubility to ensure that this appropriately reflects the gastro-intestinal fluid and that the solubility measurement is appropriate. In addition, the explicit definition of the highest dose or highest single therapeutic dose has also been highlighted. The differences that existed across different regulatory bodies prior to the introduction of the ICH harmonised guidance are provided here [2].

The 'high' solubility classification here has the potential to be much higher than the 'high' solubility goal used in drug discovery (>65 µg/mL) [3]. Only drugs where the dose is less than 15 mg would meet the BCS high solubility criteria based on measurements in drug discovery. Typically, within drug discovery, the dose is yet to be determined thus it is not always possible to provide a BCS solubility classification until later in development. Further details on solubility are presented in Chapter 4.

However, it should also be noted that solubility measurements are performed in a static system where equilibrium solubility is reported; *in vivo*, as the drug is solubilised, it will permeate the membrane thus a more dynamic situation is present in reality. The dissolution data captures the rate of solubilisation of the drug and is also a part of the BCS criteria. The dissolution criteria require that the majority of the drug is soluble within the time it takes for gastric emptying such that there is no delay to permeability within the small intestine. As with the solubility criteria, there has been some historical divergence in specific parameters to measure rapid dissolution yet these are highlighted in a review [2]. Additional details on dissolution methodology are presented in Chapter 6.

The permeability criteria relate to the extent to which a molecule crosses the intestinal membrane and reaches the systemic circulation. As stated in Chapter 5 on permeability measures of permeation can be complicated by the fact that in bioavailability studies, it is the systemic circulation that is sampled to provide data on the bioavailability of a drug. Thus, there may be pre-systemic metabolism relating to intestinal or first-pass metabolism that impacts the difference between the fraction absorbed and the overall bioavailability. Within BCS guidance a high fraction absorbed is linked to a highly permeable drug; this means that the permeation is not limiting the absorption. Typical regulatory guidelines report that greater than 85% of the drug should be absorbed for it to be classified as a highly permeable drug. There are alternative *in vitro* methods that can be used to identify whether a drug is likely to be highly permeable prior to conducting an *in vivo* study and these are described in more detail in Chapter 5 on permeability. Additionally, one *in vitro* method (Caco-2 cells) can be used to demonstrate high permeability when applying for a BCS biowaiver.

The BCS is often presented as a grid, shown in Figure 9.1.

This simple classification can be used to consider the rate-limiting steps to the absorption of orally administered drugs:

BCS 1: these are highly soluble and highly permeable agents, thus gastric emptying is usually the major rate-limiting step for absorption.

BCS 2: Low solubility may limit dissolution, which may limit the rate and extent of absorption. There are certain formulations that can improve the solubility and dissolution of drugs to improve the rate and extent of dissolution (and subsequent absorption).

BCS 3: Poor permeability may limit the absorption rate and the subsequent extent of absorption. Improving the concentration of the drug at the absorptive membrane or increasing the duration of contact would improve overall bioavailability.

BCS 4: Poor permeability and low solubility can both affect the rate and extent of absorption.

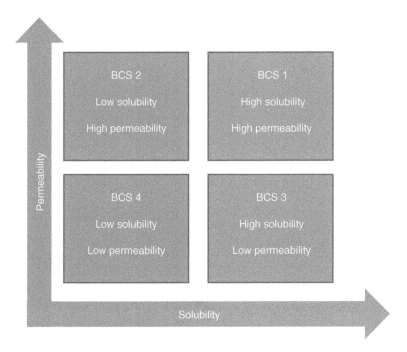

Figure 9.1 *Classification of drugs into four BCS groups based on their solubility and permeability.*

It is recognised that the BCS is a simplistic overview of the factors that control absorption. In particular, it should be recognised that the criteria that classify drugs as highly soluble and highly permeable are stringent and that those borderline compounds who just miss those targets are likely to be much more promising than those who are a long way from the target values. It should also be acknowledged that there are successful BCS 4 drugs on the market; although in much lower proportions than BCS 1, 2 and 3 with about 5% of the top 200 oral drug products being in BCS 4 [4]. A snapshot of the BCS classification of WHO Essential drugs is shown in Figure 9.2, demonstrating the distribution of these agents into the four classes.

9.3   BCS-Based Biowaivers

As stated in Chapter 1, two products can be considered to be bioequivalent when their rate and extent of absorption are the same. Bioequivalent products can be used interchangeably as the resulting pharmacokinetic profile is the same. Assessment of bioequivalence usually requires a clinical study which can be costly and time-consuming. Within the BCS framework, when certain criteria are met that provide reassurance that absorption will be rapid and complete, the *in vitro* data (e.g. solubility, dissolution and permeability data) of a drug product can be used to justify waivers of clinical bioequivalence studies (i.e. biowaivers), leading to significantly reduced cost and time associated with the human studies as well as regulatory burdens. These biowaivers can be used to provide reassurance of bioequivalence during development, for example between early clinical trial products and to-be-marketed products, or between generic and innovator products.

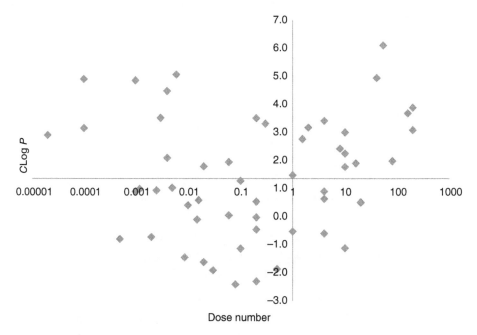

Figure 9.2 *BCS Classification of WHO Essential drugs using metoprolol as the marker for high permeability and using ClogP as the permeability marker. Source: Data from Kasim et al. [5].*

9.4 Regulatory Development of BCS-Based Biowaivers

Since its introduction, there have been some changes to the criteria definitions within the BCS and its application to biowaivers, particularly by regulatory agencies. Since the publication of the original FDA guidance in 2000 [6], several other global regulatory agencies published their own guidelines on BCS, including EMA, Health Canada and the WHO [7–9]. All of these guidelines were built along with the same BCS principles, that is classification of a drug substance according to solubility and permeability, with additional requirements regarding dissolution performance and excipient changes. However, there were subtle differences between these regional guidelines, in terms of the acceptance criteria, required testing and the supportive data required for biowaiver applications [10]. This meant that in some cases, a drug product would be eligible for a BCS-based biowaiver in some territories but not in others. Additionally, some major markets did not accept BCS-based biowaivers at all, meaning that a clinical BE study would still be required to support formulation changes in those territories even if a BCS-based biowaiver could be used elsewhere. This was a particular issue for pharmaceutical companies developing a drug product for several different global markets, as different approaches and data would be required in different regions.

9.5 International Harmonisation of BCS-Based Biowaiver Criteria – ICH M9

In 2016, the International Council for Harmonisation (ICH) endorsed the development of a new globally harmonised guideline on BCS-Based Biowaivers (ICH M9). The new

---

Box 9.1   Summary of BCS Classification and Biowaiver Requirements in the ICH M9 Guideline

- **High solubility:** highest single therapeutic dose is soluble in ≤250 mL of aqueous media across the pH range of 1.2–6.8 at 37 ± 1 °C.
- **High permeability:** preferably assessed based on data from human absolute bioavailability or mass balance studies, demonstrating >85% extent of absorption. *In vitro* permeability assessment using Caco-2 cells is also permitted. Data to demonstrate drug substance stability in the GI tract may also be required, depending on the type of data used to demonstrate high permeability.
- **Biowaivers are permitted for BCS Class 1 and BCS Class 3** drug substances, providing that the drug products meet the dissolution and excipient similarity criteria outlined in the guideline.
- **Dissolution comparisons** should be performed in at least three buffers (pH 1.2, 4.5 and 6.8);
  - For BCS 1 drug products test and reference products must be very rapidly dissolving (>85% in 15 minutes) or rapidly dissolving (>85% in 30 minutes) and similar by f2 comparison.
  - For BCS 3 drug products, both test and reference products must be very rapidly dissolving (>85% in 15 minutes).
- **Excipients:** for BCS 1, qualitative and quantitative differences in excipients are permitted, except for excipients that may affect absorption. For BCS 3, all excipients should be qualitatively the same and quantitatively similar.

Source: Bransford et al. [11] and ICH [12].

---

harmonised guideline was developed by a working group consisting of experts from global regulatory agencies and the pharmaceutical industry, spanning both new drugs and generic drugs. The final harmonised guideline was endorsed in November 2019 and will be adopted by all regulatory members of ICH, including US FDA, EMA, MHLW/PMDA Japan, Health Canada, Swissmedic, ANVISA Brazil, HSA Singapore, MFDS Republic of Korea, NMPA China, and TFDA Chinese Taipei. Previous regional guidelines will be superseded by the new globally harmonised guideline. Key points from the guideline are summarised in Box 9.1 [11, 12].

A Question and Answer document has also been developed to accompany the ICH M9 guideline [12]; this document provides additional guidance on specific applications of ICH M9.

9.5.1   Application of BCS-Based Biowaivers

The primary intent of BCS as a regulatory instrument is to enable biowaivers to be granted, that is replacing the need for a human bioequivalence study with *in vitro* data.

Initially, BCS-based biowaivers were limited to BCS 1 drugs; those with high solubility and permeability; this was expanded to encompass BCS 3 drug products in certain territories, as it was recognised that the rate-limiting factors in the absorption process from the drug product were not linked to drug substance solubility, particularly when BCS 3 compounds are available within solutions, that is dissolution is complete before gastric emptying [13]. However, it should be noted that the risks associated with BCS3 are considered to be greater than for BCS1, thus the regulations differ between these groups.

For example, a new generic drug product of a BCS Class 1 or 3 drug substance could gain marketing approval based solely on *in vitro* data, or certain formulation and process changes could be made to a marketed innovator product without the need for additional bioequivalence studies. The principles outlined in the BCS guideline can also be applied to assess bioequivalence risk at earlier stages of drug product development, even if formal bioequivalence studies would not normally be required (e.g. to assess changes made between Phase 2 and Phase 3 drug products). Specific guidance on the level of changes that can be made to immediate-release products without the need for a clinical study was first published in 1995 as the FDA Scale-Up and Post-approval Changes: Chemistry, Manufacturing, and Controls, *In Vitro* Dissolution Testing, and *In Vivo* Bioequivalence Documentation guidance [14].

The first step to determine whether a BCS biowaiver can be obtained is to determine the BCS classification of the drug substance, that is to assess solubility and permeability according to the criteria outlined in the ICH M9 guideline. It is important to note that the guideline contains specific requirements for the type of supporting data required to demonstrate high solubility/permeability, which need to be met; this may be different from the in-house assays used to assess developability and biopharmaceutic risk during development, where the emphasis is on strategy and decision making rather than the demonstration of regulatory-standard bioequivalence. If the drug substance meets the criteria for either BCS1 (high solubility, high permeability) or BCS 3 (high solubility, low permeability) classification, then it is possible to obtain BCS-based biowaivers, provided that certain additional criteria are met for the test and reference products. These are briefly outlined below; for more detail, the reader is referred to the ICH M9 guideline [12] and the associated Question and Answer document [12].

9.5.1.1   Drug Product Type

To be eligible for a BCS biowaiver, the drug product must be an immediate-release oral dosage form with systemic action. The test drug product must be the same dosage form and strength as the reference product. Drug products with buccal or sublingual absorption are not eligible, nor are orodispersible drug products administered without water.

9.5.1.2   Composition

The requirements regarding composition similarity are different for BCS 1 and BCS 3 drug products. Any excipient changes should be assessed for their potential to affect drug absorption. For BCS Class 1, changes in composition are permitted, except for excipients that can impact absorption. The excipients should be qualitatively similar, and the amount should be within ±10% of the amount in the reference product. BCS Class 3 drug substances have low permeability, and so are considered to represent a greater risk for excipient changes. Additionally, there is the less global experience in applying BCS biowaivers for BCS 3 compounds than for BCS 1 due to the historical nature of regulations that supported BCS 3 based biowaivers. Compositional similarity criteria for BCS Class 3 drug products are therefore stricter than for BCS 1. All excipients should be qualitatively and similar between the test and reference products. The definition of 'quantitatively similar' varies between different excipient classes – a table and flow chart are provided in the ICH M9 guideline which outlines the specific requirements [12]. Additionally, excipients that may affect absorption should be qualitatively similar, and the amount should be within ±10% of the amount of excipient in the reference product, as for BCS1 drug products. The ICH M9 guideline provides worked examples of permitted excipient changes and includes a decision tree for use in determining whether a proposed excipient change is acceptable. Furthermore, the guideline allows for wider excipient changes to be made, if suitable justification is provided. For any excipient change between test and reference product, it is important to mechanistically assess the potential of the change to affect drug absorption. The potential of pharmaceutical excipients to affect absorption for BCS Class 1 and 3 compounds have been reviewed by Flanagan [15].

Example excipient changes are presented within the ICH M9 guideline as case studies [12]. For example, a change in the level of sorbitol, an excipient associated with an effect on absorption [16]; where the change is within 10% of the reference product a biowaiver would still be permitted.

9.5.1.3   Dissolution Similarity

To qualify for a BCS-based biowaiver, the test and reference products should have rapid and similar dissolution in aqueous media across the physiological pH range. The dissolution tests should be performed in at least three aqueous buffers, at pH 1.2, 4.5 and 6.8, in a volume of 900 mL or less. No surfactant or organic solvent can be added to the media. The addition of enzymes to the media to overcome cross-linking for gelatin capsules is permitted. Basket or paddle apparatus should be used, at a rotation speed of 100 or 50 rpm, respectively. To demonstrate dissolution profile similarity, both drug products must be either:

- very rapidly dissolving (>85% dissolved in 15 minutes), and
- or rapidly dissolving (>85% dissolved in 30 minutes) and similar by f2 statistical comparison (permitted for BCS Class 1 only).

Further detail, including the selection of appropriate batches to perform the dissolution comparison, is provided in the guideline [12].

9.6   BCS as a Development Tool

The scientific basis of the BCS, emphasised by the regulatory endorsed BCS-based biowaivers, has led to the BCS being used within pharmaceutical development. Early identification of the BCS classification of the drug under development will inform potential absorption issues which will then inform the development strategy. More specifically, the BCS framework has laid down a foundation for pharmaceutical development from pre-formulation investigation of drug candidates, to the selection of dosage forms, formulation and associated manufacturing process. It has also provided a universal tool and language to communicate risk across the multidisciplinary teams involved in the development of a new drug product.

9.6.1   Candidate Selection

At early stages ensuring sufficient aqueous solubility and permeability indicates good systemic exposure and a greater likelihood of clinical success. Given the high cost of conducting studies on animals and humans, it is useful to generate this information as early as possible in the development pathway. It is necessary to conduct a risk-benefit analysis as sub-optimal solubility and permeability is likely to lead to a more challenging development path (e.g. more complex formulation technology, longer development time, higher cost and less certain outcome due to variability in pharmacokinetic studies). For this reason, highly soluble and highly permeable (BCS Class 1) drugs are generally most desirable. There are several formulation and processing technologies that can be applied to improve bioavailability, yet it should be noted that low permeability is much more difficult to overcome because it is an inherent biological property.

9.6.2   Solid Form Selection

If a drug candidate is known to exhibit poor solubility efforts can be directed to increasing the solubility and dissolution rate. Common considerations include the evaluation and selection of solid-state forms that are known to have better apparent solubility. This is discussed in more detail in Chapter 4. Typical examples include alternative salt forms; polymorphs of the use of an amorphous form of the drug.

9.6.3   Product Development

The principles of the BCS are widely used to guide product development and assess associated challenges and risks. During development the dose of the drug to be used can change, indeed for early ascending dose studies a wide range of doses are required. As the dose increases the likelihood of the complete dose being soluble in the recommended 250 mL of fluid decreases. Thus, solubility/dissolution enhancing strategies are required for many compounds during development. As solubility/dissolution is known to limit the overall exposure enabling formations are often used where solubility can be an issue. This is discussed in further detail in Chapter 8. During product development there is a need for appropriate *in vitro* methods that predict the *in vivo* performance; where solubility and dissolution are the rate-limiting steps, it is feasible to develop a discriminatory dissolution test that can provide an *in vivo-in vitro* correlation (IVIVC); thus, this is more feasible for BCS 2 (and some BCS 4) drugs. It can be more challenging to develop an discriminatory dissolution method for highly soluble drugs (BCS 1 and 3); however, the BCS media can be used as clinically relevant dissolution media for these compounds (refer to Chapter 10).

When formulations other than immediate-release products are considered the BCS framework will be used as part of the risk assessment. For example, an understanding of solubility and permeability, specifically regional differences in the GI tract, will influence the potential to develop a modified release dosage form. Generally, BCS class 1 drugs are suitable for modified release; certain BCS 2 drugs have potential where they have low doses with high permeability; whereas modified release is much more complex for those drugs with poor permeability (BCS 3 and 4).

FDA's Scale Up and Post-Approval Change guideline for IR products (SUPAC-IR) [14] also makes reference to the principles of BCS. The guideline specifies the data required to support post-approval changes such as changes to the formulation, manufacturing process or manufacturing scale. This includes *in vitro* dissolution comparisons of varying complexity, and sometimes clinical bioequivalence studies. The data required for some of the changes are dependent on the solubility and permeability classification of the drug substance; the extent of data required to support a given change is often stratified according to BCS class, with reduced requirements for BCS 1, 2 and 3 compounds than for BCS 4. Within these groups, BCS1 usually requires the lowest amount of data to be generated, followed by BCS 3 then BCS2. There are instances where a clinical bioequivalence study is required for BCS Class 4 compounds, but not for the other BCS classes, demonstrating an additional application of BCS to regulatory risk assessment.

9.7 Beyond the BCS

Since its introduction, there have been several adaptations and improvements proposed to the BCS. In 2003, the use of Caco-2 data as a measure of permeability was proposed as this is easier to measure than within a human mass balance study [17]. An adaptation to the BCS was proposed in 2014, where subclasses were introduced [18]. This adaptation considered the dynamic environment of the GI tract with its changing pH, buffer capacity, luminal volume, surfactant luminal conditions, permeability profile and the impact of the fasted/fed state. It is acknowledged that the dissolution of a drug is influenced by its pK_a and the pH of the media in which it is dissolving, for example, a weak acid would be insoluble in gastric conditions (hence a BCS 2 drug) but soluble within the intestine; hence the dissolution rate will increase and the absorption will be unaffected. Thus, changes could be considered low risk form a bioequivalence perspective provided that intestinal dissolution is very rapid and could be eligible for a BCS-based biowaiver. Yet, a weakly basic drug, due to limited solubility within the small intestine would have limited dissolution hence absorption can be affected. Depending on the classification of BCS 2 drugs as acids, bases, or neutral compounds, the authors proposed a change in the dissolution methodology from single media tests to a two-step dissolution setup, in which a gastric compartment is combined with the transfer into an intestinal compartment. This revised methodology better mimics the physiology and provides an understanding of the risk *in vivo*. However, this has never been incorporated into any regulatory guidelines concerning BCS and so cannot be used for biowaiver justification.

9.7.1 Biopharmaceutic Drug Disposition Classification System (BDDCS)

The BDDCS was introduced in 2005; it considers the disposition of drugs rather than just the absorption [19]. It was recognised that highly permeable (BCS 1 and 2) drugs are eliminated primarily via metabolism whereas BCS 3 and 4 drugs were primarily eliminated unchanged into urine and bile. It was also noted that transporter effects differ by BCS classification: for BCS 1 transporter effects are minimal; BCS 2 efflux transporter effects dominate; BCS 3 absorptive transporter effects dominate and for BCS 4 both efflux and absorptive transporters can be important. This classification can help in the selection of a candidate molecule and the development strategy for the compounds by improving the understanding and communication of risks associated with the compound. This greater understanding and risk assessment provided justification to expand BCS-based biowaivers beyond BCS 1 drugs [19]. Accurate assessment of permeability is complex; yet understanding disposition can support the expansion of the permeability criteria to better predict where permeability will limit overall absorption. Further advantages of the BDDCS include: predicting when transporter-enzyme effects may be clinically significant; prediction of food effects; and predicting the potential for drug–drug interaction [19]. However, this has never been incorporated into any regulatory guidelines concerning BCS, and so cannot be used for biowaiver justification.

9.7.2 Developability Classification System

The Developability Classification System (DCS) is an adaptation of the BCS, introduced in 2010 as a tool for application to formulation development [20]. The goal is to develop realistic formulation strategies by attempting to understand what critical factors are likely to control absorption. The DCS considers the biorelevant and dynamic nature of drug absorption to expand the conservative limits of the BCS. The DCS includes the use of biorelevant media for solubility and dissolution; the use of *in silico* or cell-based models to predict permeability [20]. Key aspects of the DCS are that BCS class II compounds (low solubility, high permeability) are further divided into solubility- and rate-limited dissolution. An understanding of the rate-limiting step for absorption drives the formulation strategy and guides the development of appropriate predictive tests. The DCS was revised, and a refined version was presented in 2018 [21]. The revised DCS included considerations of dose ranges and the impact of gastric solubility and precipitation to further de-risk the development for poorly soluble weak bases. The DCS is not intended to support biowaivers in a regulatory setting.

9.7.3 Fraction Absorbed Classification System

In 2015, the Fraction absorbed classification system FaCS classification was introduced by Sugano and Terada [22]. This expands upon the BCS by considering dissolution rate limited; permeability limited and solubility–permeability limited absorption. These classifications take into account the dynamic nature of drug absorption; hence if the permeability is rapid then the impact of solubility of dissolution is reduced. This system combines the dose number (Do, defined as the dose divided by the volume (250 mL) and solubility of the drug); dissolution number (Dn, defined as the ratio of mean residence time to mean dissolution time) and permeation number which combines the permeability rate and the time frame for absorption. The resulting system results in five classes: dissolution rate limited; permeability limited (subdivided into two subclasses) and solubility-permeability limited (subdivided into two subclasses). However, this has never been incorporated into any regulatory guidelines concerning BCS, and so cannot be used for biowaiver justification.

9.7.4 BCS Applied to Special Populations

The BCS was derived based on adult anatomy and physiology. However, it has been recognised that there are many populations or disease states that provide divergence from adult gastrointestinal anatomy and physiology.

There have been efforts to apply the BCS to other populations including paediatrics; where differences in the dose : solubility need to be taken into account as well as differences in permeability [23–30].

9.8 Conclusions

BCS-based biowaivers enable human bioequivalence studies to be waived for drug products which meet the criteria in the guideline, that is replaced with an *in vitro* comparison. This can reduce the cost and time associated with bringing a new medicinal product to market, and prevents unnecessary exposure of healthy volunteers to the drug product. Additionally, during development BCS classification allows scientists to anticipate the potential issues, develop strategy, and prioritise resources for pharmaceutical development.

References

[1] Amidon, G.L., Lennernas, H., Shah, V.P., and Crison, J.R. (1995). A theoretical basis for a biopharmaceutic drug classification: the correlation of *in vitro* drug product dissolution and *in vivo* bioavailability. *Pharmaceutical Research* **12** (3): 413–420.

[2] Davit, B.M., Kanfer, I., Tsang, Y.C., and Cardot, J.-M. (2016). BCS biowaivers: similarities and differences among EMA, FDA, and WHO requirements. *The AAPS Journal* **18** (3): 612–618.

[3] Kerns, E.H., Di, L., and Carter, G.T. (2008). *In vitro* solubility assays in drug discovery. *Current Drug Metabolism* **9** (9): 879–885.

[4] Takagi, T., Ramachandran, C., Bermejo, M. et al. (2006). A provisional biopharmaceutical classification of the top 200 oral drug products in the United States, Great Britain, Spain, and Japan. *Molecular Pharmaceutics* **3** (6): 631–643.

[5] Kasim, N.A., Whitehouse, M., Ramachandran, C. et al. (2004). Molecular properties of who essential drugs and provisional biopharmaceutical classification. *Molecular Pharmaceutics* **1** (1): 85–96.

[6] FDA (2000). *Guidance for Industry, Waiver of in vivo Bioavailability and Bioequivalence Studies for Immediate Release Solid Oral Dosage Forms Based on a Biopharmaceutics Classification System*. CDER/FDA.

[7] Canada, H. (2014). *Guidance Document: Biopharmaceutics Classification System Based Biowaiver*. Ottawa, ON: Health Canada.

[8] EMA (2010). Guideline on the investigation of bioequivalence. CPMP/QWP/EWP/1401/98Rev. 1. https://www.ema.europa.eu/en/documents/scientific-guideline/guideline-investigation-bioequivalence-rev1_en.pdf (accessed January 2021).

[9] WHO (2006). Proposal to waive *in vivo* bioequivalence requirements for WHO Model List of Essential Medicines immediate-release, solid oral dosage forms. WHO Technical Report Series, No. 937. Annex 8. https://www.who.int/medicines/areas/quality_safety/quality_assurance/Prop osalWaiveVivoBioequivalenceRequirementsModelListEssentialMedicinesImmediateReleaseS olidOralDosageFormsTRS937Annex8.pdf (accessed January 2021).

[10] Van Oudtshoorn, J.E., Garcia-Arieta, A., Santos, G.M.L. et al. (2018). A survey of the regulatory requirements for BCS-based biowaivers for solid oral dosage forms by participating regulators and organisations of the international generic drug regulators programme. *Journal of Pharmacy and Pharmaceutical Sciences* **21** (1): 27–37.

[11] Bransford, P., Cook, J., and Gupta, M. (2020). ICH M9 guideline in development on biopharmaceutics classification system-based biowaivers: an industrial perspective from the IQ consortium. *Molecular Pharmaceutics* **17** (2): 361–372.

[12] ICH (2019). ICH harmonised guideline M9: biopharmaceutics classification system-based biowaivers. https://database.ich.org/sites/default/files/M9_Guideline_Step4_2019_1116.pdf (accessed January 2021).

[13] Crison, J.R., Timmins, P., and Keung, A. (2012). Biowaiver approach for biopharmaceutics classification system class 3 compound metformin hydrochloride using *In Silico* modeling. *Journal of Pharmaceutical Sciences* **101** (5): 1773–1782.

[14] FDA (1995). SUPAC-IR: Immediate-release solid oral dosage forms: scale-up and post-approval changes: chemistry, manufacturing and controls, *in vitro* dissolution testing, and *in vivo* bioequivalence documentation. https://www.fda.gov/regulatory-information/search-fda-guidance-documents/supac-ir-immediate-release-solid-oral-dosage-forms-scale-and-post-approval-changes-chemistry (January 2021).

[15] Flanagan, T. (2019). Potential for pharmaceutical excipients to impact absorption: a mechanistic review for BCS Class 1 and 3 drugs. *European Journal of Pharmaceutics and Biopharmaceutics* **141**: 130–138.

[16] Adkison, K., Wolstenholme, A., and Lou, Y. (2018). Effect of sorbitol on the pharmacokinetic profile of lamivudine oral solution in adults: an open-label, randomized study. *Clinical Pharmacology & Therapeutics* **103** (3): 402–408.

[17] Rinaki, E., Valsami, G., and Macheras, P. (2003). Quantitative biopharmaceutics classification system: the central role of dose/solubility ratio. *Pharmaceutical Research* **20** (12): 1917–1925.

[18] Tsume, Y., Mudie, D.M., Langguth, P. et al. (2014). The biopharmaceutics classification system: subclasses for *in vivo* predictive dissolution (IPD) methodology and IVIVC. *European Journal of Pharmaceutical Sciences* **57**: 152–163.

[19] Wu, C.-Y. and Benet, L.Z. (2005). Predicting drug disposition via application of BCS: transport/absorption/elimination interplay and development of a biopharmaceutics drug disposition classification system. *Pharmaceutical Research* **22** (1): 11–23.

[20] Butler, J.M. and Dressman, J.B. (2010). The developability classification system: application of biopharmaceutics concepts to formulation development. *Journal of Pharmaceutical Sciences* **99** (12): 4940–4954.

[21] Rosenberger, J., Butler, J., and Dressman, J. (2018). A refined developability classification system. *Journal of Pharmaceutical Sciences* **107** (8): 2020–2032.

[22] Sugano, K. and Terada, K. (2015). Rate- and extent-limiting factors of oral drug absorption: theory and applications. *Journal of Pharmaceutical Sciences* **104** (9): 2777–2788.

[23] Batchelor, H. (2014). Paediatric biopharmaceutics classification system: current status and future decisions. *International Journal of Pharmaceutics* **469** (2): 251–253.

[24] Batchelor, H., Ernest, T., and Flanagan, T. (2016). Towards the development of a paediatric biopharmaceutics classification system: results of a survey of experts. *International Journal of Pharmaceutics* **511** (2): 1151–1157.

[25] Batchelor, H.K., Kendall, R., Desset-Brethes, S. et al. (2013). Application of *in vitro* biopharmaceutical methods in development of immediate release oral dosage forms intended for paediatric patients. *European Journal of Pharmaceutics and Biopharmaceutics* **85** (3 PART B): 833–842.

[26] Charoo, N.A., Cristofoletti, R., and Dressman, J.B. (2015). Risk assessment for extending the biopharmaceutics classification system-based biowaiver of immediate release dosage forms of fluconazole in adults to the paediatric population. *Journal of Pharmacy and Pharmacology* **67** (8): 1156–1169.

[27] Delmoral-sanchez, J.M., Gonzalez-alvarez, I., Gonzalez-alvarez, M. et al. (2019). Classification of WHO essential oral medicines for children applying a provisional pediatric biopharmaceutics classification system. *Pharmaceutics* **11** (11): 567. https://doi.org/10.3390/pharmaceutics11110567. PMID: 31683740; PMCID: PMC6920833.

[28] Gandhi, S.V., Rodriguez, W., Khan, M., and Polli, J.E. (2014). Considerations for a pediatric biopharmaceutics classification system (BCS): application to five drugs. *AAPS PharmSciTech* **15** (3): 601–611.

[29] Shawahna, R. (2016). Pediatric biopharmaceutical classification system: using age-appropriate initial gastric volume. *AAPS Journal* **18** (3): 728–736.

[30] Somani, A.A., Thelen, K., and Zheng, S. (2016). Evaluation of changes in oral drug absorption in preterm and term neonates for biopharmaceutics classification system (BCS) Class I and II compounds. *British Journal of Clinical Pharmacology* **81** (1): 137–147.

10

Regulatory Biopharmaceutics

Shanoo Budhdeo[1], Paul A. Dickinson[1] and Talia Flanagan[2]

[1] *Seda Pharmaceutical Development Services, Alderley Edge,*
Alderley Park, Cheshire, UK
[2] *UCB Pharma S.A., Avenue de l'industrie, 1420Braine l'Alleud, Belgium*

10.1 Introduction

The role biopharmaceutics plays in regulatory thinking and regulatory documentation has evolved over the last few decades. This chapter will introduce you to concepts used, guidance and tools available to the pharmaceutical industry to help during development of a new drug entity or generic drug product to provide the regulators with biopharmaceutics knowledge and evidence on drug product quality via regulatory submission documents. Biopharmaceutics is a relatively new topic in terms of regulatory documentation and historically may have been considered under the field of clinical pharmacology. It has evolved into its own scientific discipline and this was reflected in the 2000s whereby the FDA created the office of new drug quality assessment (now called Office of New Drug Products [ONDP]), and biopharmaceutic reviewers were moved from clinical pharmacology into this newly formed group.

The regulatory authority evaluates the data and information provided by drug developers in regulatory submissions to ensure drug products are safe and efficacious. This includes aspects of drug product performance that could impact safety and efficacy. Therefore, an understanding of the type of biopharmaceutic data and knowledge the regulators expect to see/are thinking about in regulatory documentation will help focus the development activities. The key focus of the biopharmaceutics reviewer is on the clinical quality of the drug

Biopharmaceutics: From Fundamentals to Industrial Practice, First Edition. Edited by Hannah Batchelor.
© 2022 John Wiley & Sons Ltd. Published 2022 by John Wiley & Sons Ltd.

product to ensure that the efficacy demonstrated in the pivotal clinical trials will also be achieved in patients taking the commercial drug product throughout its lifecycle.

Key considerations include:

- How is the proposed commercial product linked to the drug product used in pivotal safety and efficacy trials?
- Will the manufacturing process always produce drug product that has the required performance in the patient?
- Will the methods and specifications proposed detect and reject a batch that does not perform as intended in the patient?

Biopharmaceutics provides a link between the drug product quality and the clinical performance (safety and efficacy) in patients. This involves identifying attributes of the drug product that will affect the *in vivo* performance in patients and putting in place suitable controls, tests and specifications to ensure these product attributes are suitably controlled during the manufacture of the drug product, thus ensuring consistent *in vivo* performance in patients. The applicant should strive to understand and establish correlations between drug product quality attributes and clinical performance.

This chapter will introduce you to key regulatory concepts such as Quality by Design, clinically relevant specifications and BioRAM that have been introduced which build in clinical quality and put the patient at the centre of development, and where the biopharmaceutics scientist plays a key role (additional information is also within Chapters 7 and 8). Furthermore, reference to regulatory guidelines will be made throughout this chapter. These guidelines help drug developers navigate the development pathway of drug products and generate biopharmaceutic data and understanding to ensure suitable clinical performance of the drug product, and to provide evidence of this to regulatory authorities in the marketing applications.

This chapter will focus on orally administered drug products with a systemic mechanism of action. However, the same scientific principles and regulatory considerations also apply to drug products administered by other routes, although the *in vitro* tests and clinical studies used to answer these questions and demonstrate equivalence may be different.

10.2   Clinical Bioequivalence Studies

Bioequivalence is a fundamentally important concept that is used to link the performance in patients of a commercial drug product to the drug product that was used in the registration studies (traditionally controlled Phase 3 studies, although in some therapeutics areas, such as oncology conditional registration can be based on Phase 2 studies). The registration studies will provide both the efficacy and safety data that demonstrate the positive risk benefit of the drug, the dose and dosing regimen, and consequently the information that is included in the approved prescribing label.

If changes are made to the drug product between that used in the registration studies and the commercial supply then bioequivalence studies may need to be performed to demonstrate that the commercial drug product would still provide the risk-benefit profile described in the label. Bioequivalence is also the method by which generic drug product manufacturers can introduce their products without repeating the clinical safety and efficacy studies performed by the innovator (which are long and costly) but by successfully performing a clinical bioequivalence study.

Changes to a drug product can be visually obvious, such as a change to a generic product or change from capsule to tablet, or a change that may not be visually obvious, such as a change in excipients or manufacturing process. The type of changes that can occur for a drug product and whether they would require a successful bioequivalence study before the introduction of change into the supply chain are discussed in the SUPAC guidance [1] that was introduced in the BCS chapter (Chapter 9). These guidance use the risk associated with absorption as a guide: for some drug substances and drug products with low-risk absorption characteristics, a change that would ordinarily require a clinical bioequivalence study may waived based on *in vitro* evaluation (see Chapter 9, Section 9.3)

Bioequivalence has subtly different definitions based on regulatory regions; for instance, in the United States, it is defined as: 'the absence of a significant difference in the rate and extent to which the active ingredient or active moiety in pharmaceutical equivalents or pharmaceutical alternatives becomes available at the site of drug action when administered at the same molar dose under similar conditions in an appropriately designed study' (CFR 320.1 [2]).

Whereas in the EU, it is defined as: 'Two medicinal products containing the same active substance are considered bioequivalent if they are pharmaceutically equivalent or pharmaceutical alternatives and their bioavailabilities (rate and extent) after administration in the same molar dose lie within acceptable predefined limits. These limits are set to ensure comparable *in vivo* performance, i.e. similarity in terms of safety and efficacy' [3].

Recently, ICH M9 (guideline on biopharmaceutics classification system-based biowaivers) [4] defined bioequivalence as: 'Two drug products containing the same drug substance(s) are considered bioequivalent if their bioavailabilities (rate and extent of drug absorption) after administration in the same molar dose lie within acceptable predefined limits. These limits are set to ensure comparable *in vivo* performance, i.e. similarity in terms of safety and efficacy'.

However, it should be noted that bioequivalence guidance are not currently harmonised under ICH (but may be in the future under ICH M13 for immediate releases products, see Section 10.5).

These definitions are based on the precept of modern drug development that there is a direct correlation between drug concentration and activity (whether efficacy or adverse events). The EMA definition continues to say 'In bioequivalence studies, the plasma concentration time curve is generally used to assess the rate and extent of absorption'. Selected pharmacokinetic parameters and preset acceptance limits allow the final decision on bioequivalence of the tested products (see Chapter 2). AUC, the area under the concentration time curve, reflects the extent of exposure. C_{max}, the maximum plasma concentration or peak exposure, and the time to maximum plasma concentration, t_{max}, are parameters that are influenced by absorption rate.

Thus, in practice despite subtly different definitions bioequivalence is based on comparing the plasma concentration time profile of the old (reference) and new (test) drug product and if these are sufficiently similar such that the efficacy and safety is assumed to be equivalent. In this instance, plasma concentration is used as a surrogate of the drug concentration at the active site.

Simplistically, a pharmaceutical company (sponsor) would investigate plasma exposure of the test (new) formulation versus the reference (old) formulation, and asses the differences in maximum concentration (C_{max}) and overall exposure AUC (for more information on pharmacokinetic parameters see Chapter 2). T_{max}, the time to achieve C_{max} is also considered but this is a secondary outcome.

To consider whether the plasma exposure between the two products is equivalent the ratio of the geometric means of treatment: reference C_{max} and AUC are calculated along with the 90% confidence intervals. Bioequivalence is considered to have been proven if the lower and upper confidence intervals fall between 80.00 and 125.00% for both C_{max} and AUC.

Human data are associated with variability; thus, it is not appropriate to simply compare the mean data achieved for two formulations. The ratio of the geometric mean data including the 90% confidence intervals incorporates the element of intra-individual variability associated with each product; particularly where a cross over study is used.

If limits other than 80.00 and 125.00% are applied, then this study is referred to as a relative bioavailability study and the metrics can be used to compare formulations that may not be intended to be bioequivalent, for example, to measure the relative performance of a solid dosage form compared to a non-precipitating oral solution. Often during clinical testing, a non-precipitating oral solution will be used in early clinical studies for proof of concept of drug safety and efficacy whilst this would not be a suitable commercial dosage form. A non-precipitating oral solution would be expected to have the best possible absorption and thus determining relative bioavailability versus the oral solution allows the drug product developer and regulators to understand how close to optimum performance the solid dosage form performance is.

Understanding the relative performance of two formulations is commonly called 'bridging'. For drug product changes prior to marketing, for example, at early stages of the development pathway, a less stringent standard than bioequivalence limit may be appropriate for introducing a new formulation into the clinical programme. Formal bioequivalence is only required once pivotal registration data have been generated and final dose/exposure has been selected to prove the risk-benefit. However, a comparison of C_{max} and AUC provides a useful means to compare formulations during development to ensure that the safety (and efficacy) data that have been generated to that point in development are still relevant for the new formulation is still applicable or whether a dose change is required to control risks to subjects in subsequent clinical studies.

10.3   Design of Clinical Bioequivalence (BE) Studies

BE studies are generally performed as single-dose studies with a randomised, two-period cross-over design in the fasted state (as drug pharmacokinetics are most sensitive to changes in absorption in the fasted state) in healthy subjects that are representative of the general population, unless safety considerations prohibit such a population, as shown in Figure 10.1. For the introduction of a generic drug product, bioequivalence in the fed state may be required. Regulatory guidance (eg [3, 5]) also specify other aspects of the study design such as:

- doses to be investigated,
- nature of fed or fasted period,
- % of AUC to be characterised and allowed carry over in the pre-dose sample of the second dose as a percentage of second dose C_{max},
- number of plasma samples to be taken and
- Selection and retention of dosage form dosed.

The US FDA has also drafted compound specific guidance for generic drug development [6].

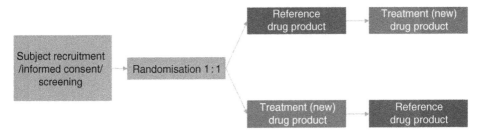

Figure 10.1 *Schematic of 'standard' clinical bioequivalence study design.*

10.4   Implication of Bioequivalence Metrics

For the most common bioequivalence study design shown in Figure 10.1, the confidence that a study will return for the treatment to reference ratio and thus associated confidence intervals depends predominately on two factors:

1. the within subject variability on AUC and C_{max}, e.g. how much C_{max} and AUC vary on repeated administration of the same formulation as single doses to the same subject and
2. the number of subjects included.

Figure 10.2 shows potential outcomes for a bioequivalence with test/reference (T/R) ratio and 90% confidence interval plotted as a Forest plot (both C_{max} and AUC needs to be considered but for simplicity only a single indicative metric is plotted). Case A shows a study that successfully demonstrates bioequivalence with both the ratio and upper and lower 90% confidence limits contained within the bioequivalence metrics of 80.00 to 125.00%. In this case, the T/R ratio is close to 100%. Case B also successfully demonstrates bioequivalence; however, the T/R ratio is ~113% and the lower confidence interval does not cross 100%. As the lower confidence interval does not cross 100%, this indicates that the test treatment produces greater exposure than the reference formulation (based on the upper confidence interval up to 23% greater exposure than the reference formulation). In this case, the study demonstrates bioequivalence because within-subject variability is low and/or a large number of subjects have been dosed reducing the confidence interval of the T/R ratio to within acceptable limits even though on average the test formulation gives higher exposure than the test. In Case C, the test exhibits lower exposure than the reference and the lower confidence interval is less than 80.00% so bioequivalence is not proven. For Case D, bioequivalence is also not proven as both the lower and higher confidence intervals are outside the accepted levels; however, the T/R ratio is ~100% suggesting the formulations are similar and the study 'failed' due to high within-subject variability/insufficient subjects.

Some drugs are demonstrated to be highly variable as their intersubject variance is greater than expected. It can be more challenging for these drugs to demonstrate bioequivalence due to this higher than typical intersubject variability. However, there can be cases where despite the high intersubject variance the wide therapeutic index of the drug means that it is safe and effective despite having a different exposure. For highly variable drugs, other designs than that shown in Figure 10.1 maybe be more appropriate, such as a replicate design or calculation scaled-average BE approaches. For further information, please

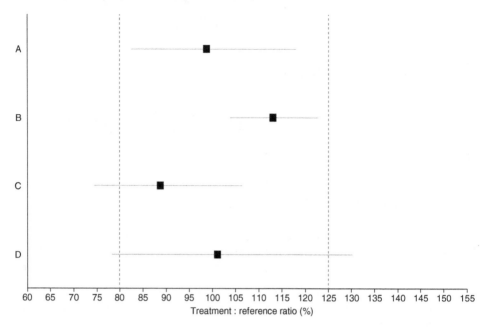

Figure 10.2 *Forest plot representing potential outcomes of the T/R results from clinical bioequivalence study.*

see these references [7, 8]. These approaches are currently allowed by some regulatory agencies (US FDA, EMA and Health Canada), although the metrics to be applied to assess BE for these drugs differ between these regions.

10.5   Bioequivalence Regulatory Guidelines

This chapter has discussed the high-level concepts contained within bioequivalence regulatory guidance and used predominantly current European, US and ICH guidance to illustrate the main concepts. Currently, each territory and in some cases individual countries may have their own bioequivalence guidance(s) covering different aspects of bioequivalence studies and data presentation; thus, the reader should consult each regulatory authorities' websites to understand their particularly requirements for bioequivalence. The science and societal expectations underpinning these guidance are dynamic and ever evolving and as such the guidance are periodically updated.

While the fundamental concept of bioequivalence is globally accepted, there are some differences between the different national/regional guidelines in terms of how the bioequivalence study should be performed and the metrics used to interpret the data. An ICH guideline is currently under development on the topic of bioequivalence studies (ICH M13: Bioequivalence for Immediate-Release Solid Oral Dosage Forms). This harmonisation initiative will result in a single globally harmonised guideline on the design of bioequivalence studies, that can be applied in all ICH territories. This will help streamline drug product development, by providing a common standard for study design and reducing the number of 'duplicate' bioequivalence studies that are performed to meet the different requirements of various territories. Up-to-date information on the progress of this guideline can be found on the ICH website (https://www.ich.org/page/multidisciplinary-guidelines).

10.6 Biowaivers

The discussion above describes the conduct of clinical bioequivalence studies conducted in human subjects. While this type of data remains the 'gold standard' from a regulatory perspective to demonstrate formulation bioequivalence, many territories now allow the requirement for clinical bioequivalence data to be waived for some low-risk drug product changes, and *in vitro* data to be used as evidence of bioequivalence (a biowaiver). These are usually applied to drug products where the API is considered to be low biopharmaceutics risk according to specific regulatory criteria, and/or the nature of the manufacturing process or formulation change is low risk. BCS-based biowaivers are discussed in detail in the BCS chapter (Chapter 9).

10.7 Biopharmaceutics in Quality by Design

During the 2000s, changes in regulatory thinking were stimulated by the US FDA critical path initiative which identified a lack of innovation in pharmaceutical manufacturing and led to the introduction of ICH Q8 and the concept of Quality by Design [9]. Previously, the perception was that the pharmaceutical manufacturing industry had focussed on 'testing quality in' rather than designing drug products and manufacturing processes that were fit for their intended purpose. Typically, this meant testing products upon finalisation of manufacture so that each batch would pass or fail the release specification criteria and only batches that passed would be released to the commercial market, while this approach ensured quality it is inefficient and hampered innovation and continual improvement.

Thus, there was reconsideration of pharmaceutical quality and ICHQ8 introduced the concept of the quality target product profile (QTPP), which is defined as: 'a prospective summary of the quality characteristics of a drug product that ideally will be achieved to ensure the desired quality, taking into account safety and efficacy of the drug product'. The QTPP specifically referenced the clinical performance of the product (safety and efficacy) and aimed to put the patient at the centre of pharmaceutical product development by asking formulators and manufacturers to explicitly think about the *in vivo* performance required and what this meant for formulation and process design and quality testing (see Section 10.8). ICH Q8 states that the QTPP 'could include therapeutic moiety release or delivery and attributes affecting pharmacokinetic characteristics (e.g. dissolution, aerodynamic performance) appropriate to the drug product dosage form being developed'.

This notion of placing the patient at the centre of drug product design was developed further in a series of papers introducing the Biopharmaceutics Risk Assessment Roadmap 'BioRAM' as a concept for optimising clinical drug product performance and integrating thinking across all disciplines involved in drug development. Biopharmaceutics sits at the interface between formulation science and clinical pharmacology and is therefore ideally placed to create an integrated pathway for product development.

An initial publication focussed on integrating knowledge across pharmaceutical and other functions and especially clinical science using biopharmaceutics and pharmacokinetics thinking as the vehicle [10]. However, later papers very much focussed on multi-stakeholder systems thinking for decision making to optimise drug development.

The first BioRAM paper [10], discussed four drug delivery scenarios/PK profiles that cover many of the potential PK considerations for therapy-driven product performance (see Figure 10.3).

1. Scenario 1: Rapid therapeutic onset,
2. Scenario 2: Multi phasic delivery,
3. Scenario 3: Delayed therapeutic onset (e.g. chronotherapy) and
4. Scenario 4: Maintenance of target exposure.

These scenarios were not intended to be exhaustive or for classification of the drug/drug product but rather to serve as analogues, which are learning tools, for those trying to implement BioRAM and therapy-driven product development. The use of the scenarios provided context for patient-centred design and integrated thinking to cover key concepts in the development of new products. The differences in the pharmacokinetic profiles of each of the scenarios ensured clear communication between clinical and pharmaceutical development scientists to design an integrated approach to development.

The second BioRAM paper [11] introduced the BioRAM scoring grid made up of 12 key elements spanning the patient population, pharmacology and pharmaceutical quality with

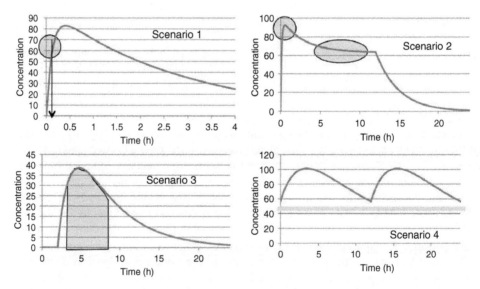

Figure 10.3 *Four drug delivery scenarios depicted as drug-concentration time profiles. Source: From Selen et al. [10] / With permission of Elsevier.*

development stage specific questions posed for each these elements. The scoring grid is used to drive a cross-disciplinary risk assessment discussion, identifying known risks and knowledge gaps that could prevent the required drug delivery profile from being reliably achieved in the patient population. Additional benefits of the scoring grid include:

- Facilitates cross-disciplinary thinking across all disciplines involved in drug development,
- Translates uncertainty into patient-focused action,
- Facilitates phase-appropriate actions and knowledge generation,
- Enables customised development plans for each specific project and
- Encourages an entrepreneurial mindset.

Thus, using the BioRAM approach and the scoring grid, the product development plan is driven by the critical knowledge needed and focuses on areas of potential risk identified in the scoring tool. It will focus the cross-functional development group on what they need to learn from a particular study or set of experiments, rather than mapping the project to a standard development plan or collecting data without fully assessing its relevance. A common theme across the BioRAM papers is an emphasis on considering the use of novel tools and approaches, rather than relying upon typical or historical approaches to address the problem/question.

The third BioRAM paper [12] describes the implementation of BioRAM to embed systems thinking during drug product development, and to drive timely and risk-based decision-making. Developing a shared understanding across different scientific disciplines enables risks to be identified and addressed in a balanced manner across all of the different aspects of drug product development. Three examples are described illustrating the application of the BioRAM approach.

Collectively, the BioRAM demonstrates that the biopharmaceutics discipline can act a bridge across numerous scientific disciplines involved in drug development to enable patient-centric design and development of drugs/drug products. The concepts described in the BioRAM publications focus on defining the needs of the patients and identifying critical information to move forward. Because the roadmap and scoring grid is hypothesis based, this supports innovation rather than dependence of habitual approaches and techniques.

10.8 Control of Drug Product and Clinically Relevant Specifications

The advent of quality by design had a significant impact on industrial biopharmaceutics. One area, as described above, was the increased emphasis on proactive design of the drug product to meet the patients' needs, including a suitable *in vivo* release profile to ensure safety and efficacy. These 'design intent' aspects are captured in the QTPP. Having designed a drug product with the required clinical performance, it is necessary to ensure that this is maintained during routine batch manufacture and on stability. In other words, the drug product should have the desired *in vivo* performance in the patient every time they take the dosage form, regardless of any intentional or unintentional variation in the drug product manufacturing process or input material properties.

This is the purpose of the control strategy. ICH Q8 R2 defines the control strategy as:

'A planned set of controls, derived from current product and process understanding that ensures process performance and product quality. The controls can include parameters and attributes related to drug substance and drug product materials and components, facility and equipment operating conditions, in-process controls, finished product specifications, and the associated methods and frequency of monitoring and control.'

Put simply, the control strategy is a combination of manufacturing controls, in-process and end product tests, that systematically control all relevant sources of risk to drug product manufacturability and product quality as defined in the QTPP, including the elements that are related to *in vivo* performance of the drug product. It is therefore essential to understand the **clinical relevance** of the tests, parameter limits and specifications that comprise the control strategy. Dissolution is widely recognised as the main quality test to assess drug product performance in the patient, linking the CMC elements of development to the clinic – it is usually identified as a **critical quality attribute** (CQA) for oral drug products. A clinically relevant dissolution method and specification ensures that every batch released during commercial manufacture has the required performance in the patient, and is therefore a key element of the control strategy.

10.9   Establishing Clinically Relevant Dissolution Methods and Specifications

Key advances were made in linking dissolution to clinical performance using the QbD approach. There was also emphasis on the development of appropriate knowledge during development for the lifecycle of a product such that future formulations could be developed and launched using more flexible regulatory approaches. A key example presented to highlight the value of QbD in biopharmaceutics established clinical boundaries based on product design and manufacturing processes to demonstrate the link between dissolution and clinical performance. This example is presented in detail in this reference [13]. In brief, a series of tablets were produced that included the highest product and process risk factors for performance and these tablets were tested using both dissolution and within a clinical study. The tablet variants selected included a standard tablet, a tablet with a large API particle size, an over-granulated tablet and a tablet with increased binder and reduced disintegrant. All variants showed much reduced dissolution compared to the standard tablet. However, when tested *in vivo* all tablets exhibited similar pharmacokinetics. This result showed that there was a wide range in acceptable dissolution profiles whereby clinical performance was unaffected. This finding provides confidence that small process or product changes are unlikely to affect the *in vivo* performance and highlights that the dissolution test may over discriminate between tablet variants.

To establish a clinically relevant dissolution method and specification, it is necessary to develop understanding of the relationship between *in vitro* dissolution and the *in vivo* performance of the drug product. There are currently no regulatory guidelines describing how this should be done; however, a recent IQ Consortium White Paper set out a harmonized

---

Box 10.1   Five step process to establishing clinically relevant specifications
described by Dickinson et al. [13, 15]

Step 1	Conduct Quality Risk Assessment,
Step 2	Develop Appropriate CQA tests,
Step 3	Understand the *in vivo* importance of changes,
Step 4	Establish Appropriate CQA limits and
Step 5	Use the product knowledge in subsequent QbD steps.

Sources: Based on Dickinson et al. [13, 15]. © John Wiley & Sons.

---

industry position on the topic [14]. Dickinson et al. [13, 15] described a five-step process which is illustrated in Box 10.1.

The first step is to perform a risk assessment to identify the manufacturing process and formulation attributes that can potentially influence dissolution *in vivo*. This should be based on a mechanistic assessment of the drivers for dissolution and the biopharmaceutics properties of the API, for example, considering the effect of parameters such as the particle size of the API and compression force used for tablets on dissolution. Next, formulation and process variants incorporating the highest risk factors are produced (e.g. using altered input materials, composition and/or processing parameters), and *in vitro* dissolution tests which can detect these changes are identified. The next step in the process involves a key strategic decision for the project team – **is it necessary to generate clinical bioavailability data to understand the impact of these *in vitro* changes on *in vivo* performance?** For drug products containing high solubility (BCS Class 1 and 3) APIs, the BCS biowaiver dissolution conditions and acceptance (refer to Chapter 9) can be considered to provide clinical relevance, as the principal of BCS biowaivers is that the drug products meeting these conditions will be bioequivalent *in vivo*. For drug products containing low solubility (BCS Class 2 and 4) APIs, if development teams want to develop clinically relevant specifications, they will need to perform a clinical relative bioavailability study to compare the pharmacokinetics of process and formulation variants with the standard clinical drug product or leverage data on different formulation from clinical studies performed during development. Detailed examples of this approach are provided in the following references: [13, 15–17].

The next step is to interpret the data from the clinical study and use this to select a clinically relevant dissolution test and specification for routine batch release. This interpretation will involve comparing the *in vitro* dissolution data to the clinical data to assess the impact that changes in *in vitro* dissolution had on *in vivo* performance. There are four potential ways that the outcomes from such a clinical study can be translated into a clinically relevant specification; these are described in detail in the IQ White Paper [17], and summarized in Table 10.1.

Table 10.1 *Potential relationships between dissolution and clinical performance, and translation into a clinically relevant specification.*

Study outcome	Description	Impact on specification setting
Level A IVIVC[a]	Changes to *in vitro* dissolution result in corresponding changes to *in vivo* performance, and a mathematical relationship can be developed which enables the PK profile to be predicted using the *in vitro* dissolution profile.	Specification set using IVIVC model to define boundary for equivalent *in vivo* performance.
Level C IVIVC[a]	Changes to *in vitro* dissolution result in corresponding changes to *in vivo* performance. The entire dissolution profile cannot be used to predict the PK profile; however, a mathematical correlation can be developed between a single dissolution timepoint and a PK parameter (e.g. between dissolution at 30 minutes and C_{max}).	Specification set using IVIVC model to define boundary for equivalent *in vivo* performance.
Safe space	Changes to *in vitro* dissolution have no impact on PK across the range tested in the clinical study. This enables a dissolution 'safe space' to be defined, within which batches will be clinically equivalent (usually based on bioequivalence criteria).	Specification set within the region where equivalent *in vivo* performance has been demonstrated in the clinical study.
In silico IVIVe	A link between *in vitro* dissolution data and clinical PK is established using an *in silico* physiologically-based biopharmaceutics model (PBBM). This approach can be applied to scenarios where any dissolution change leads to a corresponding PK change (mechanistic IVIVC), or where changes in PK are seen for some dissolution profiles but not for all of them (mixed safe space/IVIVC). It is more difficult to apply this approach in the regulatory setting when a 'safe space' outcome is achieved, as it can be difficult to validate that the model can appropriately predict the point at which dissolution starts to impact PK.	Specification set using *in silico* PBBM model to define boundary for equivalent *in vivo* performance.

[a]*For regulatory definitions of different levels of IVIVC, please refer to the FDA IVIVC guideline [18].*
Source: Based on Hermans et al. [14].

10.10   Application of *In Silico* Physiologically Based Biopharmaceutics Modelling (PBBM) to Develop Clinically Relevant Specifications

The use of *in silico* PBBM modelling to establish clinically relevant dissolution specifications is becoming more widespread, with numerous examples presented by both industry and regulators. For example, Pepin et al. [19] described the application of PBBM to establish dissolution and particle size specifications for lesinurad, a BCS Class two weak acid [19]. A PBBM was established using a combination of *in vitro* and clinical PK data, and was used to interpolate between bioequivalent and non-bioequivalent batches to define a dissolution safe space. Compared to safe space or traditional IVIVC, this approach offers enhanced mechanistic insight into:

- Factors influencing absorption,
- The level of biopharm risk,
- The *in vitro* and *in vivo* data generated for a drug product and
- How results from healthy volunteer studies can be extrapolated to special populations or disease states.

Biopharmaceutics absorption modelling is discussed in Chapter 12. It should be noted that there is a higher level of expectation on these models when they are used in a regulatory setting compared to a development setting, for example, in terms of the data inputs used and level of validation required. For a recent discussion of these considerations, the reader is referred to Pepin et al. [20].

10.11   Additional Considerations for Establishing Dissolution Methods and Specifications

While ensuring suitable drug product performance in the patient is the primary consideration, a number of additional factors need to be taken into account when establishing a dissolution test and release specification. These include method robustness and suitability of the method for use in a routine QC environment. It is also important to consider the discriminatory power of the dissolution method in the context of the wider control strategy, for example, the dissolution method does not need to detect all potential dissolution risks as some of them may already be adequately controlled by the manufacturing process controls or input material specifications. Flanagan and Mann [21] describe the different factors that need to be balanced when establishing a clinically relevant dissolution test and specification, including a tool to help project teams work through this process (the Dissolution Universal Strategy Tool, DUST) [21].

The global regulatory environment for development and acceptance of clinically relevant dissolution tests and specifications is rapidly (and non-homogenously) evolving, including the acceptability of the modern biopharmaceutics toolkit, such as non-standard *in vitro* tests and PBBM. This is an area of active discussion between industry and regulators in recent years – for further detail, the reader is referred to the following publications: [17, 22, 23].

10.12 Common Technical Document (CTD)

The common technical document (CTD) is the format used by all major markets including Europe, United States and Japan to assemble and organise the quality, safety and efficacy information required in an application for the registration of a new drug product intended for human use. The CTD was introduced in the 2000s to harmonise the authoring and review of the regulatory application, and reduce the time and resources needed to compile the application, review (by the regulators) and communication with the applicant. The overall structure of the CTD is detailed in the ICH M4 guidelines [24]. The CTD is divided into five main modules each of which contain multiple sub-sections providing a structure and layout of relevant information required to enable the regulators to review the CTD efficiently. The five modules of the CTD are as follows:

- Module 1 – Administrative Information and prescribing information,
- Module 2 – Overviews and summaries of Modules 3–5,
- Module 3 – Quality,
- Module 4 – Nonclinical reports and
- Module 5 – Clinical reports.

Biopharmaceutic knowledge and data are included in multiple parts of the CTD including: Module 3 – Quality; and in Module 2 – overview and summaries of Modules 3–5, specifically in the 'Clinical Summary'. Module 3 – Quality presents the chemistry, manufacturing and controls information for the drug product and is spilt into two sections such as drug substance and drug product (see Figure 10.4). The drug substance and drug product sections contain a series of sub-sections whereby biopharmaceutics information is incorporated into the drug product section in two main sub-sections '3.2.P.2 Pharmaceutical Development' and '3.2.P.5 Control of Drug Product'.

A summary of biopharmaceutic information is provided including:

- API properties (i.e. BCS classification),
- formulation development,
- dissolution methods used during development, including justification of methods used and link to *in vivo* performance and
- a bridging narrative of formulations used during clinical development.

Module 2 contains overviews and summaries of Modules 3–5 and contains seven sections (Figure 10.4). Each of the seven sections is further broken down into sub-sections guiding the applicant on the information required. Additional biopharmaceutics information is included in Module 2.7 clinical summaries (particularly 2.7.1. Summary of Biopharmaceutic Studies and Associated Analytical Methods). The information provided is similar to Module 3; however, the focus is on the *in vivo* including clinical study data versus the CMC and quality aspects of the drug product focussed on in Module 3. A summary of all the biopharmaceutics (i.e. BE/BA studies) and clinical pharmacology studies plus a narrative on cross-study understanding of the drug product performance is provided in this section. Furthermore, details of the drug product used in each study and how the *in vitro* dissolution performance links to the *in vivo* performance should also be provided.

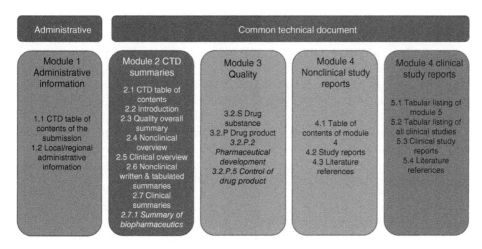

Figure 10.4 *Submission structure and Common Technical Document Module Content. Subsections with significant biopharmaceutics input are highlighted in italics.*

10.13   Other Routes of Administration and Locally Acting Drug Products

This chapter has focussed on regulatory biopharmaceutics considerations for drug products administered orally with a systemic mechanism of action (i.e. plasma concentrations can reasonably be considered as a surrogate for efficacy). Special consideration is needed for drug products administered via non-oral routes for systemic action, and for locally acting drug products (administered via oral or non-oral routes). The same high-level principles apply as described earlier in the chapter, i.e. the commercial drug product must have equivalent performance to that used in pivotal safety and efficacy studies, and the control strategy must be clinically relevant, i.e. able to guarantee suitable clinical performance from all batches manufactured and released. However, the design of the *in vivo* and *in vitro* studies used to assess this may be different. For drug products administered via alternative routes for systemic action, similar principles can be applied for the demonstration of bioequivalence, i.e. a clinical study to demonstrate statistical similarity of plasma concentration vs. time profiles. However, for drug products intended for local action, alternative study designs to demonstrate equivalent therapeutic effect may be needed. Regarding control strategy, the *in vitro* tests applied are likely to be different for non-oral drug products. The design of the dissolution/drug release test may be different, and there may be an increased emphasis on other drug product attributes that relate to clinical performance (e.g. particle size for inhaled products, device aspects for inhaled and injectable products). Only a small number of specific regulatory bioequivalence guidelines are available for many of these product types, while others do not have specific guidance and therefore default to standard bioequivalent guidelines.

10.14 Conclusion

Biopharmaceutics is a key area of focus for regulators as it is the critical link between the drug product and clinical safety and efficacy. If the link is not made to the satisfaction of the regulators or is not proven then the safety and efficacy of the commercially supplied drug product cannot be ascertained and thus there will be a delay to marketing approval for that product.

The application of biopharmaceutics thinking to enable patient centric drug product design is creating an environment, where novel or innovative biopharmaceutics tools and approaches may be applied in a regulatory setting to support registration/marketing approval for drug products resulting direct benefit to patients.

References

[1] FDA (2000). Guidance for industry, waiver of *in vivo* bioavailability and bioequivalence studies for immediate release solid oral dosage forms based on a biopharmaceutics classification system. https://www.fda.gov/media/70963/download (accessed January 2021).

[2] FDA (2019). Title 21 – food and drugs Part 320 – bioavailability and bioequivalence requirements. https://www.govinfo.gov/content/pkg/CFR-2019-title21-vol5/xml/CFR-2019-title21-vol5-part320.xml (accessed January 2021).

[3] EMA (2010). Committee for Medicinal Products for Human Use (CHMP) Guideline on the Investigation of Bioequivalence https://www.ema.europa.eu/en/documents/scientific-guideline/guideline-investigation-bioequivalence-rev1_en.pdf (accessed January 2021).

[4] ICH (2020). ICH M9 guideline on biopharmaceutics classification system-based biowaivers. https://www.ema.europa.eu/en/documents/scientific-guideline/ich-m9-biopharmaceutics-classification-system-based-biowaivers-step-5_en.pdf (accessed January 2021).

[5] Food and Drug Administration, Centre for Drug Evaluation and Research (CDER), US Department of Health, and Human Services (2003). Bioavailability and bioequivalence studies for orally administered drug products—general considerations. https://www.fda.gov/files/drugs/published/Guidance-for-Industry-Bioavailability-and-Bioequivalence-Studies-for-Orally-Administered-Drug-Products---General-Considerations.PDF (accessed August 2021)

[6] FDA (2013). Guidance for industry. Bioequivalence studies with pharmacokinetic endpoints for drugs submitted under an ANDA (draft guidance). https://www.fda.gov/media/87219/download (accessed January 2021).

[7] Haidar, S.H., Davit, B., Chen, M.-L. et al. (2008). Bioequivalence approaches for highly variable drugs and drug products. *Pharmaceutical Research* **25** (1): 237–241.

[8] Endrenyi, L. and Tothfalusi, L. (2019). Bioequivalence for highly variable drugs: regulatory agreements, disagreements, and harmonization. *Journal of Pharmacokinetics and Pharmacodynamics* **46** (2): 117–126.

[9] ICH (2006). ICH guideline Q8 (R2) on pharmaceutical development. https://www.ema.europa.eu/en/documents/scientific-guideline/international-conference-harmonisation-technical-requirements-registration-pharmaceuticals-human-use_en-11.pdf (accessed January 2021).

[10] Selen, A., Dickinson, P.A., Müllertz, A. et al. (2014). The biopharmaceutics risk assessment roadmap for optimizing clinical drug product performance. *Journal of Pharmaceutical Sciences* **103** (11): 3377–3397.

[11] Dickinson, P.A., Kesisoglou, F., Flanagan, T. et al. (2016). Optimizing clinical drug product performance: applying biopharmaceutics risk assessment roadmap (BioRAM) and the BioRAM scoring grid. *Journal of Pharmaceutical Sciences* **105** (11): 3243–3255.

[12] Selen, A., Müllertz, A., Kesisoglou, F. et al. (2020). Integrated multi-stakeholder systems thinking strategy: decision-making with biopharmaceutics risk assessment roadmap (BioRAM) to optimize clinical performance of drug products. *The AAPS Journal* **22** (5): 97.

[13] Dickinson, P.A., Lee, W.W., Stott, P.W. et al. (2008). Clinical relevance of dissolution testing in quality by design. *The AAPS Journal* **10** (2): 380–390.

[14] Hermans, A., Abend, A.M., Kesisoglou, F. et al. (2017). Approaches for establishing clinically relevant dissolution specifications for immediate release solid oral dosage forms. *The AAPS Journal* **19** (6): 1537–1549.

[15] Dickinson, P., Flanagan, T., Holt, D., and Stott, P. (2017). Chapter 15 clinically relevant dissolution for low-solubility immediate-release products: dissolution and drug release. In: *Poorly Soluble Drugs: Dissolution and Drug Release* (eds. G.K. Webster, R.G. Bell and D.J. Jackson), 511–552. Pan Stanford Publishing.

[16] Abend, A., Xiong, L., Zhang, X. et al. (2019). Biowaiver applications in support of a polymorph during late-stage clinical development of verubecestat—current challenges and future opportunities for global regulatory alignment. *The AAPS Journal* **22** (1): 17.

[17] Suarez-Sharp, S., Cohen, M., Kesisoglou, F. et al. (2018). Applications of clinically relevant dissolution testing: workshop summary report. *The AAPS Journal* **20** (6): 93.

[18] FDA (1997). Guidance for industry extended release oral dosage forms: development, evaluation, and application of *in vitro/in vivo* correlations. https://www.fda.gov/media/70939/download (accessed January 2021).

[19] Pepin, X.J., Flanagan, T.R., Holt, D.J. et al. (2016). Justification of drug product dissolution rate and drug substance particle size specifications based on absorption PBPK modeling for lesinurad immediate release tablets. *Molecular Pharmaceutics* **13** (9): 3256–3269.

[20] Pepin, X.J.H., Parrott, N., Dressman, J. et al. (2021). Current state and future expectations of translational modeling strategies to support drug product development, manufacturing changes and controls: a workshop summary report. *Journal of Pharmaceutical Sciences* **10** (2): 584–593.

[21] Flanagan, T. and Mann, J. (2019). Dissolution universal strategy tool (DUST): a tool to guide dissolution method development strategy. *Dissolution Technologies* **26**: 6–16.

[22] Lennernäs, H., Lindahl, A., Van Peer, A. et al. (2017). *In vivo* predictive dissolution (IPD) and biopharmaceutical modeling and simulation: future use of modern approaches and methodologies in a regulatory context. *Molecular Pharmaceutics* **14** (4): 1307–1314.

[23] McAllister, M., Flanagan, T., Boon, K. et al. (2020). Developing clinically relevant dissolution specifications for oral drug products—industrial and regulatory perspectives. *Pharmaceutics* **12** (1): 19.

[24] ICH (2004). ICH topic M 4 common technical document for the registration of pharmaceuticals for human use – organisation CTD. https://www.ema.europa.eu/en/documents/scientific-guideline/ich-m-4-common-technical-document-registration-pharmaceuticals-human-use-organisation-ctd-step-5_en.pdf (accessed January 2021).

11

Impact of Anatomy and Physiology

Francesca K. H. Gavins[1], Christine M. Madla[1], Sarah J. Trenfield[1],
Laura E. McCoubrey[1], Abdul W. Basit[1] and
Mark McAllister[2]

[1] *Department of Pharmaceutics, UCL School of Pharmacy,*
University College London, London,
United Kingdom
[2] *Pfizer Drug Product Design, Sandwich, United Kingdom*

11.1 Introduction

The oral route of administration remains the preferred route of administration for small molecules. Meeting the demands of fast-moving clinical programmes requires the formulation and biopharmaceutics scientist to have a good understanding of the impact of critical material attributes (e.g. solid form, particle size distribution) and critical process parameters (e.g. granulation or lubrication parameters) on dosage form performance. This is often assessed with conventional *in vitro* dissolution approaches, but to answer questions on the clinical performance of a product, e.g. what is the impact of administering the proposed formulation with food, a knowledge of the product and its interactions with the luminal environment of the GI tract is essential.

Our fundamental understanding of the physiological factors (e.g. pH, motility, fluid volume and composition), which are critical for dosage form performance has increased significantly over recent years and was a core research area for the Innovative Medicines Initiative OrBiTo programme [1]. Data gained through novel human clinical studies using

Biopharmaceutics: From Fundamentals to Industrial Practice, First Edition. Edited by Hannah Batchelor.
© 2022 John Wiley & Sons Ltd. Published 2022 by John Wiley & Sons Ltd.

advanced imaging, telemetry and intubation approaches has enriched our understanding of the GI tract and key aspects of anatomy and physiology which are critical for GI processing of orally administered formulations.

The dosage form journey begins in the oral cavity when an oral drug product is swallowed with a glass of water, transiting the oesophagus in a few seconds (although it should be noted that in the elderly and some disease states, dysphagia can impact swallowing [2]) before entering the stomach via the oesophageal sphincter. If the dosage form is administered in the fasted state, residual volumes of gastric fluids in the stomach are small and the bulk of the gastric volume is provided by the fluid swallowed from the glass of co-administered water. For immediate-release dosage forms, the process of release begins upon interaction with the gastric fluids. Wetting, disintegration/disaggregation are initiated and if gastric residence time permits, dissolution begins. Gastric emptying in the fasted state is a first-order process with a half-life of approximately 10–15 minutes after which the disintegrated dosage form, excipients and potentially dissolved API are emptied into the first region of the small intestine, the duodenum [3]. Here, intestinal bicarbonate secretions rapidly neutralise the acidic secretions from the stomach and the absorption process begins with permeation across the duodenal membrane. However, the duodenum is a relatively short segment of the small intestine and the bulk of absorption will occur along the length of the jejunum and for some compounds or modified-release dosage form later in the ileum and colon. Transit through the small intestine takes around three to four hours before the dosage form or undissolved/unabsorbed API reaches the ileo-caecal junction (ICJ). After passage through the ICJ, our dosage form enters the final stage of its journey along the GI tract with transit through the large intestine and colon. This chapter will provide a detailed review of the anatomy and physiology of the GI tract which a dosage encounters during its journey after oral dosing and highlight the importance of key aspects which are critical for drug release and absorption.

11.2   Influence of GI Conditions on Pharmacokinetic Studies

The impact of food on drug absorption has been reported previously [4–8]. The fed state changes the volume and composition of gastrointestinal media and also has effects on the motility/mixing and transit times. Meals that are high in total calories and fat content are likely to result in a larger effect on the absorption of a drug substance or drug product as they show the greatest effects on GI physiology. A high-fat and high-calorie meal has been proposed by the FDA as an example of a worst-case fed effect: two eggs fried in butter, two strips of bacon, two slices of toast with butter, four ounces of hash brown potatoes and eight ounces of whole milk [9].

For oral drug absorption studies, fasted and fed studies are conducted after an overnight fast of approximately 8–10 hours and no food is allowed for at least four hours post-dose. Subjects in the fasted state will take the drug with a standardised volume of water (250 mL) [9, 10]. Whereas, subjects in the fed state will take the drug with a standardised volume of water 30 minutes after the start of the consumption of a high-fat, high-caloric (800–1000 kcal) meal.

11.3   The Stomach

11.3.1   Gastric Anatomy

From a drug development perspective, the anatomy of the stomach (see Figure 11.1) is often oversimplified and considered as a simple one-compartment organ. However, this overlooks the complex physiological function of the stomach in terms of initiating digestion. If we consider the anatomy of the stomach, there are distinct anatomical regions, which serve different functions in the digestion process. The fundus and proximal corpus regions function as a flexible reservoir for food intake with the distal corpus and antral regions controlling mixing and emptying [11].

Gastric absorption is limited by the relatively low epithelial surface area (approximately $0.1\,m^2$, see Table 11.1). In addition, the structure of the stomach epithelium is dominated by surface mucosal cells rather than absorptive cells. However, the lack of a significant absorptive function does not mean the stomach is simply a *waiting room* for the small intestine, its physiological function is critical for processing of foods and dosage forms [18]. It is often noted that the stomach is not a major absorption organ although it is capable of absorbing some non-ionized lipophilic molecules of moderate size, especially weakly acidic drugs.

The stomach mucosa is folded into ridged structures called rugae which contain acid secreting parietal cells located in the mucosal glands. Distribution of the H+ secreting parietal cells is not uniform with a high concentration is present in the body of the stomach,

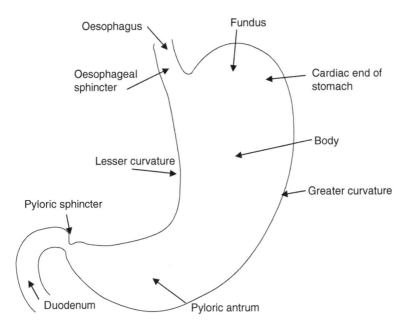

Figure 11.1   Stomach anatomy.

Table 11.1 *The relative anatomical lengths and surface areas of various regions of the GI tract in adult humans.*

Region of the intestinal tract	Average length (cm)	Average diameter (cm)	Absorbing surface area (m²)
Stomach	20–30	10	~0.5
Duodenum	25–30	4	30–40
Jejunum	250	3	
Ileum	300	3	
Caecum	6	7.5	~0.05
Colon	110–190	5	~2
Rectum	15–19	2.5	~0.2

Sources: Based on Ritschel et al. [12], Helander et al. [13], Collins et al. [14], Shroyer and Kocoshis [15], Sadahiro et al. [16], and Ferrua and Singh [17].

relatively few in the fundus and none in the antrum. Surface mucosal cells secrete a thick mucus layer which protects the stomach lining from the effects of high acid levels and digestive enzymes [19]. The adherent mucus layer has been reported to have a mean thickness of 180 µm, in the range of 50–450 µm [3, 20, 21], which upon hydration forms two distinct layers on the gastric epithelium [19]. Gastric mucus is composed of the gel-forming protein MUC5AC [22], with other glycosylated mucins found in adjacent crypts [23]. A detailed overview of the structural biology and of stomach mucosa and regional differences is provided in a review by Vertzoni et al. [3].

11.3.2   Gastric Motility and Mixing

Gastric motility is characterized by tonic contraction in the proximal areas and peristaltic muscular contractions in the distal region of the stomach. The result of these muscular contractions is to churn the ingested contents towards the closed pyloric sphincter and generate hydrodynamic conditions, which serve to mix gastric contents and reduce the size of ingested materials. Intragastric pressures are highest in the antral area around the pyloric sphincter and may be sufficient to disrupt the mechanical integrity of hydrophilic matrix tablets if they have not been designed appropriately [24]. Meal components are reduced in size to around 2–5 mm (or 1–2 mm according to Vrbanac [25]) before being passed through the pyloric sphincter and into the duodenum, where the digestion of nutrients begins. The rate of emptying of food materials is based on calorific load with emptying controlled at rate (1–4 kcal/min) which can maintain effective digestion in the small intestine [26].

Between meals, fasted state GI motility is controlled by the enteric nervous system through the Interdigestive Migratory Motor Complex – the IMMC. The IMMC is a peristaltic contraction starting from the gastric midcorpus region and subsequently migrates along the gastric wall, increasing in propagation velocity and amplitude to the pylorus. The IMMC is a triphasic, continuous cyclical pattern, each phase differing in contraction frequency and amplitude.

Under fasting conditions, phases of minimal contractility or quiescence (IMMC phase I) alternate with phases of irregular low amplitude contractions (IMMC phase II) and strong, high-frequency bursts of contractile activity (IMMC phase III) [27]. This is shown schematically in Figure 11.2. IMMC phase III contractions, also known as the 'housekeeper

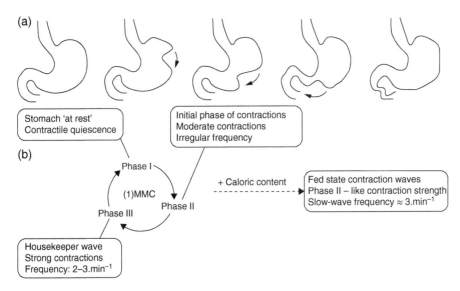

Figure 11.2 *Schematic representation of (a) peristaltic contractions waves in the stomach and (b) gastric motility features under fasted and fed state conditions. Source: Modified from Van Den Abeele et al. [18].*

wave', have the physiological function of transferring remaining content from the stomach to the small intestine [28, 29]. The housekeeper wave is also responsible for the gastric emptying of non-disintegrating dosage forms such as extended-release formulations or enteric-coated dosage forms. Gastric residence times for this type of dosage form will be largely dictated by the phase of IMMC at the time of administration. Housekeeper wave contractions have a frequency of 2–3 and 11–12 contractions/min in the antrum and duodenum respectively. They occur only in the fasted state and the intake of food immediately disrupts the IMMC and induces the digestive myoelectric motor activity.

The motility pattern in the fed state is characterised by a very different pattern of contractility with constantly propagating waves of antral contraction at a frequency of around 3 waves/min [30–32]. Gastric motility may have implications for pharmacokinetic studies. For fasted state studies, there is growing evidence that when drug administration coincides with phase 1 or phase II IMMC cycle's differences in resulting pharmacokinetics may be observed. Studies with paracetamol, atazanavir and fosamprenavir suggest that dosing study volunteers who are in phase I of the IMMC result in a wider range of T_{max} values than those dosed in a phase II cycle of the IMMC [29].

11.3.3 Gastric Emptying

Gastric emptying time depends on the size of the administered dosage form, size and type of meal, posture and chronological effects. Generally, emptying in the fasted state is reported to be a first-order process driven by the gastro-duodenal pressure gradient with a glass of water emptying with a half-life of 10–15 minutes [19, 25]. MRI studies by Mudie et al. [33] have shown that after ingestion of a glass of water and it takes around 30–45 minutes for the gastric volume to return to pre-ingestion levels.

In contrast to non-caloric liquids, emptying of the meal volume is largely dictated by calorific content with the pylorus opening periodically to allow the passage of chyme into the duodenum. This process is controlled by a complex neurohumoral feedback loop from duodenal receptors and results in gastric emptying being controlled to rate which does not breach the absorptive capacity of the small intestine [26].

The stomach road, or Magenstrasse, has been shown to be important for drug release and absorption. The Magenstrasse is not a new phenomenon [30], being first described more than 100 years ago [34–36], but a series of more recent studies demonstrate its importance for the performance of oral dosage forms administered in the fed state. The Magenstrasse mechanism describes the conditions under which liquid may be rapidly emptied from the fed stomach [3, 25, 30, 37–40]. For oral dosage forms administered in the fed state, the Magenstrasse can have a significant impact. In some cases, it may facilitate an early exit from the stomach and enable rapid absorption to occur in the small intestine under fed conditions. For a drug administered as an immediate-release dosage form, early onset of absorption may occur as the dosage form or disintegrated particles (containing API) bypass the ingested meal. This effect will contribute to increased inter-individual variability in PK profiles obtained from food-effect studies. The disintegration, dissolution and physico-chemical properties of the drug molecule will be factors which contribute to the sensitivity of different drug products to the Magenstrasse effect. Further details of the impact of the Magenstrasse on PK profiles can be found in a paper by Schick et al [40].

11.3.3.1 Gastric Fed State

The FDA/EMA standardised high-fat breakfast (comprising 800–1000 kcal of which 50% is derived from lipid) is widely utilised in food effect studies to shift GI physiology to a perceived extreme state to allow the maximal potential of food on dosage form performance and oral drug absorption to be assessed. It has been shown that after consumption of such a meal, gastric volumes rise to around 500–628 mL in the period immediately following ingestion [39, 41]. The gastric volume is derived not only from the volume of the ingested meal, but also from salivary and gastric secretions both of which are significantly stimulated as a result of food intake [3]. Gastric volumes during the initial phase of gastric emptying in the fed state are a result of the balance achieved between zero-order calorie controlled emptying output rate and the inputs from gastric and salivary secretions. After around 60–90 minutes, the gastric volume profile is dominated by emptying with an overall half-life for gastric emptying of a high-fat meal reported in the range of 15–180 minutes [42]. Whilst this general assessment of gastric volumes and emptying and in the fed state is widely accepted and supported by data across many different studies, it is also important to note dynamic effects associated with water intake following meal ingestion, which can be important for dosage form performance.

11.3.4 Gastric Fluid Volume

Gastric fluid volumes, composition and pH are important for dosage form disintegration and dissolution. As noted by Vertzoni et al. [3], the fluid environment of the stomach is directly affected by the ingestion of fluids and food. MRI studies [33, 43, 44] have shown that the volume of gastric fluids in the fasted state in adults is typically less than 100 mL. In a recent study by de Waal [45], the proton pump inhibitor (PPI) esomeprazole was

Table 11.2 *Gastric fluid volumes reported in healthy subjects as determined by magnetic resonance imaging (MRI) under fasting conditions.*

Publication	Number of volunteers	Volume reported (mL) in the fasted stomach		
		Min	Max	Mean (s.d.)
Schiller [43]	12	13	72	45 (18)
Steingoetter [47]	12	103	149	122
Fruehauf [48]	8			153 (41)
	8			129 (46)
Mudie [33]	12			35 (7)
Grimm [49]	6			23 (36)
	6			24 (19)
	6			17 (12)
	6			26 (24)
Grimm [44]	8			15 (8)
	8			33 (23)
	120			25 (18)

Sources: Based on Mudie et al. [33], Schiller et al. [43], and Grimm et al. [44].

reported to decrease the gastric and duodenal fluid volume by 41 and 44%, respectively. The findings suggest that the dissolution and subsequent absorption of poorly soluble drugs may be affected in patients using PPIs. The reader is directed towards an excellent review on drugs administered for GI diseases and their effects on the pharmacokinetics of co-administered drugs [46]. Table 11.2 shows the reported values of gastric fluid present in the fasted state from published studies.

11.3.5 Gastric Temperature

The impact of a glass of water on the temperature profile of the gastric contents has been quantified in several studies. Garbacz et al. [50], estimated an initial intragastric temperature of 25–30 °C after intake of 240 mL of water at 20 °C. It was shown that temperature effects can be relevant for the disintegration and dissolution behaviour of capsules and tablets. Koziolek et al. [41] reported an exponential increase of the temperature inside the human stomach from approximately 23–36 °C within approximately 10 minutes, after the intake of an Intellicap® together with a glass of water.

11.3.6 Gastric Fluid Composition

In the fasted state, stomach content is composed of saliva, gastric secretions (hydrochloric acid, digestive enzymes including pepsin and gastric lipase), dietary food and refluxed fluids from the duodenum [51].

11.3.6.1 Gastric pH

It has been reported that the median pH of gastric fluids in the fasted state of healthy adults is in the range of 2–3 [52–57], although in some cases the extremes of pH reported can vary between pH 1 and 6. However, the exact pH can be hard to define as its values fluctuate on a minute-to-minute basis and can be dependent on the sampling technique [58, 59]. Regional differences can be observed in the stomach with less acidic values between pH 2 and 4 noted in the fundal area, which is related to the lower density of acid-secreting parietal cells in this anatomical region.

The ingestion of a typical glass water with a dosage form in the fasted state will cause a shift in the pH values of gastric fluids and large inter-individual differences can be observed (due to differences in the volume of gastric secretions at the point of dosing) [30, 41, 60, 61].

The pH value of gastric fluids after food ingestion varies between pH 3 and pH 7. The pH value and rate of change with time is strongly related to the type and size of the meal [59]. The pH of gastric fluids returns to the fasted state range within two to three hours due to stimulation of acid secretion on meal ingestion. In the fed state, peak acid secretion is around 42 ± 22 mmol/h, which is around ten times higher than the basal acid output rate [62].

Whilst fasted gastric conditions are consistently regarded as acidic, certain populations with specific health related conditions such as achlorhydria can have gastric fluid pH values above 6 [59, 63].

11.3.6.2   Gastric Bile Salt Composition and Concentration

Surface tension in gastric fluids varies between 35 and 46 mN/m [61, 64], which indicates presence of surface active agents such as lecithin and lysolecithin [65]. According to Solvang and Finholt, the levels of these surfactants *in vivo* are below the critical micelle concentration (CMC) [66]. In a review of several studies which measured bile salts levels in the stomach, Vertzoni et al. concluded that bile salt reflux into the stomach occurs sporadically and that concentrations in the stomach are very low compared with those in the small intestine [58].

11.4   Small Intestine

11.4.1   Small Intestinal Anatomy

The small intestine (SI) is a long, narrow and highly convoluted tube extending from the stomach to the large intestine and provides the region within the gastrointestinal (GI) tract where most digestion and absorption of food and nutrients takes place. In adults, the small intestine is around 3–8 metres long and is contained in the central and lower abdominal cavity.

Three successive regions of the small intestine are defined as duodenum, jejunum, and ileum as shown in Figure 11.3. These regions form one continuous tube, and, despite each area exhibiting distinct characteristics, there are no obviously marked separations between them. The first region, the duodenum, is characteristically C-shaped and is positioned adjacent to the stomach; it is only 25–30 cm long and has the widest diameter [67].

The duodenojejunal flexure, an arc finishing with a sharp bend, indicates the junction of duodenum with the jejunum [68]. The jejunum, is around 2.5 m in length and is located in the central section of the abdomen [14]. The colour of the jejunum is deep red because of its extensive blood supply. No anatomical feature separates the jejunum from the ileum.

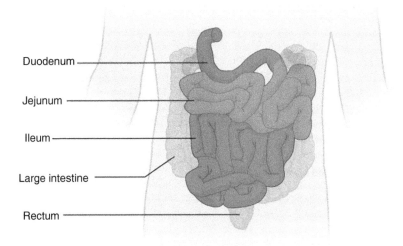

Duodenum

Jejunum

Ileum

Large intestine

Rectum

Figure 11.3 *Diagram of the small intestine with the duodenum, jejunum and ileum marked. Source: Anatomy & Physiology, 2013, Connexions Web site. CC BY 4.0*

The ileum, is around 3 m long and located in the lower abdomen. Its walls are narrower and thinner than in the previous sections in addition blood supply is more limited. The jejunum and ileum lie on a mesentery, which anchors the small intestines to the abdominal wall and contains blood vessels and lymphatic drainage that is supported by fatty connective tissue [14]. The ileum joins the caecum (the first part of the large intestine) at the ileo-caecal sphincter.

The wall of the small intestine is composed of the same four layers (the serosa, muscularis, submucosa and mucosa) typically present in the GI tract. However, certain features of the mucosa and submucosa in the small intestine are unique, including circular folds, villi and microvilli. These adaptations function to increase the absorptive surface area of the small intestine more than 600-fold and are most abundant in the proximal two-thirds of the small intestine.

From the duodenum to the ileum, there is a gradual decrease in diameter, thickness of the wall and number of circular folds with no folds in the distal ileum. Within the circular folds are small, finger-like vascularised projections called villi that extend into the small intestinal lumen. Villi are approximately 0.5–1.5 mm in length depending on their location in the intestinal tract. There are about 20–40 villi per square millimetre, responsible for increasing the surface area and hence absorptive capacity of the small intestine. Each villus contains a capillary bed composed of one arteriole, one venule and a lymphatic capillary [69]. As their name suggests, microvilli are much smaller (1 μm in length, 0.9 μm in diameter) compared with villi. They are apical surface extensions of the plasma membrane of the mucosa's epithelial cells, which are supported by microfilaments. Although their small size makes it difficult to see each microvillus, their combined microscopic appearance suggests a mass of bristles, which is termed the brush border. There are an estimated 200 million microvilli per square millimetre of small intestine, greatly expanding the surface area of the plasma membrane and thus greatly enhancing absorption.

Interspersed in the GI mucosa are goblet (mucus producing) cells secreting mucus when mechanically or chemically stimulated. The mucus is rich in mucin glycoproteins (MUC-2) and produces a single, thin, unattached and discontinuous layer allowing the absorption of nutrients [3]. The mucus in the duodenum is less viscoelastic than in the stomach and colon allowing particles to become entrapped and driven down the gut [70].

Embedded in the lamina propria of the gut is mucosa-associated lymphoid tissue (MALT). MALT exists as single isolated nodules or lymphoid nodes. In the ileum, lymph nodes called Peyer's Patches exist in the mucosa and submucosa to prevent bacteria from entering the bloodstream.

11.4.2 Small Intestinal Motility and Mixing

The residence time and the strength of contractions in the small intestine will influence the rate of absorption for drug products. The frequency of basal contractile waves declines distally and the duodenal frequency was reported to be 11.8 cycles/min, compared with an ileal frequency of 8 cycles/min in healthy adults [71], encouraging the distal movements.

11.4.3 Small Intestinal Transit Time

Small intestinal transit time is frequently quoted at three to four hours, comparable in the fasted and fed state [72] and is reported to be similar for tablets, pellets and liquid formulations [73, 74], however this figure hides the considerable inter- and intra-subject variability for small intestinal transit times of multiple- and single-unit systems, with values typically ranging from between 1 and 9.5 hours [73, 75]. It has also been observed that pellets tend to move as a bolus instead of distribute widely within the small intestine [76]. Most oral dosage forms also spend the majority of the total transit time at rest within the small intestine [74, 77].

11.4.4 Small Intestinal Volume

The availability of fluids within the small intestine plays a major role in dissolution and the distribution of the drug within the intestinal lumen and subsequent absorption. Under fasted conditions, dosage forms are not always in contact with intestinal water [43]. Fluid volumes are found to be highly dynamic. In addition, fluid flow across the mucosa cannot be fully understood and often unpredictable. As such, pharmaceutical scientists should exercise caution when using intestinal fluid volumes for *in vitro* and *in silico* models [3].

Schiller and co-workers measured the fluid volumes in the small intestine of adults in the fasted and fed state using MRI [43]. In the fasted state, small intestinal fluid volumes were reported as highly variable, ranging from 45 to 319 mL. On average, the small intestinal fluid volume was 105 ± 72 mL. The fluid was shown to be distributed as 'fluid pockets' along the length of the small intestine. The largest fluid volume is usually present in the terminal ileum as fluids rests at the ileo-caecal junction before transfer to the caecum. A summary of all reported data on the volume of fluids in the small intestine is presented in Figure 11.4.

After the intake of a high-caloric breakfast (one hour), the fluid volume was found to be 54 ± 41 mL, significantly lower than the preprandial volume, with various distinct 'fluid pockets'. A further study by Marciani and colleagues studied the change of intestinal fluid

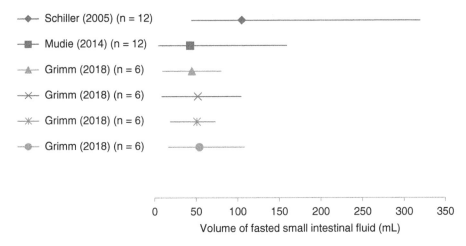

Figure 11.4 *Literature data on reported small intestinal fluid volumes in a fasted healthy adult. Sources: Based on Mudie [33], Schiller et al. [43], and Grimm et al. [44].*

volumes over a period of over eight hours following ingestion of a meal; after one hour, a decrease in fluid volume was observed. Although another study, found that the fluid volume began to increase after 90 minutes with a maximum fluid reached after approximately 3 hours, only slightly higher than the fasted state values [78].

Péronnet and colleagues used D_2O as a marker to understand the rate of water absorption from the small intestine. Pharmacokinetic modelling found a rate of 3–3.5 mL/min for intestinal water absorption after intake of 300 mL of water [79]. Although, other researchers reported rates of up to 25 mL/min with significant regional differences. Water absorption was found to be higher in proximal parts than in distal parts [80]. Shi and Passe reported that the net water flux across the intestinal membrane is influenced by the secretion of fluids and the osmolality and the electrolyte and the nutrient composition [81].

11.4.5 Small Intestinal Fluid Composition

Fluid sampling can be conducted using a perfusion tube (Loc-I-Gut) positioned in the intestine. Studies have mainly characterised the luminal contents in the fasted state of healthy adults due to the practical difficulties in sampling the upper small intestine in the fed state. To investigate the fed state, studies have been conducted by administering liquid meals with a similar caloric content and type of calories similar with those in meals, which are typically administered in oral drug absorption studies [82].

Ducts from the liver, gallbladder, and pancreas provide digestive fluids to the duodenum that neutralise acids coming from the stomach and help to digest carbohydrates, proteins and fats. The liver produces approximately 600–800 mL of bile per day, a fluid that is critical for the digestion and absorption of fats and fat-soluble vitamins, as well as the clearance of waste products in the faeces. Bile is comprised of water (~95%), in which are dissolved several constituents including bile salts, bilirubin, amino acids, fats, inorganic salts, cholesterol amongst others [83]. The sphincter of Oddi, located around two-thirds along the descending duodenum, controls the flow of bile and pancreatic juices.

Inter-individual variability in buffer capacity, osmolality and surface tension, and to a lesser extent pH, was reported in pooled intestinal fluids [84]. In addition, high variability was found in the drug solubility of two model drugs, furosemide and dipyridamole, known to have highly variable oral bioavailabilities. Furthermore, the luminal fluid composition and properties can vary greatly between the different regions of the small intestine [85]. To understand drug release, particularly for modified release preparations, it is important to understand how the GI luminal environment changes. Biorelevant media such as fasted-state human intestinal fluid (FaSSIF) and fed-state human intestinal fluid (FeSSIF) are used as dissolution media to simulate *In vivo* conditions. Their composition is designed to reflect the surface tension, osmolality, buffer capacity and ionic strength of the human intestinal fluid [86]. *In vitro* and *in silico* models that replicate the complex GI physiology are of value in predicting absorption under a range of GI conditions [87].

11.4.5.1   Small Intestinal pH

In the fasted state, the duodenal pH is slightly acidic with median values from various studies ranging between 5.6 and 7.0 with a median of 6.3 [88]. In the fed state, after the administration of homogenous or heterogeneous liquid meals, decreased duodenal pH values are found with medians ranging between 4.8 and 6.5 during the first three hours post meal administration [3]. In the jejunum, the mean fasted pH was reported to be 7.5 and the fed value 6.1 [89]. The pH of the jejunum is less variable than in the duodenum, where the emptying of the gastric contents and secretion of bile, pancreatic and intestinal secretion takes place and can cause fluctuations in the pH profile. Finally, the mean pH of the ileum has been reported to be 6.5 [90] in the fasted state and 7.5 in the fed state [51, 91].

11.4.5.2   Small Intestinal Buffer Capacity

In the duodenum, median buffer capacity value of $5.6 \, mmol \, L^{-1} \, pH^{-1}$ have been reported in the fasted state [61]. In the jejunum, a mean value of $3.23 \, mmol \, L^{-1} \, pH^{-1}$ was found in the fasted state [51, 85, 92] and in the ileum, a higher buffer capacity of $6.4 \, mmol \, L^{-1} \, pH^{-1}$ has been reported [85]. Values in the fed state are raised and on average more than double [51, 92]. The values are relatively low compared with the buffer capacity of the stomach (ranging from 7 to $18 \, mmol \, L^{-1} \, pH^{-1}$ in the fasted state and in the fed state a range $14–28 \, mmol \, L^{-1} \, pH^{-1}$ [61]).

11.4.5.3   Small Intestinal Surface Tension

Duodenal surface tension appears to be slightly higher in the fasted state; reported median values range from 32.7 to 35.3 mN/m in the fasted state and from 30.2 mN/m (at 3 hours) to 35.1 mN/m (at 0.5 hour) in the fed state [3]. Decreased surface tension was shown in the jejunum 28 mN m/m in the fasted state and 27 mM in the fed state [89]. This is significantly lower than the surface tension of water (72 mN/m), primarily due to the pancreatic secretions and gall bladder secretions of bile salts.

11.4.5.4   Small Intestinal Osmolality

Duodenal contents are hypo-osmotic in the fasted state (median range: 124–266 mOsm/kg) [93] and are mostly hyper-osmotic in the fed state 250–367 mOsm kg^{-1} [93] median values at various times during the first three hours after meal administration range from 215 mOsm/kg (at three hours) to 423 mOsm/kg (at two hours) [3, 61, 94, 95]. Fasted jejunal osmolality was found to be 271 ± 15 mOsm kg^{-1} [53], with less variable values than in the duodenum. No studies of the osmolality of fed jejunal fluids have been reported to date.

11.4.5.5 Bile Salt Composition and Concentration

The overall mean bile salt concentration in the fasted-state human intestinal fluid is approximately 3.3 mM in the duodenum [61, 92, 93, 96–98] and slightly lower, 3 mM in the jejunum [53, 85, 89, 92, 96]. Bile salt levels are highly variable and range from 0.3 to 9.6 mM. Concentrations above 6 mM are uncommon, although may result after the gall bladder empties into the duodenum [96]. Total bile salt concentrations are slightly higher in the fed state, overall median values during the first three hours, after the administration of a liquid meal, range from 3.7 mM (at three hours) to 18.2 mM (at one hour) with the increased bile secretion [3]. Jejunal fluids show bile salt concentrations of 4.5 and 8.0 mM in the fed state [89].

Approximately 70–75% of the total bile salt concentration in the human small intestine was found to be composed of the three main bile salts; taurocholate, glycocholate and glycochenodeoxycholate [92, 96, 97, 99]. Other bile salts detected are glycodeoxycholate, glycoursodeoxycholate, tauroursodeoxycholate and taurodeoxycolate. Lithocholates have been found in low quantities of less than 1%. Each bile salts show different pK_a values and molecular aggregation number resulting in different colloidal aggregations [100, 101].

Phospholipids are also key natural surfactants, amphiphilic in structure [96]. The overall mean phospholipid concentration in fasted HIF is highly variable and ranges between 0.003 and 2.7 mM with a mean of 0.32 mM [89, 93, 97–99]. The concentrations are about 4.5 times higher in the duodenum (0.53 mM), compared with the jejunum (0.13 mM) [96]. In the fed state, duodenal concentrations range from 1.2 to 6.0 mM (median of 2.15 mM) and 2.0–3.0 mM in the jejunum [88, 89].

Other components exist in human intestinal fluid (HIF). Cholesterol was observed to be present at concentration of about 0.18 mM [99]. Free fatty acids can affect the surface tension and colloidal nature of the fasted HIF. Monoglycerides and free fatty acids are surface-active molecules from the digestion of triglycerides [88], therefore will be related to the fat content of the meal. Free fatty acids have been reported at concentrations of 0.1 mM in the fasted state, from the hydrolysis of lecithin [89]. In the fed state, monoglyceride concentration was reported as 5.9 and 8.1 mM and the free fatty acid as 39.4 and 52 mM in the duodenum [95, 102]. In the fed jejunum, reports found 2.2 nM and 13.2 mM for the monoglycerides and free fatty acids, respectively [89]. The lower concentration of monoglycerides and free fatty acids in the jejunum suggest absorption may occur in the duodenum [88]

11.5 The Colon/Large Intestine

The terms colon and large intestine are used interchangeably in general use and within this section. The large intestine receives chyme after most of the nutrients have been absorbed. The role of the ascending colon is to absorb remaining water and key vitamins as well as producing the faeces. The descending colon will store faeces before defecation [103].

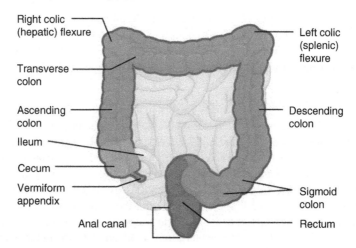

Figure 11.5　*Diagram of the large intestine. Source: Anatomy & Physiology, 2013, Connexions Web site. CC BY 4.0*

11.5.1　Large Intestine Anatomy

At the distal end of the small intestine, thickened smooth muscle termed the ileo-caecal valve separates the ileum from the colon and reduces reflux from the colon. The large intestine is divided into four sections; caecum, colon, rectum and anal canal as shown in Figure 11.4. The human colon is further divided into four sections; ascending (or proximal) colon, transverse colon, descending (or distal) colon, and sigmoidal colon.

The length of the large intestine length is up to 1.5 m, compared with the small intestine length of 6 m [104, 105]. The colon is a reservoir and has a wider lumen and significantly smaller surface area of around 1.3 m², compared with 200 m² in the small intestine [104, 106].

The colonic mucosa is made up of a simple columnar epithelium composed of absorptive colonocytes, goblet cells, and enteroendocrine cells. Enterocytes absorb water, salts and metabolites from the intestinal microbiota. The colon has no villi, although microvilli are found on the transverse furrows on the colon for chloride secretion and/or sodium absorption and passive diffusion of ions and water to maintain osmolality. Microvilli are thought to increase the surface area of the colon by ~6.5 times, facilitating absorption, digestion, and transport of water and nutrients across the mucosa [13].

The proportion of mucus-producing goblet cells in the distal colon is approximately 16% [23]. There is a thick, two-layer, protective mucus system, in the range of 20–52.5 μm, which eases the movement of faeces [107].

11.5.2　Large Intestinal Motility and Mixing

The colon is mostly fixed in position, with high amplitude peristaltic contractions propelling the contents (mass movement) and segmental mixing movements of the gut contents with enzymes. Motility is controlled by autonomic nervous inputs from the splanchnic and vagal nerves, by local responses and extrinsic loops from the stomach [108].

Retrogade contractions of the proximal colon can prolong residence time in the caecum and ascending colon. Non-continuous movement was observed, with movements in intervals during magnetic marker monitoring studies. In the ascending colon bidirectional movements were seen and in the transverse colon, the dosage forms were found to be often at rest [109]. Colonic motility and mass transport are believed to be related to the caloric content of ingested food, with fat and carbohydrates with the ability to stimulate colonic motility [110].

11.5.3　Large Intestinal Transit Time

Colonic transit times are considerably longer and more variable than the small intestine, and highly dependent on individual bowel habit. The proximal colon has shown a mean transit time of 11 hours, with a standard deviation of 3 hours. Although transit times of less than one hour have also been reported [108].

11.5.4　Large Intestinal Volume

The primary factor hindering drug dissolution in the colon is the lack of fluid [111]. During colonic transit, the volume of free fluid decreases, whilst the viscosity of the contents increases, causing implications for the dissolution of drugs in the colon. High intra- and inter-individual variability are found for colonic fluid volumes. A study by Schiller and colleagues using MRI imaging found the luminal fluid in the range of 1–44 mL with a mean of 13 ± 12 mL in the fasted state. Fluid is not homogenously distributed, instead it tends to congregate into discrete fluid pockets in the ascending and descending colon [43]. These fluid pockets have been found to be very small, mostly <0.5 mL, and can be primarily grouped in a single region of the colon [112]. In the distal parts of the small intestine, slurry or fluid pockets are rarely present. In the fed state 30 mL free water can be recovered from the caecum [113].

After food intake, the number of liquid pockets increase but not the volume of the pockets [43]. Although, endoscopic investigations in the distal ileum and proximal colon five hours after administration of the FDA reference meal, did not find distinct water pockets but instead high water content in the mucosa [82, 113–115]. The colonic environment is moist where most water in the colon is not freely available and instead is associated with bacteria or biomass [112]. Disintegration and dissolution of colon-targeted medicines may be highly dependent on contact with fluid pockets. The low free water content in the colon could partly explain why solubility of drugs in the small intestine is often higher and suggests that medicines targeted to the colon should be uniquely formulated with this environment in mind [116].

11.5.5　Large Intestinal Fluid Composition

The luminal environment is composed of a semi-solid mass of faeces, volatile short chain fatty acids, gases, minerals and salts and water. The gases are in the form of nitrogen, carbon dioxide, methane and hydrogen, with a volume of approximately 200–300 mL and are confined to the ascending and transverse colon [104, 117].

11.5.5.1　Large Intestinal pH

In the proximal colon, the pH is 7.8 in the fasted state and approximately 6 in the fed state, due to the bacterial fermentation of polysaccharides in food, producing of short chain fatty acids (SCFAs) [3, 113, 118]. The pH is approximately 6.4 in the ascending colon, with influence from SCFA. The upwards trend continues from 6.6 in the transverse colon to 7.0 in the descending colon, although large intra- and inter-individual variability exists [110].

11.5.5.2   Large Intestinal Buffer Capacity

Aspirates from the distal ileum observed that the buffer capacity was not significantly different between the fasted (8.9 mmol/L/ΔpH) and fed states (15.2 mmol/L/ΔpH) [114]. A significantly higher buffer capacity was found in the proximal colon compared with the ileal contents in the fed state. In the caecum, the buffer capacity was 19.2 mmol/L/ΔpH in the fasted state and 33.6 mmol/L/ΔpH [114]. The buffer capacity was similar in ascending colon to the caecum; 21.5 mmol/L/ΔpH in the fasted state and 37.7 mmol/L/ΔpH in the fed state [113, 114].

11.5.5.3   Large Intestinal Surface Tension

The surface tension is the colon is higher than in the upper small intestine, most likely due to the nearly complete re-absorption of bile acids due to passive re-uptake in the jejunum and active transport in the ileum [119] and significantly lower than water. The fasted state ascending colon showed with a mean of 39.2 and 42.7 mN/m in the fed state colon [113].

11.5.5.4   Large Intestinal Osmolality

In both prandial states, the lower intestinal fluids are hypo-osmotic, with readings lower than in the upper intestines [82]. In the ileum, the mean osmolality was reported as 60 mOsmol/Kg in the fasted state and 252 mOsmol/Kg in the fed state [113]. The osmolality of the caecum and ascending colon did not appear to differ between the fasted and fed state [113, 114]

11.5.5.5   Bile Salt Composition and Concentration

Bile reabsorption occurs in the distal ileum and therefore, bile acid concentrations are lower in the large intestine compared with the upper intestine in the both prandial states [82]. The bile salt concentration in the ascending colon is higher in the fed state than the fasted state [113, 114]. In the fed state, there are higher concentrations of cholic acid and chenodeoxycholic acid (primary bile acids) than deoxycholic acid and lithocholic acid (secondary bile acids), the opposite is found in the fasted state. In the post-prandial state, the conversion of primary bile acids to secondary bile acids may become saturated, resulting in more primary bile acids entering the ascending colon [113]. The type of meal has an influence on the bile acid concentration in the distal ileum [120].

Protein contents were higher in the ileum in the fasted state (5.1 mg/mL) than the fed state (3.4 mg/mL). A similar trend was found in caecal samples, the protein content was 6.2 mg/mL in the fed state and higher in the fasted state, 10.2 mg/mL, which may be due to mucosal cell regeneration [114]. The lower values in the fed state may be because the digestion and subsequent absorption of protein is almost complete in the ileum, and the liquid portion of the chyme may dilute residing proteins [114]. The carbohydrate content was significantly lower in the fasted than in the fed state in the ileum and caecum. In the fasted state, the carbohydrate content increase from the distal ileum to the caecum then to the ascending colon, a trend not observed in the fed state.

The total short chain fatty acid (SCFA) concentration is higher in the caecum, than the distal ileum in both prandial states, attributed to the SCFA [114]. A significant food effect was reported in the ascending colon [113]. It is believed that the bacteria in the ascending colon have a greater efficiency in fermentation than those in the distal ileum and caecum [114]. Acetate was the most prevalent species followed by propionate and butyrate with carporate in trace amounts in the distal ileum, caecum and ascending colon [113, 114].

Cholesterol levels do not differ significantly in the large intestine, although higher values are found in the caecum compared with the distal ileum [114]. This is expected as higher concentration of the SCFA acetate is found in the caecum which is the primary substrate for cholesterol synthesis [114]. Similar concentration of total phospholipids (phosphatidyl-choline and lysophosphatidylcholine) were found in the ileum in the fasted and fed states. In the fasted state, an increase was seen in the caecum, whereas in the fed state, an increase was seen in the ascending colon. The free fatty acid concentration (oleic, linoleic and stea-tic acid) in the proximal colon was quantified as twice that of the distal ileal contents in both prandial states. Triglycerides were below the limit of detection and monoglycerides and diglycerides were found in the distal ileum and caecum, higher in the fed state [114].

11.5.6   Impact of Microbiome on Oral Drug Delivery

The density of micro-organisms rises significantly towards the distal gut, with a substantial increase in the colon [121]. The realisation that imbalances in the microbiome can influ-ence the onset of both local and systemic diseases has altered the concept of a pharmaceutical-microbiome relationship. The gut microbiome is relatively stable, however, changes in the structure and diversity of the microbiome are also associated with large array of pathologi-cal states such as inflammatory bowel diseases [122, 123], metabolic diseases such obesity and diabetes [124], atherosclerosis [125] and cardiovascular disease [126].

Complex interactions can occur between drugs and the gut microbiome in a bidirectional manner [127]. In one direction, the composition and/or function of the gut microbiome may be influenced by a range of factors such as diet [128, 129], probiotics [130] and drugs [131], especially antibiotics [132]. In the other direction, the gut microbiome is capa-ble of the biotransformation of drugs [133], which may enhance or inhibit the clinical response. To date, a portfolio of drugs and diet-derived bioactive compounds have been reported to undergo direct microbial modification and their number is still growing. In almost all cases, the exact mechanism, the specific reaction and the responsible microbial species remain unknown. However, several mechanisms have been proposed and demon-strated. Recent studies have shown that the gut microbiota can influence the pharmacoki-netics of orally administered drugs and, thus, may have significant implications for their oral bioavailability [127, 134, 135].

Oral delivery of drugs to the colon has several advantages for the delivery of small molecules and biopharmaceuticals to treat both local and systemic diseases [105, 136]. For example, the reduced expression of drug efflux pumps. However, limitations exist in traditional colon-targeting technologies, dependent on pH and transit time, due to intra- and inter-individual variabilities [137]. To combat such limitations, a colon-specific, dual-mechanism coating (Phloral™) was designed, as a single line matrix, with two independent release mechanisms; a pH trigger (Eudragit® S dissolving a pH 7) and a microbiota-trigger (resistant starch digestible by bacterial enzymes) [137, 138]. Recently, the OPTICORE™ coating technology has been developed by Tillots utilising Phloral™ in the outer layer and DuoCoat™, a release acceleration mechanism, in the inner layer [139]. These technologies offer significant benefits for colonic drug delivery, compared with conventional enteric coating, by exploiting the activity of the microbiota.

11.6 Conclusions

The gastrointestinal tract's physiology can influence drug absorption. The composition of the GI fluid has a significant impact on the solubility, dissolution and absorption of an orally administered drug. An in-depth knowledge of the human gastrointestinal fluid is required to understand the intricacies of *in vivo* drug product performance, design drug delivery systems, and to optimise the biorelevant media used in drug product performance studies.

References

[1] Abrahamsson, B., McAllister, M., Augustijns, P. et al. (2020). Six years of progress in the oral biopharmaceutics area – a summary from the IMI OrBiTo project. *Eur. J. Pharm. Biopharm.* **152**: 236–247.

[2] Stillhart, C., Vucicevic, K., Augustijns, P. et al. (2020). Impact of gastrointestinal physiology on drug absorption in special populations—an UNGAP review. *Eur. J. Pharm. Sci.* **147**: 105280.

[3] Vertzoni, M., Augustijns, P., Grimm, M. et al. (2019). Impact of regional differences along the gastrointestinal tract of healthy adults on oral drug absorption: an UNGAP review. *Eur. J. Pharm. Sci.* **134**: 153–175.

[4] Cheng, L. and Wong, H. (2020). Food effects on oral drug absorption: application of physiologically-based pharmacokinetic modeling as a predictive tool. *Pharmaceutics* **12** (7): 1–18.

[5] O'Shea, J.P., Holm, R., O'Driscoll, C.M., and Griffin, B.T. (2019). Food for thought: formulating away the food effect – a PEARRL review. *J. Pharm. Pharmacol.* **71** (4): 510–535.

[6] Koziolek, M., Kostewicz, E., and Vertzoni, M. (2018). Physiological considerations and *in vitro* strategies for evaluating the influence of food on drug release from extended-release formulations. *AAPS PharmSciTech* **19** (7): 2885–2897.

[7] Koziolek, M., Alcaro, S., Augustijns, P. et al. (2019). The mechanisms of pharmacokinetic food-drug interactions – a perspective from the UNGAP group. *Eur. J. Pharm. Sci.* **134**: 31–59.

[8] Varum, F.J.O., Hatton, G.B., and Basit, A.W. (2013). Food, physiology and drug delivery. *Int. J. Pharm.* **457** (2): 446–460.

[9] FDA (2002). Guidance for industry: food-effect bioavailability and fed bioequivalence studies. https://www.fda.gov/files/drugs/published/Food-Effect-Bioavailability-and-Fed-Bioequivalence-Studies.pdf (14 October 2020).

[10] EMA (2010). Guideline on the investigation of bioequivalence. https://www.ema.europa.eu/en/documents/scientific-guideline/guideline-investigation-bioequivalence-rev1_en.pdf (accessed March 2021).

[11] Imai, Y. et al. (2012). Antral recirculation in the stomach during gastric mixing. *Am. J. Physiol. Liver Physiol.* **304**: G536–G542.

[12] Ritschel, W.A. (1991). Targeting in the gastrointestinal tract: new approaches. *Methods Find. Exp. Clin. Pharmacol.* **13** (5): 313–336.

[13] Helander, H.F. and Fändriks, L. (2014). Surface area of the digestive tract – revisited. *Scand. J. Gastroenterol.* **49** (6): 681–689.

[14] Collins, J.T., Nguyen, A., and Badireddy, M. (2020). Anatomy, abdomen and pelvis, small intestine. In: *StatPearls*. Treasure Island (FL): StatPearls Publishing LLC. Copyright © 2020.

[15] Shroyer, N.F. and Kocoshis, S.A. (2011). Anatomy and physiology of the small and large intestines. In: *Pediatric Gastrointestinal and Liver Disease*, 4e (eds. R. Wyllie and J.S. Hyams), 324–336.e2. Saint Louis: W.B. Saunders.

[16] Sadahiro, S., Ohmura, T., Saito, T., and Suzuki, S. (1991). Relationship between length and surface area of each segment of the large intestine and the incidence of colorectal cancer. *Cancer* **68** (1): 84–87.

[17] Ferrua, M.J. and Singh, R.P. (2011). Understanding the fluid dynamics of gastric digestion using computational modeling. *Proc. Food Sci.* **1**: 1465–1472.

[18] Van Den Abeele, J., Brouwers, J., Tack, J., and Augustijns, P. (2017). Exploring the link between gastric motility and intragastric drug distribution in man. *Eur. J. Pharm. Biopharm.* **112**: 75–84.

[19] Van Den Abeele, J., Rubbens, J., Brouwers, J., and Augustijns, P. (2017). The dynamic gastric environment and its impact on drug and formulation behaviour. *Eur. J. Pharm. Sci.* **96**: 207–231.

[20] Newton, J.L., Jordan, N., Pearson, J. et al. (2000). The adherent gastric antral and duodenal mucus gel layer thins with advancing age in subjects infected with *Helicobacter pylori*. *Gerontology* **46** (3): 153–157.

[21] Newton, J., Jordan, N., Oliver, L. et al. (1998). *Helicobacter pylori in vivo* causes structural changes in the adherent gastric mucus layer but barrier thickness is not compromised. *Gut* **43** (4): 470–475.

[22] Johansson, M.E., Sjövall, H., and Hansson, G.C. (2013). The gastrointestinal mucus system in health and disease. *Nat. Rev. Gastroenterol. Hepatol.* **10** (6): 352–361.

[23] Kim, Y.S. and Ho, S.B. (2010). Intestinal goblet cells and mucins in health and disease: recent insights and progress. *Curr. Gastroenterol. Rep.* **12** (5): 319–330.

[24] Hens, B., Corsetti, M., Spiller, R. et al. (2017). Exploring gastrointestinal variables affecting drug and formulation behavior: methodologies, challenges and opportunities. *Int. J. Pharm.* **519** (1–2): 79–97.

[25] Vrbanac, H., Trontelj, J., Berglez, S. et al. (2020). The biorelevant simulation of gastric emptying and its impact on model drug dissolution and absorption kinetics. *Eur. J. Pharm. Biopharm.* **149**: 113–120.

[26] Goelen, N., de Hoon, J., Morales, J.F. et al. (2020). Codeine delays gastric emptying through inhibition of gastric motility as assessed with a novel diagnostic intragastric balloon catheter. *Neurogastroenterol. Motil.* **32** (1): e13733.

[27] Braeckmans, M., Brouwers, J., Masuy, I. et al. (2020). The influence of gastric motility on the intraluminal behavior of fosamprenavir. *Eur. J. Pharm. Sci.* **142**: 105117.

[28] Paixão, P., Bermejo, M., Hens, B. et al. (2018). Linking the gastrointestinal behavior of ibuprofen with the systemic exposure between and within humans—part 2: fed state. *Mol. Pharm.* **15** (12): 5468–5478.

[29] Hens, B., Masuy, I., Deloose, E. et al. (2020). Exploring the impact of real-life dosing conditions on intraluminal and systemic concentrations of atazanavir in parallel with gastric motility recording in healthy subjects. *Eur. J. Pharm. Biopharm.* **150**: 66–76.

[30] Koziolek, M., Grimm, M., Schneider, F. et al. (2016). Navigating the human gastrointestinal tract for oral drug delivery: uncharted waters and new frontiers. *Adv. Drug Deliv. Rev.* **101**: 75–88.

[31] Madsen, J.L. and Dahl, K. (1990). Human migrating myoelectric complex in relation to gastrointestinal transit of a meal. *Gut* **31** (9): 1003–1005.

[32] Kwiatek, M.A., Steingoetter, A., Pal, A. et al. (2006). Quantification of distal antral contractile motility in healthy human stomach with magnetic resonance imaging. *J. Magn. Reson. Imaging* **24** (5): 1101–1109.

[33] Mudie, D.M., Murray, K., Hoad, C.L. et al. (2014). Quantification of gastrointestinal liquid volumes and distribution following a 240 mL dose of water in the fasted state. *Mol. Pharm.* **11** (9): 3039–3047.

[34] Scheunert, A. (1912). Über denMagenmechanismus des Hundes bei der Getränkaufnahme. *Pflugers Arch. Gesamte Physiol. Menschen Tiere* **144**: 569–576.

[35] Jefferson, G. (1915). The human stomach and the canalis gastricus (Lewis). *J. Anat. Physiol.* **49**: 165–181.

[36] Baastrup, C. (1924). Roentgenological studies of the inner surface of the stomach and of the movements of the gastic contents. *Acta Radiol.* **3**: 180–204.

[37] Pal, A.B., G, J., and Abrahamsson, B. (2006). A stomach road or "magenstrasse" for gastric emptying. *J. Biomech.* **40**: 1202–1210.

[38] Koziolek, M., Gorke, K., Neumann, M. et al. (2014). Development of a bio-relevant dissolution test device simulating mechanical aspects present in the fed stomach. *Eur. J. Pharm. Sci.* **57**: 250–256.

[39] Koziolek, M., Grimm, M., Garbacz, G. et al. (2014). Intragastric volume changes after intake of a high-caloric, high-fat standard breakfast in healthy human subjects investigated by MRI. *Mol. Pharm.* **11** (5): 1632–1639.

[40] Schick, P., Sager, M., Wegner, F. et al. (2019). Application of the gastroduo as an *in vitro* dissolution tool to simulate the gastric emptying of the postprandial stomach. *Mol. Pharm.* **16** (11): 4651–4660.

[41] Koziolek, M., Grimm, M., Becker, D. et al. (2015). Investigation of pH and temperature profiles in the GI tract of fasted human subjects using the Intellicap(®) system. *J. Pharm. Sci.* **104** (9): 2855–2863.

[42] Gentilcore, D., Chaikomin, R., Jones, K.L. et al. (2006). Effects of fat on gastric emptying of and the glycemic, insulin, and incretin responses to a carbohydrate meal in Type 2 diabetes. *J. Clin. Endocrinol. Metab.* **91** (6): 2062–2067.

[43] Schiller, C., Frohlich, C.-P., Giessmann, T. et al. (2005). Intestinal fluid volumes and transit of dosage forms as assessed by magnetic resonance imaging. *Aliment. Pharmacol. Ther.* **22** (10): 971–979.

[44] Grimm, M., Koziolek, M., Kuhn, J.P., and Weitschies, W. (2018). Interindividual and intraindividual variability of fasted state gastric fluid volume and gastric emptying of water. *Eur. J. Pharm. Biopharm.* **127**: 309–317.

[45] de Waal, T., Rubbens, J., Grimm, M. et al. (2020). Exploring the effect of esomeprazole on gastric and duodenal fluid volumes and absorption of ritonavir. *Pharmaceutics* **12** (7): 670. https://doi.org/10.3390/pharmaceutics12070670.

[46] Litou, C., Effinger, A., Kostewicz, E.S. et al. (2019). Effects of medicines used to treat gastrointestinal diseases on the pharmacokinetics of coadministered drugs: a PEARRL review. *J. Pharm. Pharmacol.* **71** (4): 643–673.

[47] Steingoetter, A., Fox, M., Treier, R. et al. (2006). Effects of posture on the physiology of gastric emptying: a magnetic resonance imaging study. *Scand. J. Gastroenterol.* **41** (10): 1155–1164.

[48] Fruehauf, H., Goetze, O., Steingoetter, A. et al. (2007). Intersubject and intrasubject variability of gastric volumes in response to isocaloric liquid meals in functional dyspepsia and health2. *Neurogastroenterol. Motil.* **19** (7): 553–561.

[49] Grimm, M., Koziolek, M., Saleh, M. et al. (2018). Gastric emptying and small bowel water content after administration of grapefruit juice compared to water and isocaloric solutions of glucose and fructose: a four-way crossover MRI pilot study in healthy subjects. *Mol. Pharm.* **15** (2): 548–559.

[50] Garbacz, G., Cade, D., Benameur, H., and Weitschies, W. (2014). Bio-relevant dissolution testing of hard capsules prepared from different shell materials using the dynamic open flow through test apparatus. *Eur. J. Pharm. Sci.* **57**: 264–272.

[51] Mudie, D.M., Amidon, G.L., and Amidon, G.E. (2010). Physiological parameters for oral delivery and *in vitro* testing. *Mol. Pharm.* **7** (5): 1388–1405.

[52] Pedersen, B.L., Mullertz, A., Brondsted, H., and Kristensen, H.G. (2000). A comparison of the solubility of danazol in human and simulated gastrointestinal fluids. *Pharm. Res.* **17** (7): 891–894.

[53] Lindahl, A., Ungell, A.-L., Knutson, L., and Lennernäs, H. (1997). Characterization of fluids from the stomach and proximal jejunum in men and women. *Pharm. Res.* **14** (4): 497–502.

[54] Dressman, J.B., Berardi, R.R., Dermentzoglou, L.C. et al. (1990). Upper gastrointestinal (GI) pH in young, healthy men and women. *Pharm. Res.* **7** (7): 756–761.

[55] Pedersen, P.B., Vilmann, P., Bar-Shalom, D. et al. (2013). Characterization of fasted human gastric fluid for relevant rheological parameters and gastric lipase activities. *Eur. J. Pharm. Biopharm.* **85**: 958–965.

[56] Press, A.G., Hauptmann, I.A., Hauptmann, L. et al. (1998). Gastrointestinal pH profiles in patients with inflammatory bowel disease. *Aliment. Pharmacol. Ther.* **12**: 673–678.

[57] Feldman, M. and Barnett, C. (1991). Fasting gastric pH and its relationship to true hypochlorhydria in humans. *Dig. Dis. Sci.* **36**: 866–869.

[58] Vertzoni, M., Dressman, J., Butler, J. et al. (2005). Simulation of fasting gastric conditions and its importance for the *in vivo* dissolution of lipophilic compounds. *Eur. J. Pharm. Biopharm.* **60** (3): 413–417.

[59] Dressman, J.B., Amidon, G.L., Reppas, C., and Shah, V.P. (1998). Dissolution testing as a prognostic tool for oral drug absorption: immediate release dosage forms. *Pharm. Res.* **15** (1): 11–22.

[60] Petrakis, O., Vertzoni, M., Angelou, A. et al. (2015). Identification of key factors affecting the oral absorption of salts of lipophilic weak acids: a case example. *J. Pharm. Pharmacol.* **67** (1): 56–67.

[61] Kalantzi, L., Goumas, K., Kalioras, V. et al. (2006). Characterization of the human upper gastrointestinal contents under conditions simulating bioavailability/bioequivalence studies. *Pharm. Res.* **23** (1): 165–176.

[62] Weinstein, D.H., deRijke, S., Chow, C.C. et al. (2013). A new method for determining gastric acid output using a wireless pH-sensing capsule. *Aliment. Pharmacol. Ther.* **37**: 1198–1209.

[63] Ogata, H., Aoyagi, N., Kaniwa, N. et al. (1984). Development and evaluation of a new peroral test agent GA-test for assessment of gastric acidity. *J. Pharm.* **7** (9): 656–664.

[64] Efentakis, M. and Dressman, J.B. (1998). Gastric juice as a dissolution medium: surface tension and pH. *Eur. J. Drug Metab. Pharmacokinet.* **23** (2): 97–102.

[65] Gibaldi, M. and Feldman, S. (1970). Mechanisms of surfactant effects on drug absorption. *J. Pharm. Sci.* **59** (5): 579–589.

[66] Solvang, S. and Finholt, P. (1970). Effect of tablet processing and formulation factors on dissolution rate of the active ingredient in human gastric juice. *J. Pharm. Sci.* **59** (1): 49–52.

[67] Lopez, P.P., Gogna, S., and Khorasani-Zadeh, A. (2020). *Anatomy, Abdomen and Pelvis, Duodenum, in StatPearls*. Treasure Island (FL): StatPearls Publishing LLC. Copyright © 2020.

[68] Saenko, V.F., Markulan, L., Belianski, L.S. et al. (1989). The role of the duodenojejunal flexure in regulating the motor evacuatory function of the duodenum. *Klin. Khir.* **2**: 24–27.

[69] Schofield, P.F., Haboubi, N.Y., and DF, M. (1993). The small intestine: normal structure and function. In: *Highlights in Coloproctology*. London: Springer.

[70] Taherali, F., Varum, F., and Basit, A.W. (2018). A slippery slope: on the origin, role and physiology of mucus. *Adv. Drug Deliv. Rev.* **124**: 16–33.

[71] Kerlin, P. and Phillips, S. (1982). Variability of motility of the ileum and jejunum in healthy humans. *Gastroenterology* **82** (4): 694–700.

[72] Fadda, H.M., McConnell, E.L., Short, M.D., and Basit, A.W. (2009). Meal-induced acceleration of tablet transit through the human small intestine. *Pharm. Res.* **26** (2): 356–360.

[73] Davis, S.S., Hardy, J.G., and Fara, J.W. (1986). Transit of pharmaceutical dosage forms through the small intestine. *Gut* **27** (8): 886–892.

[74] McConnell, E.L., Fadda, H.M., and Basit, A.W. (2008). Gut instincts: explorations in intestinal physiology and drug delivery. *Int. J. Pharm.* **364** (2): 213–226.

[75] Yuen, K.H. (2010). The transit of dosage forms through the small intestine. *Int. J. Pharm.* **395** (1–2): 9–16.

[76] Varum, F.J., Merchant, H.A., and Basit, A.W. (2010). Oral modified-release formulations in motion: the relationship between gastrointestinal transit and drug absorption. *Int. J. Pharm.* **395** (1–2): 26–36.

[77] Weitschies, W., Blume, H., and Mönnikes, H. (2010). Magnetic marker monitoring: high resolution real-time tracking of oral solid dosage forms in the gastrointestinal tract. *Eur. J. Pharm. Biopharm.* **74** (1): 93–101.

[78] Marciani, L., Cox, E.F., Hoad, C.L. et al. (2010). Postprandial changes in small bowel water content in healthy subjects and patients with irritable bowel syndrome. *Gastroenterology* **138** (2): 469–477.e1.

[79] Péronnet, F., Mignault, D., du Souich, P. et al. (2012). Pharmacokinetic analysis of absorption, distribution and disappearance of ingested water labeled with D2O in humans. *Eur. J. Appl. Physiol.* **112** (6): 2213–2222.

[80] Lambert, G., Chang, R.-T., Xia, T. et al. (1997). Absorption from different intestinal segments during exercise. *J. Appl. Physiol.* **83** (1): 204–212.

[81] Shi, X. and Passe, D.H. (2010). Water and solute absorption from carbohydrate-electrolyte solutions in the human proximal small intestine: a review and statistical analysis. *Int. J. Sport Nutr. Exerc. Metab.* **20** (5): 427–442.

[82] Pentafragka, C., Symillides, M., McAllister, M. et al. (2019). The impact of food intake on the luminal environment and performance of oral drug products with a view to *in vitro* and *in silico* simulations: a PEARRL review. *J. Pharm. Pharmacol.* **71** (4): 557–580.

[83] Boyer, J.L. (2013). Bile formation and secretion. *Comprehens. Physiol.* **3** (3): 1035–1078.

[84] Rabbie, S.C., Flanagan, T., Martin, P.D., and Basit, A.W. (2015). Inter-subject variability in intestinal drug solubility. *Int. J. Pharm.* **485** (1): 229–234.

[85] Fadda, H.M., Sousa, T., Carlsson, A.S. et al. (2010). Drug solubility in luminal fluids from different regions of the small and large intestine of humans. *Mol. Pharm.* **7** (5): 1527–1532.

[86] Klein, S. (2010). The use of biorelevant dissolution media to forecast the *in vivo* performance of a drug. *AAPS J.* **12** (3): 397–406.

[87] Basit, A.W., Madla, C.M., and Gavins, F.K.H. (2020). Robotic screening of intestinal drug absorption. *Nat. Biomed. Eng.* **4** (5): 485–486.

[88] Bergström, C.A., Holm, R., Jørgensen, S.A. et al. (2014). Early pharmaceutical profiling to predict oral drug absorption: current status and unmet needs. *Eur. J. Pharm. Sci.* **57**: 173–199.

[89] Persson, E.M., Gustafsson, A.S., Carlsson, A.S. et al. (2005). The effects of food on the dissolution of poorly soluble drugs in human and in model small intestinal fluids. *Pharm. Res.* **22** (12): 2141–2151.

[90] Youngberg, C.A., Berardi, R.R., Howatt, W.F. et al. (1987). Comparison of gastrointestinal pH in cystic fibrosis and healthy subjects. *Dig. Dis. Sci.* **32** (5): 472–480.

[91] Bown, R.L., Sladen, G.E., Clark, M.L., and Dawson, A.M. (1971). The production and transport of ammonia in the human colon. *Gut* **12** (10): 863.

[92] Perez de la Cruz Moreno, M., Oth, M., Deferme, S. et al. (2006). Characterization of fasted-state human intestinal fluids collected from duodenum and jejunum. *J. Pharm. Pharmacol.* **58** (8): 1079–1089.

[93] Clarysse, S., Tack, J., Lammert, F. et al. (2009). Postprandial evolution in composition and characteristics of human duodenal fluids in different nutritional states. *J. Pharm. Sci.* **98** (3): 1177–1192.

[94] Litou, C., Vertzoni, M., Goumas, C. et al. (2016). Characteristics of the human upper gastrointestinal contents in the fasted state under hypo- and A-chlorhydric gastric conditions under conditions of typical drug–drug interaction studies. *Pharm. Res.* **33** (6): 1399–1412.

[95] Vertzoni, M., Markopoulos, C., Symillides, M. et al. (2012). Luminal lipid phases after administration of a triglyceride solution of danazol in the fed state and their contribution to the flux of danazol across Caco-2 cell monolayers. *Mol. Pharm.* **9** (5): 1189–1198.

[96] Fuchs, A. and Dressman, J.B. (2014). Composition and physicochemical properties of fasted-state human duodenal and jejunal fluid: a critical evaluation of the available data. *J. Pharm. Sci.* **103** (11): 3398–3411.

[97] Brouwers, J., Tack, J., Lammert, F., and Augustijns, P. (2006). Intraluminal drug and formulation behavior and integration in *in vitro* permeability estimation: a case study with amprenavir. *J. Pharm. Sci.* **95** (2): 372–383.

[98] Kossena, G.A., Charman, W.N., Wilson, C.G. et al. (2007). Low dose lipid formulations: effects on gastric emptying and biliary secretion. *Pharm. Res.* **24** (11): 2084–2096.

[99] Psachoulias, D., Vertzoni, M., Goumas, K. et al. (2011). Precipitation in and supersaturation of contents of the upper small intestine after administration of two weak bases to fasted adults. *Pharm. Res.* **28** (12): 3145–3158.

[100] Söderlind, E., Karlsson, E., Carlsson, A. et al. (2010). Simulating fasted human intestinal fluids: understanding the roles of lecithin and bile acids. *Mol. Pharm.* **7** (5): 1498–1507.

[101] Kossena, G.A., Charman, W.N., Boyd, B.J. et al. (2004). Probing drug solubilization patterns in the gastrointestinal tract after administration of lipid-based delivery systems: a phase diagram approach. *J. Pharm. Sci.* **93** (2): 332–348.

[102] Kalantzi, L., Persson, E., Polentarutti, B. et al. (2006). Canine intestinal contents vs. simulated media for the assessment of solubility of two weak bases in the human small intestinal contents. *Pharm. Res.* **23** (6): 1373–1381.

[103] Azzouz, L.L. and Sharma, S. (2019). *Physiology, Large Intestine*. Treasure Island (FL): StatPearls Publishing.

[104] Basit, A.W. (2005). Advances in colonic drug delivery. *Drugs* **65** (14): 1991–2007.

[105] Bak, A., Ashford, M., and Brayden, D.J. (2018). Local delivery of macromolecules to treat diseases associated with the colon. *Adv. Drug Deliv. Rev.* **136–137**: 2–27.

[106] Rampton, D. (1994). Colonic drug absorption and metabolism. *Gut* **35** (3): 432.

[107] Strugala, V., Dettmar, P., and Pearson, J. (2008). Thickness and continuity of the adherent colonic mucus barrier in active and quiescent ulcerative colitis and Crohn's disease. *Int. J. Clin. Pract.* **62**: 762–769.

[108] Wilson, C.G. (2010). The transit of dosage forms through the colon. *Int. J. Pharm.* **395** (1-2): 17–25.

[109] Weitschies, W., Kosch, O., Mönnikes, H., and Trahms, L. (2005). Magnetic marker monitoring: an application of biomagnetic measurement instrumentation and principles for the determination of the gastrointestinal behavior of magnetically marked solid dosage forms. *Adv. Drug Deliv. Rev.* **57** (8): 1210–1222.

[110] Roldo, M., Barbu, E., Brown, J.F. et al. (2007). Azo compounds in colon-specific drug delivery. *Exp. Opin. Drug Deliv.* **4** (5): 547–560.

[111] McConnell, E.L., Liu, F., and Basit, A.W. (2009). Colonic treatments and targets: issues and opportunities. *J. Drug Target.* **17** (5): 335–363.

[112] Murray, K., Hoad, C.L., Mudie, D.M. et al. (2017). Magnetic resonance imaging quantification of fasted state colonic liquid pockets in healthy humans. *Mol. Pharm.* **14** (8): 2629–2638.

[113] Diakidou, A., Vertzoni, M., Goumas, K. et al. (2009). Characterization of the contents of ascending colon to which drugs are exposed after oral administration to healthy adults. *Pharm. Res.* **26** (9): 2141–2151.

[114] Reppas, C., Karatza, E., Goumas, C. et al. (2015). Characterization of contents of distal ileum and cecum to which drugs/drug products are exposed during bioavailability/bioequivalence studies in healthy adults. *Pharm. Res.* **32** (10): 3338–3349.

[115] Vertzoni, M., Goumas, K., Söderlind, E. et al. (2010). Characterization of the ascending colon fluids in ulcerative colitis. *Pharm. Res.* **27** (8): 1620–1626.

[116] Tannergren, C., Bergendal, A., Lennernäs, H., and Abrahamsson, B. (2009). Toward an increased understanding of the barriers to colonic drug absorption in humans: implications for early controlled release candidate assessment. *Mol. Pharm.* **6** (1): 60–73.

[117] Macfarlane, G., Gibson, G., and Cummings, J. (1992). Comparison of fermentation reactions in different regions of the human colon. *J. Appl. Bacteriol.* **72** (1): 57–64.

[118] Koziolek, M., Schneider, F., Grimm, M. et al. (2015). Intragastric pH and pressure profiles after intake of the high-caloric, high-fat meal as used for food effect studies. *J. Control. Release* **220**: 71–78.

[119] Segregur, D., Flanagan, T., Mann, J. et al. (2019). Impact of acid-reducing agents on gastrointestinal physiology and design of biorelevant dissolution tests to reflect these changes. *J. Pharm. Sci.* **108** (11): 3461–3477.

[120] Ladas, S.D., Isaacs, P.E., Murphy, G.M., and Sladen, G.E. (1984). Comparison of the effects of medium and long chain triglyceride containing liquid meals on gall bladder and small intestinal function in normal man. *Gut* **25** (4): 405–411.

[121] Ranmal, S.R., Yadav, V., and Basit, A.W. (2017). Targeting the end goal: opportunities & innovations in colonic drug delivery. *ONdrugDelivery Mag.* **77**: 22–26.

[122] Ferreira, C.M., Vieira, A.T., Vinolo, M.A. et al. (2014). The central role of the gut microbiota in chronic inflammatory diseases. *J. Immunol. Res.* **2014**: 689492.

[123] Yadav, V., Varum, F., Bravo, R. et al. (2016). Inflammatory bowel disease: exploring gut pathophysiology for novel therapeutic targets. *Transl. Res.* **176**: 38–68.

[124] Forslund, K., Hildebrand, F., Nielsen, T. et al. (2015). Disentangling type 2 diabetes and met-formin treatment signatures in the human gut microbiota. *Nature* **528** (7581): 262–266.

[125] Koeth, R.A., Wang, Z., Levison, B.S. et al. (2013). Intestinal microbiota metabolism of L-carnitine, a nutrient in red meat, promotes atherosclerosis. *Nat. Med.* **19** (5): 576–585.

[126] Singh, V., Yeoh, B.S., and Vijay-Kumar, M. (2016). Gut microbiome as a novel cardiovascular therapeutic target. *Curr. Opin. Pharmacol.* **27**: 8–12.

[127] Weersma, R.K., Zhernakova, A., and Fu, J. (2020). Interaction between drugs and the gut microbiome. *Gut* **69** (8): 1510.

[128] Scott, K.P., Gratz, S.W., Sheridan, P.O. et al. (2013). The influence of diet on the gut micro-biota. *Pharmacol. Res.* **69** (1): 52–60.

[129] Coombes, Z., Yadav, V., McCoubrey, L.E. et al. (2020). Progestogens are metabolized by the gut microbiota: implications for colonic drug delivery. *Pharmaceutics* **12** (8).

[130] Stojancevic, M., Bojic, G., Salami, H.A., and Mikov, M. (2014). The influence of intestinal tract and probiotics on the fate of orally administered drugs. *Curr. Issues Mol. Biol.* **16**: 55–68.

[131] Sousa, T., Paterson, R., Moore, V. et al. (2008). The gastrointestinal microbiota as a site for the biotransformation of drugs. *Int. J. Pharm.* **363** (1–2): 1–25.

[132] Maurice, C.F., Haiser, H.J., and Turnbaugh, P.J. (2013). Xenobiotics shape the physiology and gene expression of the active human gut microbiome. *Cell* **152** (1–2): 39–50.

[133] Wang, J., Yadav, V., Smart, A.L. et al. (2015). Stability of peptide drugs in the colon. *Eur. J. Pharm. Sci.* **78**: 31–36.

[134] McCoubrey, L.E., Elbadawi, M., Orlu, M. et al. (2021). Harnessing machine learning for development of microbiome therapeutics. *Gut Microb.* **13**: 1–20. https://doi.org/10.1080/1949 0976.2021.1872323.

[135] Flowers, S.A., Bhat, S., and Lee, J.C. (2020). Potential implications of gut microbiota in drug pharmacokinetics and bioavailability. *Pharmacother. J. Hum. Pharmacol. Drug Therapy* **40** (7): 704–712.

[136] Lee, S.H., Bajracharya, R., Min, J.Y. et al. (2020). Strategic approaches for colon targeted drug delivery: an overview of recent advancements. *Pharmaceutics* **12** (1).

[137] Varum, F., Freire, A.C., Fadda, H.M. et al. (2020). A dual pH and microbiota-triggered coating (Phloral™) for fail-safe colonic drug release. *Int. J. Pharm.* **583**: 119379.

[138] Ibekwe, V.C., Fadda, H.M., McConnell, E.L. et al. (2008). Interplay between intestinal pH, transit time and feed status on the *in vivo* performance of pH responsive ileo-colonic release systems. *Pharm. Res.* **25** (8): 1828–1835.

[139] Varum, F., Freire, A.C., Bravo, R., and Basit, A.W. (2020). OPTICORE™, an innovative and accurate colonic targeting technology. *Int. J. Pharm.* **583**: 119372.

12

Integrating Biopharmaceutics to Predict Oral Absorption Using PBPK Modelling

Konstantinos Stamatopoulos

Biopharmaceutics, Pharmaceutical Development, PDS, MST, RD Platform Technology and Science, GSK, Ware, Hertfordshire, United Kingdom

12.1 Introduction

Initially, biopharmaceutical physiologically based models were static systems, for example they did not account for the dynamic changes in the lumen of the GI tract, e.g. fluid volumes and pH. However, the drug absorption process is dynamic where fluid composition can change depending on location, food digestion, time and the scale of the investigation; there is potential for interactions with endogenous and exogenous materials as well as metabolism of the drug compound. It is very difficult for static models to account for these dynamic aspects.

Pharmacokinetic models can have different levels of complexity, starting from being 'exploratory' and empirical and/or semi-mechanistic to complex physiologically based pharmacokinetic (PBPK) models. The complexity is conditional and depends on the goal of the modelling as well as the amount and quality of the available data. Pharmacokinetic models can be either built based mainly on the clinical data ('top down' approach) or based on our general knowledge of the human physiology and its mechanisms ('bottom up' approach).

Biopharmaceutics: From Fundamentals to Industrial Practice, First Edition. Edited by Hannah Batchelor.
© 2022 John Wiley & Sons Ltd. Published 2022 by John Wiley & Sons Ltd.

Physiologically based pharmacokinetic (PBPK) models are built using a mathematical framework. The mathematical framework uses known physiology and divides the body into a series of compartments that correspond to the different organs or tissues in the body. These compartments are connected by flow rates that parallel the circulating blood system. Many of the early models were focused on the liver and metabolism of drug with inclusion of absorption parameters coming somewhat later. Preliminary models used first-order oral uptake constants to mimic ingestion then several studies used multiple compartments within the GI tract, finally the compartmental absorption and transit model was introduced in 1996 to include the dynamic GI transit aspects that influence absorption [1]. Current commercial software uses nine compartments within the gastrointestinal tract [2].

Advances in computing power have enabled software to be developed to predict the uptake of drugs based on human physiology; commercial software packages include PKSim, Gastroplus and SimCYP. These are physiologically based pharmacokinetic (PBPK) software packages that use physiological inputs to build models that replicate the complexity of the gastrointestinal tract (as well as skin and pulmonary systems) to predict the absorption of drugs based on their chemical properties.

The mathematical equations that underpin PBPK were first reported in the 1920s to simulate the concentration of the noxious volatile anaesthetic ethyl ether in the blood after inhalation exposure [3]. In the 1930s, mathematical equations to simulate ADME processes were reported [4]. Advances in PBPK modelling were made based on defining and combining body tissues based on their blood perfusion rates, grouping highly and poorly perfused tissues. The link between kinetics of inhaled compounds and tissue volume, perfusion rate and partition coefficient was reported in 1951 [5].

More recently, the use of PBPK modelling to support regulatory approval of drugs is increasingly common with PBPK guidance being issues by both the FDA [6] and the EMA [7]. PBPK has reduced the burden of both animal and clinical testing during drug development.

Fundamental to PBPK models is high-quality input data on human physiology as well as robust pharmacokinetic data that can be used to verify and validate these models.

Biopharmaceutics-related applications of PBPK modelling include (i) prediction of oral drug absorption by integrating drug permeability, dissolution, particle size and controlled/modified-release rates and formulation selection based on the model-predicted absorption; (ii) prediction of the effects of food on drug absorption and (iii) demonstration of bioequivalence of formulations through numerical or mechanistic IVIVC to support biowaivers. Furthermore, PBPK modelling can be used to extrapolate drug exposure observed in healthy adults to other populations to better understand how physiological and anatomical differences may impact drug pharmacokinetics (see Chapter 13).

This chapter will consider key parameters relevant to biopharmaceutics to demonstrate how these are incorporated into PBPK models. PBPK models that predict oral absorption need to include and link all processes that occur following the administration of an oral drug product.

12.2   Mechanistic Models

It is not easy to simultaneously explore and investigate in depth all the different parameters (and their relationships) that affect oral drug absorption. PBPK modelling has been utilised to explore all these phenomena together by describing mathematically the human

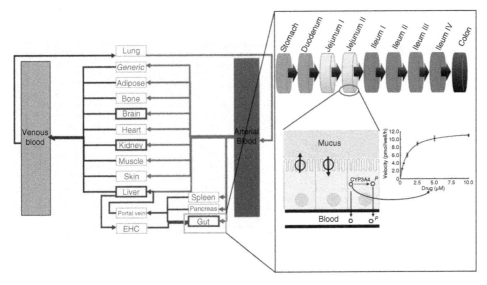

Figure 12.1 *Structure of the compartmentalised gut model into the whole body PBPK model.*

physiology, drug properties and formulation behaviour. Every part of the human body is treated as a compartment (think about a box with input and output and inside the box a function/equation that describe a mechanism, e.g. drug metabolism by CYP's enzymes). The equations that describe each process within each compartment combine drug parameters (e.g. solubility, diffusion coefficient, etc.) and physiological conditions (e.g. pH, bile micelle concentration, transit time, etc.). Thus, a PBPK platform is a bunch of compartments connected to each other reproducing the connection of the internal organs through the blood circuits (Figure 12.1). Each compartment can be further compartmentalised adding more complexity, reflecting regional differences in the gut (Figure 12.1).

As biopharmaceutics of orally administered drugs are the focus of this chapter; further exploration of the processes that follow oral administration of a solid dosage form is presented in Figure 12.1. This provides an illustration of the many processes that occur and need to be modelled as inputs for a PBPK model.

Figure 12.2 shows the sequence of processes which take place in the human GI tract after the oral administration of a dosage form. When a dosage form enters the GI tract, the disintegration/erosion of the formulation starts (based on different triggering mechanisms, e.g. pH, pressure forces, osmolality and so on) resulting in the release of the drug particles to the luminal fluid (step 1). Then the released drug particles are dissolved (step 2). Around the particle, there is an unstirred layer, so-called diffusion layer, where the detached drug molecules from the solid surface must pass through in order to reach the bulk media. Precipitation of the dissolved drug molecules can occur if the concentration of the drug in the media is above the amount of the drug that can be dissolved (step 3). The precipitated drug particles can be resolubilised (step 4) following dissolution. How fast or to what extent these precipitated particles will be resolubilised depends on the solid state of the precipitant, i.e. crystalline or amorphous, and the concentration of the drug in the bulk media. Depending upon pKa of the drug, ionization can take place and based on the lipophilicity of the drug, partitioning into the micelles can occur, further contributing to the solubilisation of the drug.

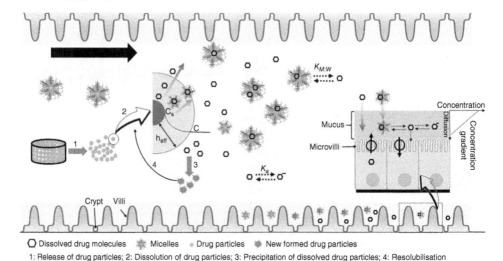

O Dissolved drug molecules   ✳ Micelles   ○ Drug particles   ✳ New formed drug particles
1: Release of drug particles; 2: Dissolution of drug particles; 3: Precipitation of dissolved drug particles; 4: Resolubilisation

Figure 12.2   *Schematic representation of the phenomena which takes place in the luminal environment of the human GHI tract affecting the performance of the dosage form and the absorption of the drug molecules; K_a ionization constant, $K_{M:W}$: micelle-to-water partition coefficient.*

Free and bound to the micelles, drug molecules should diffuse the mucus layer that is lying on the surface of the enterocytes. Within the mucus, when bound to the micelles, drug molecules should be released before the drug will passively or actively (i.e. involvement of transporters) permeate the apical membrane of the enterocytes. Thus, it is assumed that only the free drug monomers are taken up by the enterocytes. Ionized drug molecules can be taken up by passive diffusion through the junction zone (the space between the enterocytes) which is controlled by the molecular weight of the drug and its charge relative to the charge of the proteins located in the junction zones. Thus, positively charged molecules cannot pass through the junction zones due to the repulsion forces from the also positively charged proteins.

Within this complex process of oral absorption, it can be important to conceptually understand what the rate-limiting step for absorption is likely to be for a given drug/drug product; then modelling and simulation can be used to better understand sensitivities to each aspect. For example, if the solubility is limiting the absorption then there is a risk of reaching saturated solubility at low concentrations where the concentration gradient will not drive permeation.

12.3   Solubility Inputs

Solubility was the focus of Chapter 4. This chapter revisits certain concepts to underpin the detail required to develop a PBPK model. Solubility is normally referred to as equilibrium solubility which is a condition where the dissolved and undissolved drug molecules are in equilibrium. Quantitatively, solubility describes the amount of drug that can be dissolved in a given amount of solvent. The rate at which the drug molecule is transferred from a solid state into solution is called dissolution (see Chapter 6). Thus, solubility and dissolution are fundamental terms that both describe the process of solvation and are linked by the Noyes–Whitney equation (Eq. 12.1).

$$\frac{dC}{dt} = \frac{DS}{h}(C_s - C) \tag{12.1}$$

The left term of the Eq. (12.1) (dC/dt) refers to the dissolution rate, S is the contact surface area of solute with the solvent, C is the bulk concentration, C_s is the concentration at the drug particle surface (Figure 12.3), D is the diffusion coefficient and h is the diffusion layer thickness.

Ionizable drugs, e.g. weak acids and weak bases, exhibit pH-dependent solubility as described in the solubility chapter (Chapter 4). This is because the charged form of the drug molecule is more soluble. Thus, the pH of the aqueous solution determines whether the drug molecule exists in its neutral or ionized form. The dependency of the solubility of different classes of the ionizable drugs by the pH of the solution is described by the Henderson–Hasselbalch equations as follows:

$$\log S_{\text{total}}^{\text{acid}} = \log S_o + \log\left(1 + 10^{\text{pH}-\text{p}K_a}\right) \text{monoprotic acids} \tag{12.2}$$

$$\log S_{\text{total}}^{\text{base}} = \log S_o + \log\left(1 + 10^{\text{p}K_a-\text{pH}}\right) \text{monoprotic base} \tag{12.3}$$

$$\log S_{\text{total}}^{\text{ampholytes}} = \log S_o + \log\left(1 + 10^{\text{pK}-\text{p}K_a} + 10^{\text{pK}_b-\text{pH}}\right) \text{ampholytes} \tag{12.4}$$

Where S_o is the intrinsic solubility.

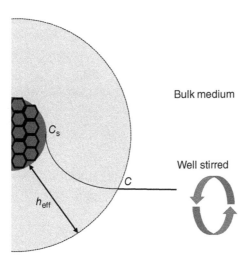

Figure 12.3 *Schematic representation of a dissolving drug particle and concentration gradient in the diffusion layer. C is bulk concentration, C_s is the concentration at the drug particle surface and h_{eff} is the diffusion layer thickness.*

The pH-solubility profile of a drug can be measured in a simple buffer without solubilisers (i.e. surfactants) and it is controlled by the pK_a, intrinsic solubility (S_0), and solubility product (Ksp) of a drug, as well as the pH and the common ion concentration in the fluid. To develop a PBPK model, the pK_a of the drug needs to be known as well as the intrinsic solubility.

The solubility of the drug in the luminal environment of the human GI tract is also affected by the presence of bile salts and other surfactants like lysophospholipids. The extent of bile salt-mediated solubility enhancement depends on the 'ability' of the drug to partition into the micelle which is directly related to the lipophilicity of the drug. Solubility experiments in biphasic systems (e.g. octanol/water) are normally conducted to obtain $\log P$ value which is a parameter that determines the lipophilicity of the drug. The higher the $\log P$ is, the more lipophilic the drug is with potential implications in the solubility as well as potential food effect. The last scenario, i.e. food effect, is relevant due to the fact that in the fed state the concentration of the bile salts and other surfactants, e.g. lysophospholipids, are in higher concentration compared to fasted state resulting in forming micelles. The partitioning of the drug in those micelles will increase its solubility and potentially its absorption. In PBPK modelling, the following equations are used to calculate the solubility of the different drug species in the luminal environment:

$$S_{dissolv} = S_o * \left(1 + \frac{K_{m:w}[BS]}{[Water]}\right) \text{ undissociated drug molecules} \qquad (12.5)$$

$$S_{dissolv} = S_o * \left(1 + \frac{K_a}{H^+} + \frac{K_{m:w_o}[BS]}{[Water]} + \frac{K_a}{H^+} + \frac{K_{m:w_-}[BS]}{[Water]}\right) \text{monoprotic acid} \quad (12.6)$$

$$S_{dissolv} = S_o * \left(1 + \frac{H^+}{K_a} + \frac{K_{m:w_o}[BS]}{[Water]} + \frac{H^+}{K_a} + \frac{K_{m:w_+}[BS]}{[Water]}\right) \text{monoprotic base} \quad (12.7)$$

Where $S_{dissolv}$ is the solubility of drug at the present of bile salt, S_o is the intrinsic solubility, $K_{m:w}$ is the micelle-to-water partition coefficient with $K_{m:w_o}$, $K_{m:w_-}$ and $K_{m:w_+}$ referring to undissociated molecule, monoprotic anion and monoprotic cation, respectively, K_a is the dissociation constant, [BS] is the bile salt concentration and [Water] is the water concentration.

The micelle-to-water partition coefficient ($K_{m:w}$) can either calculated based on Eq. (12.8) [8] or from $\log P$ Eq. (12.9) [9].

$$K_{m:w} = \frac{[Drug - BS]/[BS]}{[Drug]/[water]} \qquad (12.8)$$

$$\text{Log } K_{m:w_o} = 0.74 * \log P_{oct} + 2.29 \qquad (12.9)$$

The corresponding micelle-to-water partition coefficient for anion and monocation (i.e. $K_{m:w_-}$ and $K_{m:w_+}$) can be estimated using Eqs. (12.10) and (12.11) [10, 11].

$$\text{Log } K_{m:w_-} \approx \log K_{m:w_o} - 2 \tag{12.10}$$

$$\text{Log } K_{m:w_+} \approx \log K_{m:w_o} - 1 \tag{12.11}$$

However, Eqs. (12.10) and (12.11) should be drug specific as the structure alongside the functional group (i.e. monoprotic acid or monoprotic base) determines the partition of the ionized species into the micelles (Figure 12.4) [12]. The Eqs. (12.10) and (12.11) need to be adjusted based on experimental data for each drug. The equations for ibuprofen and ketoprofen are shown as examples to demonstrate the adjustments made:

$$\text{Ibuprofen} \log K_{m:w_-} \approx \log K_{m:w_o} - 2.1$$

$$\text{Ketoprofen} \log K_{m:w_-} \approx \log K_{m:w_o} - 1.6$$

The key measurements required to develop a PBPK model that relate to solubility are:

- Basic
  - Physicochemical properties of the compound type
  - Type: neutral, acid and base
  - pK_a (not applied on neutral APIs)
  - Molecular weight
  - Aqueous (intrinsic) solubility
  - Lipophilicity ($\log P$)
- Advanced
  - Biorelevant solubility
  - pH-dependent solubility (measured data)
  - Micelle-to-water partition coefficient ($\text{Log} K_{m:w}$)
  - Particle surface solubility
  - Precipitation/supersaturation

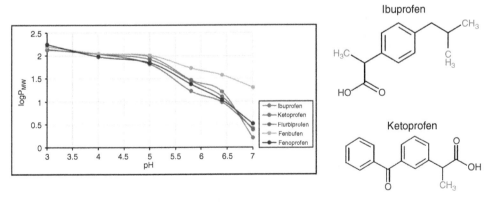

Figure 12.4   *Log* P_{MW} *of five drugs at various mobile phase pH values. Source: From Shahzad [12].*

12.4   Dissolution Inputs

Dissolution was introduced in Chapter 6; a brief overview is provided here in the context of PBPK modelling. To understand dissolution phenomena, it is important to understand diffusion and convection, the two factors govern mass-transfer phenomena. Diffusion is a random 'walk' process, but the driving force behind the net movements of the drug molecules is the concentration gradient. Drug molecules move/diffuse from high concentration regions to low concentration regions. Convection is simply the motion, flow, of the solvent. This motion contributes to mass transfer. A simple example of convection is when stirring is applied to dissolution media to reach homogeneity of the dissolved mass faster. Diffusion kinetics are described by Fick's law whereas convection is described by the Navier–Stokes equation. The mass-transfer equation is derived from the Fick's laws of diffusion and the Navier–Stokes equation; for more information, see Chapters 1–7 of the book 'Mass Transfer: Basics and Application' by Kohichi Asano [13].

There are two approaches on how to integrate *in vitro* dissolution data into PBPK models. One is to use the dissolution profile as it is: in this case, the PBPK model assumes that the drug is completely dissolved, and any supersaturation and precipitation phenomena are unimportant. The other approach is to use the *in vitro* data to estimate the parameters included in the so-called mechanistic diffusion layer model (DLM). As can be seen in Eq. (12.12), the DLM incorporates the particle size, the microenvironment around the particle (i.e. diffusion layer), the diffusivity of the detached drug particles from the solid surface and the concentration gradient $(S_{surf}(t) - C_b(t))$ which is the driving force for dissolution (12.8). However, even this model needs to be adjusted, on occasion, to capture *in vivo* data using an empirical scalar.

$$DR(t) = -NS \frac{D}{h(t)} 4\pi\alpha(t)(\alpha(t) + h(t))(S_{surf}(t) - C_b(t)) \qquad (12.12)$$

Where N is the number of particles (in a given particle size bin for polydispersed formulations), S is the empirical scalar, D is the effective diffusion coefficient, h is the thickness of diffusion layer, α is the particle radius at time t, S_{surf} is the concentration of drug at the particle surface at time t; C_b is the concentration of the drug in the bulk solution at time t and $(S_{surf}(t) - C_b(t))$ is the driving force for dissolution.

Thus, the DLM model considers the drug properties (particle size, lipophilicity, pK_a and pH-solubility) and the physiological aspects of GI tract (regional and between-subject differences related to pH, bile salt concentration, buffer capacity and fluid volume dynamics). The factors that are related to these factors and the relevant equations are listed in Table 12.1.

The DLM is an approach that facilitates a mechanistic translation of *in vitro* experiments to *in vivo* environment [14]. Proper translation of the *in vitro* data to *in vivo* can be conducted in stepwise process starting from generating data for the aqueous solubility, biorelevant solubility and dissolution experiments. In each step, a parameter is obtained to inform DLM. In step 1, the intrinsic solubility of the drug is determined with aqueous solubility experiments. In step 2, the solubility of the drug in the presence of bile salts is determined and the $LogK_{m:w}$ can be estimated. This will give information about the fraction of the drug being bound in micelles which has a direct effect on D_{eff} (Figure 12.2) and hence on DR as the diffusion of the drug bound to micelles will be slower within the diffusion

Table 12.1 *Summary of key parameters within the DLM and the equations that relate to these factors.*

Parameter	Where this is incorporated	Which equation
Particle size	N is the number of particles (in a given particle size bin for polydispersed formulations) a is the particle radius at time t	Equation (12.12)
Lipophilicity	S_{surf} is the concentration of drug at the particle surface at time t C_b is the concentration of the drug in the bulk solution at time t	Equations (12.5), (12.6) and, (12.7)
pK_a pH-solubility	S_{surf} is the concentration of drug at the particle surface at time t C_b is the concentration of the drug in the bulk solution at time t	Equations (12.2), (12.3), and (12.4)

layer. In step 3, dissolution experiments will help to estimate the disintegration rate which is important especially for immediate release formulations. If the drug is a weak base, transfer dissolution experiments can help to estimate supersaturation and precipitation parameters with direct effect on DR. Figure 12.5 shows the stepwise process as reported by Pathak et al. [14].

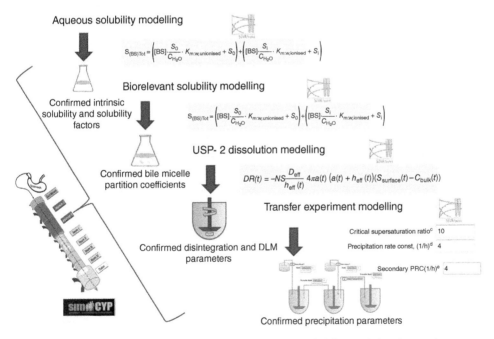

Figure 12.5 *Sequential workflow for integrating in vitro solubility and dissolution data into PBPK modelling (Source: From Pathak et al. [14] / American Chemical Society / CC BY 4.0).*

12.4.1   Fluid Dynamics and Dissolution

The dissolution rate is affected by the hydrodynamics around a dissolving particle. This can be seen in Eq. (12.12) where the radius of the particle and the diffusion layer thickness are both proportionally linked to the dissolution rate. To understand the fluid dynamics model, we will start from the main parameter that links the DLM with the hydrodynamics, i.e. h is the thickness of diffusion layer.

$$h = \frac{2a(t)}{Sh} \tag{12.13}$$

Where Sh is the Sherwood number and a is the particle's radius.

Sherwood number (Sh) is linked to Reynolds and Schmitt number (Sc) via Eq. (12.14) which is based on Prandtl's boundary layer.

$$Sh = 2 + Re^{1/2} * Sc^{1/3} \tag{12.14}$$

Schmidt number (Sc) is defined as the ratio of kinematic viscosity to the diffusion coefficient.

$$Sc = \frac{v}{D_{eff}} = \frac{\mu_f}{\rho_f D_{eff}} \tag{12.15}$$

Where μ_f is the fluid's viscosity, ρ_f is the fluid's density and D_{eff} is the effective diffusion coefficient.

And finally, the Reynolds number (Re) is defined as

$$Re = \frac{\rho_f * 2a(t)}{\mu_f} |\vec{U}_f - \vec{U}_p|(4), \tag{12.16}$$

Where $\vec{U}_f, \vec{U}_p$ is representative axial and tangential velocities of fluid and particle, respectively, which can be derived via complex CFD modelling of compendial dissolution apparatuses (e.g. USP II, μDIss).

It should be pointed out that the luminal contents possess viscoelastic properties and hence the viscosity will depend on the shear rates which will in turn affect the calculation of these parameters.

The complexity of applying these mathematical models is increased when the time-dependent change in viscosity within the GI tract as well as the discontinuous motility patterns are considered. Complex computational fluid dynamics are required to better understand *in vivo* fluid dynamics and to reflect these in PBPK models.

The key measurements required to develop a DLM model include:

- fluid viscosity,
- particle density and
- agitation strength (i.e. impeller speed (rpm).

12.5   Permeability Inputs

Following dissolution, the dissolved drug particles should permeate enterocytes to reach the systemic circulation. Currently, the gold standard to measure drug's permeability is the Loc-I-Gut method as described in (Chapter 5). In this method, a region of the small intestine is isolated and drug solution is infused. The disappearing rate (of the drug) is calculated providing the permeability of the drug. The available *in vivo* data reflect mainly jejunal permeability as this is the region where most of the Loc-I-Gut studies have been conducted. However, jejunal permeability is not representative of the entire GI tract. This is due to regional differences in the GI tract, such as surface area, fluid volumes, enzymes/transporters abundances and luminal pH. Moreover, Loc-I-Gut studies are expensive and time consuming, and are not a routine part of the drug development process. Existing *in vivo* permeability data have been used to build a correlation with *in vitro* data derived either from cell lines (e.g. CaCo2, MDCK etc) or from artificial permeable membranes (e.g. PAMPA); this was shown in Chapter 5.

A simple linear relationship has been extensively used in biopharmaceutics. However, a more mechanistic approach is needed which takes into account species differences, regional anatomical differences in GI tract (e.g. villi dimensions, mucus layer, etc.) as well as regional transporter abundances.

The general equation describing the mechanistic permeability of the drug is given below (Eq. 12.17).

$$P_{eff} = \left(\left(\left(P_{Trans,0} * f_{neutral,pH} + P_{para} \right) * ACC * MVE * f_{u_{UBL}} \right)^{-1} + \left(P_{UBL} \right)^{-1} \right)^{-1} * FE_p \quad (12.17)$$

Where P_{eff} is the effective intestinal membrane permeability, $P_{Trans,0}$ is the membrane passive intrinsic transcellular permeability of unionized drug molecules, $f_{neutral,pH}$ is the fraction of drug molecules neutral at a given pH, P_{para} is the paracellular pore permeability (this parameter is a function of molecular size, charge and pore size, ACC is the effective surface area scalar, MVE is the maximum villous surface expansion scalar, $f_{u_{UBL}}$ is the fraction of drug unbound in the unstirred boundary layer (UBL), P_{UBL} is the permeability across the unstirred boundary layer and FE_p is the fold expansion (Plicae circulares) scalar.

Practically, $P_{Trans,0}$ can be determined either by modelling $CaCo_2$ and MDCK cell lines [15] or can be predicted by correlation with $\log P_{o:w}$ or $\log P_{PAMPA,0}$ [16].

Equation (12.18) is used to calculated P_{UBL} which is directly related to the effective diffusion coefficient (D_{eff}) and the regional mucus thickness ($h_{eff,UBL}$).

$$P_{UBL} = D_{eff} / h_{eff_{UBL}} \quad (12.18)$$

Where D_{eff} is calculated based on Eq. (12.19).

$$D_{eff} = f_{unbound} * D_{mono} + \left(1 - f_{unbound} \right) * D_{micelle} \quad (12.19)$$

Where $f_{unbound}$ is the fraction of the unbound drug to micelles, D_{mono} is the diffusion coefficient of drug monomers and $D_{micelle}$ is the diffusion coefficient of micelles.

Thus, as it can been seen also from Figure 12.2, D_{eff} is linked to the diffusion of the free and bound to micelle drug within the diffusion layer, around a dissolved drug particle, affecting dissolution rate as well as to the permeability within UBL affecting the overall permeability. P_{UBL} is important for drug where the permeability in UBL will be the upper limit of the drug's overall permeability.

Equation (12.20) shows the link between permeability and absorption rate.

$$\frac{dA_n}{dt} = A_{diss_n} * 2 \frac{P_{eff_n}}{R_{SI_n}} \quad (12.20)$$

Where A_{diss_n} is the amount of drug dissolved in the luminal fluids of the nth GI tract segment and R_{SI_n} is the regional small intestinal radius.

The key measurements required to develop a PBPK model that relate to permeability are:

- Basic
  - $\log P$
  - Apparent permeability data from an *in vitro* system (derived from artificial membranes like PAMPA as minimum requirement)
- Advanced
  - Passive permeability derived from cell lines experiments inhibiting any active efflux and/or uptake transport
  - Mucus permeability using co-cultured cell lines (e.g. $CaCo_2$-HT29)
  - Diffusion coefficient of drug monomers and drug–micelle complex in mucus
  - Permeability experiments performed using biorelevant media

12.6   Incorporation of Modelling and Simulation into Drug Development

Modelling and simulation brings together mathematical models and physical measurements that both represent the real situation; these can be used to explore a hypothesis and to generate simulated data. For example, to predict the solubility of a drug over a pH range using the pK_a as described in Eqs. (12.6) and (12.7). These predicted values can then be checked using a physical experiment to understand the hypothesis and the effect on observed data. It is essential that this checking step is undertaken to ensure that modelling and simulation enhance knowledge from both mathematical and physical experiments. In cases where the observed data are very different to the predicted data, this can highlight complex issues that are not predicted using simple mechanistic understanding; thus, further research may be required to understand why the model is not predicting the observed data.

Modelling and simulation are used throughout the drug development process; in the early stages, to identify the risk factors that may limit drug absorption and explore the formulation variants and their consequences (e.g. particle size) on pharmacokinetics; finally, the models can be used to better predict food effects or drug–drug interactions. They are used in regulatory submissions to demonstrate mechanistic understanding and can also be used in biowaiver submissions to replace clinical studies.

12.6.1   Understanding the Effect of Formulation Modifications on Drug Pharmacokinetics

Robust models can be used to understand how small changes in a formulation may affect pharmacokinetics; for example, the sensitivity of particle size on pharmacokinetics can be predicted. This approach was reported by Parrot et al. [17] who compared the dissolution and pharmacokinetics of bitopertin of a range of particle sizes and could then use the model to better understand the impact of particle size and thus set a limit on the particle size of drug substance to be used in a product. The use of modelling eliminates the need for trial and error, and reduces the extent of clinical testing required to understand formulation attributes.

Changes in drug release rate can be used to mitigate certain factors, for example, improve exposure from a CYP 3A4 substrate where a high extraction ratio in the gut wall can limit exposure. Midazolam is a CYP3A4 substrate. An immediate release (IR) formulation will release most of the drug at the proximal part of the GI tract where CYP3A4 is present at the highest abundance. A change to a formulation that delays release to the distal part of the intestine where CYP3A4 enzymes have a lower abundance may be considered an attractive option. However, this needs to be balanced with the overall permeability as this also decreased along the GI tract, so there may be less extraction in the gut wall but there may also be less overall absorption. A PBPK model can be utilised to understand how changing the release rate of the drug from the dosage form will affect its PK profile and simultaneously consider all factors. Further information on how controlling the rate of drug release can affect the bioavailability of CYP3A4 substrates using PBPK models was reported by Olivares-Morales et al. [18].

12.6.2 Model Verification/Validation

The use of a PBPK model to understand clinical scenarios is depended upon the confidence within the model. Typically, a learn-and-confirm cycle is used where the model is developed to predict a clinical scenario and the accuracy of the model is evaluated using *in vivo* data. These data may come from preclinical animal studies in early development and from phase 1 clinical studies in humans.

Typically, the model is developed using this early clinical data to refine inputs based on the estimated sensitive or critical parameters. The model can subsequently be verified as the additional clinical data becomes available. If modelled simulations do not match this new clinical data then the model is refined to fit the observations and further clinical data sets are required to verify the model. Due to the inherent variability associated with clinical data, a limit of predicted data being within twofold of the observed is considered to be a sufficiently predictive model. Further descriptions on PBPK model verification and validation are available [19].

12.6.3 Using Modelling to Understand Bioequivalence

Formulation development occurs in parallel with clinical evaluation. Therefore, the formulation will change from phase 1 studies prior to development of the final commercial formulation. A robust PBPK model can be used to understand the impact of formulation changes on the pharmacokinetics to determine whether the products are likely to be bioequivalent. This can support bridging of clinical exposure within studies when a new formulation is introduced.

In supporting bridging of formulations, PBPK models are used to establish an *in vitro–in vivo* correlation to dissolution and permeation models that are mechanistically informed based on the drug substance and drug product under evaluation. The advantage of PBPK correlations rather than dissolution is the ability to incorporate non-linearity or regional-specific permeability. Further details on the use of PBPK models to manage bridging can be found [20].

12.7 Conclusions

Biopharmaceutical models are useful to better understand drug and formulation design, and to predict a human pharmacokinetic profile from the limited data available. As the drug development programme proceeds and clinical human data become available, the validity of the model can be improved and fine-tuned to best understand the pharmacokinetic output.

Hence, these models can be used to investigate scenarios for future development; for example, what are the effects in special population based on our knowledge in a healthy population? They can also help to understand formulation variants; for example, in developing a rapid-dissolving or slow-release product.

References

[1] Lawrence, X.Y., Crison, J.R., and Amidon, G.L. (1996). Compartmental transit and dispersion model analysis of small intestinal transit flow in humans. *International Journal of Pharmaceutics* **140** (1): 111–118.

[2] Agoram, B., Woltosz, W.S., and Bolger, M.B. (2001). Predicting the impact of physiological and biochemical processes on oral drug bioavailability. *Advanced Drug Delivery Reviews* **50** (Suppl 1): S41–S67.

[3] Haggard, H.W. (1924). The absorption, distribution and elimination of ethyl ether. *Journal of Biological Chemistry* **59**: 753–770.

[4] Teorell, T. (1937). Kinetics of distribution of substances administered to the body, I: the extravascular modes of administration. *Archives Internationales de Pharmacodynamie et de Therapie* **57**: 205–225.

[5] Kety, S.S. (1951). The theory and applications of the exchange of inert gas at the lungs and tissues. *Pharmacological Reviews* **3** (1): 1–41.

[6] FDA (2018). Physiologically based pharmacokinetic analyses—format and content guidance for industry. https://www.fda.gov/regulatory-information/search-fda-guidance-documents/physiologically-based-pharmacokinetic-analyses-format-and-content-guidance-industry (accessed January 2021).

[7] EMA (2018). Guideline on the reporting of physiologically based pharmacokinetic (PBPK) modelling and simulation. https://www.ema.europa.eu/en/documents/scientific-guideline/guideline-reporting-physiologically-based-pharmacokinetic-pbpk-modelling-simulation_en.pdf (accessed January 2021).

[8] Takagi, T., Ramachandran, C., Bermejo, M. et al. (2006). A provisional biopharmaceutical classification of the top 200 oral drug products in the United States, Great Britain, Spain, and Japan. *Molecular Pharmaceutics* **3** (6): 631–643.

[9] Glomme, A., Marz, J., and Dressman, J.B. (2006). Predicting the intestinal solubility of poorly soluble drugs. In: *Pharmacokinetic Profiling in Drug Research* (eds. B. Testa, S.D. Krämer, H. Wunderli-Allenspach and G. Folkers), 259–280. Wiley.

[10] Avdeef, A., Box, K.J., Comer, J.E.A. et al. (1998). pH-metric log *P* 10. Determination of liposomal membrane-water partition coefficients of ionizable drugs. *Pharmaceutical Research* **15** (2): 209–215.

[11] Miyazaki, J., Hideg, K., and Marsh, D. (1992). Interfacial ionization and partitioning of membrane-bound local anaesthetics. *Biochimica et Biophysica Acta (BBA) – Biomembranes* **1103** (1): 62–68.

[12] Shahzad, Y. (2013). *Micellar Chromatographic Partition Coefficients and their Application in Predicting Skin Permeability*. University of Huddersfield.

[13] Asano, K. (2007). *Mass Transfer: From Fundamentals to Modern Industrial Applications*. Wiley.

[14] Pathak, S.M., Ruff, A., Kostewicz, E.S. et al. (2017). Model-based analysis of biopharmaceutic experiments to improve mechanistic oral absorption modeling: an integrated *in vitro in vivo* extrapolation perspective using ketoconazole as a model drug. *Molecular Pharmaceutics* **14** (12): 4305–4320.

[15] Avdeef, A. (2012). *Absorption and Drug Development: Solubility, Permeability, and Charge State*. Wiley.

[16] Pade, D., Jamei, M., Rostami-Hodjegan, A., and Turner, D.B. (2017). Application of the MechPeff model to predict passive effective intestinal permeability in the different regions of the rodent small intestine and colon. *Biopharmaceutics & Drug Disposition* **38** (2): 94–114.

[17] Parrott, N., Hainzl, D., Scheubel, E. et al. (2014). Physiologically based absorption modelling to predict the impact of drug properties on pharmacokinetics of bitopertin. *The AAPS Journal* **16** (5): 1077–1084.

[18] Olivares-Morales, A., Kamiyama, Y., Darwich, A.S. et al. (2015). Analysis of the impact of controlled release formulations on oral drug absorption, gut wall metabolism and relative bioavailability of CYP3A substrates using a physiologically-based pharmacokinetic model. *European Journal of Pharmaceutical Sciences* **67**: 32–44.

[19] Rostami-Hodjegan, A. (2018). Reverse translation in PBPK and QSP: going backwards in order to go forward with confidence. *Clinical Pharmacology and Therapeutics* **103** (2): 224–232.

[20] Stillhart, C., Pepin, X., Tistaert, C. et al. (2019). PBPK absorption modeling: establishing the *in vitro–in vivo* link—industry perspective. *The AAPS Journal* **21** (2): 19.

13

Special Populations

Christine M. Madla, Francesca K. H. Gavins, Sarah J. Trenfield and Abdul W. Basit

Department of Pharmaceutics, UCL School of Pharmacy, University College London, London, United Kingdom

13.1 Introduction

During the development of new oral pharmaceutical products, the behaviour of drug compounds is usually studied in healthy adults. However, the morphology and physiology of the gastrointestinal (GI) tract are influenced by multiple factors including the sex of the individual, ethnicity, diet, age, the state of pregnancy and disease. These factors can significantly alter the kinetics of drug absorption, the total amount of drug absorbed amongst other processes and as such, elicit differing drug response to the observed drug behaviour in healthy adults. Consequently, in order for populations to receive appropriate pharmaceutical therapy, differences in GI tract physiology should be considered at the start of formulation development for orally administered products.

The main physiological parameters that are often altered in patients when compared to healthy populations are gastric emptying rates, pH, transit times, intestinal surface area, epithelial permeability and intestinal enzyme and transporter expression. Recent research has also explored the differences in the colonic microbiome and its interplay with health and disease, and drug response. Adding further to the complexity of optimised drug therapy is the possible interaction of co-administered drugs (otherwise known as polypharmacy) as well as drug-mediated effects on GI physiology.

This chapter will focus on special populations often overlooked in drug development to better understand why drug absorption may change in these populations.

Biopharmaceutics: From Fundamentals to Industrial Practice, First Edition. Edited by Hannah Batchelor.
© 2022 John Wiley & Sons Ltd. Published 2022 by John Wiley & Sons Ltd.

13.2   Sex Differences in the Gastrointestinal Tract and Its Effect on Oral Drug Performance

It is commonly accepted that males and females respond differently to medicines. Following a review of 10 drugs that were withdrawn from the market from 1997 to 2000, it was identified that the withdrawal of eight of the ten drugs were due to greater risks of adverse effects in women [1]. In addition, out of 67 new molecular entities approved by the US FDA between 2000 and 2002, 25 compounds demonstrated significant sex-specific differences in pharmacokinetics (PK) and efficacy [2]. For decades, the default human model subject was a '70 kg Caucasian male' [3]. The Food and Drug Administration (FDA) and the National Institute of Health (NIH) mandated the inclusion of women in clinical trials in the United States in 1993 [4] to address the disparity between the sexes in clinical research. A recent review highlighted major sex differences in factors that affect drug absorption [5, 6]; these are summarised in Figure 13.1 [7].

Drug dissolution is often the rate-limiting step in drug absorption. Studies have shown that fluid volumes in the stomach and small intestine (following the normalisation of body weight) were higher in men than in women [8, 9]. In terms of gastric and small intestinal fluid volumes, males have been reported to have higher volumes than

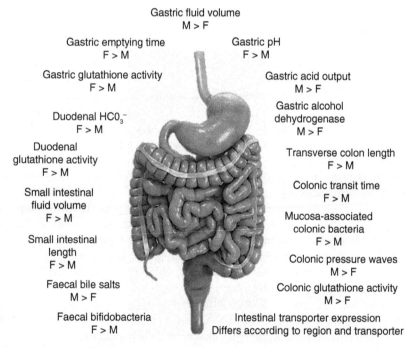

Figure 13.1   *Key sex differences at the level of the gastrointestinal tract that impact oral drug delivery and bioavailability, (M = male; F = female). Source: From Freire et al. [7] / With permission of Elsevier.*

females [8] which may affect the extent of drug dissolution. Average fasted gastric pH is significantly higher in females (2.79 ± 0.18; $n = 133$) than in males (2.16 ± 0.09; $n = 252$) ($p < 0.05$) which may be attributed to reduced acid secretion and the smaller stomach size seen in females [10]. Lowered gastric acid secretion may influence drug ionisation and solubility of pH-sensitive drugs, thereby impairing absorption and consequently, oral drug bioavailability [11].

With regard to motility, females have a significantly longer gastric emptying time (GET) for solids and calorific liquids (118.0 ± 8.1 min) than males (91.4 ± 7.5 min) however, GET decreases in post-menopausal women (97.9 ± 7.6 min) similar to that in men [12]. Variabilities in drug PK can be attributed to differences in GET; for example, peak plasma concentration of orally administered carbidopa was achieved 22 minutes later in women than men due to longer GET [13]. Sex differences in the oral bioavailability of a gastroresistant ketoprofen formulation have also been demonstrated. Males showed a higher C_{max}/AUC than females (0.468 ± 0.094 vs $0.361 \pm 0.087\,h^{-1}$) and a significantly lower T_{max} (3–5 versus 5–10 hours), respectively. Such differences were attributed to the faster small intestinal transit time in males allowing for ketoprofen to reach the appropriate intestinal environment for dissolution and absorption to occur more rapidly [14].

In terms of colonic transit time (CTT), it was identified that longer transverse and descending CTT, but shorter rectosigmoid transit time was associated with females than males [15]. The longer GI residence time for sustained-release dosage forms may facilitate enhanced drug absorption in women, as demonstrated with diltiazem which is sensitive to GI transit time [16]. This, however, may be further implicated by the regulation of intestinal membrane transporters and metabolising enzymes located in the GI mucosa.

Cytochrome P450 (CYP) enzymes are responsible for the metabolism of a number of drug substrates of which CYP2C and 3A are most commonly expressed in the small intestine. For example, the oral bioavailability of verapamil (a CYP3A substrate) was higher in women than in men [17] due to higher intestinal CYP3A expression and activity [18]. It is, however, also important to note that significant sex differences are observed in the expression of hepatic drug metabolising enzymes which contribute variabilities in clinical drug performance [19]. Drugs may also compete for intestinal membrane transporters into enterocytes which affect the downstream metabolism or availability of the drug at its target site. For example, the OATP1B1 transporters are responsible for the transport of oestrogens including oestrone-3-sulfate and oestradiol 17-beta-D-glucoronide. Statins, however, are also transported by OATPB1. As such, competitive inhibition can occur if multiple substrates are present; several studies have found sex-specific effect of SLCO1B1 genetic variants which compromise the efficacy of statin treatment [20].

In addition, there is an increasing body of literature evidence that report the inherent sex-specific expression of a number of uptake and efflux intestinal transporters [21] that elicit differential treatment outcomes. For example, a recent study showed that cimetidine, a drug substrate of the intestinal efflux transporter P-glycoprotein (P-gp) was significantly higher in females than males due to the innately higher expression of P-gp in the male proximal small intestine [22, 23]. Adding further to the complexity in formulation response in males and females is the sex-specific influence of excipients. In the presence of the solubilising excipient polyethylene glycol (PEG) 400, cimetidine bioavailability significantly increased, although only in males. No such enhancement was seen in females [22]. In addition, when co-formulated with PEG 400, the bioavailability of ranitidine significantly increased in male subjects but decreased in females when orally administered with the same formulations [24]. In a rat model, other solubilising excipients including PEG 2000 Cremophor RH 40, Poloxamer 188 and Tween 80 significantly enhanced ranitidine bioavailability in males but not females. Span 20 was also studied on its influence on ranitidine; although this excipient was able to increase oral drug bioavailability, such effects were in a non-sex-dependent manner. Sex-specific effects may be attributed to the presence of a polyoxyethylated group in PEG 2000, Cremophor RH 40, Poloxamer 188 and Tween 80, but not for Span 20 [25].

Research invested in the understanding of differences between the sexes and the clinical performance of drugs continues to be very limited. It is clear that males and females respond differently to medicines due to the dynamic interplay of GI physiology, drug PK itself and contributions from other associated organs. A single PK parameter cannot be considered as the only rate-limiting step as this may occur in a drug-by-drug basis. For a better understanding of the basic mechanisms of sex differences, future studies should be designed with this primary focus in mind to determine the extent these differences may have on clinical management.

13.3 Ethnic Differences in the Gastrointestinal Tract

Ethnicity can influence the variability in PK, pharmacodynamics, safety and efficacy of a drug. Ethnicity can be linked to be intrinsic and extrinsic factors: intrinsic factors include genetics, physiology and disease whereas extrinsic factors include diet, culture, medical practice, environment and socio-economic status [26, 27]. When a candidate drug product is developed in a new region in bridging studies (studies that compare the exposure in a test population compared to the reference), candidate drug products are advised by the International Council on Harmonization (ICH) to be classified as 'ethnically insensitive' or 'sensitive' depending on certain drug properties (chemical class, bioavailability, therapeutic dose range, intersubject variation, likelihood of co-medications, PK, pharmacodynamics, metabolic pathway, protein binding and pharmacologic class) [26, 27]. To further understand the influence of ethnicities on the PK of drugs, mechanistic studies into genetic variants of drug transporters and metabolising enzymes are needed to understand their contribution to drug product performance. Not only should differences in PK be investigated but also pharmacodynamics and dose responses. In addition, ethnic minorities, which are currently underrepresented, should be included in clinical drug studies and evaluated as a separate group.

There are limited findings for ethnic differences in GI physiological parameters such as gastric emptying, intestinal motility and GI pH. Such differences may be attributed to extrinsic factors such as diet and culture. However, lower gastric acidity has been shown in the Japanese population, with a far higher incidence of achlorhydria than other ethnicities [28].

Absorption of most drugs is by passive diffusion and ethnic differences in absorption would not be expected, with no examples found in the literature [29]. On the other hand, for drugs that are substrates of active transport and/or metabolic pathways, there is potential for variability in drug bioavailability.

Pharmacogenetics and multiallelic genetic polymorphisms, which strongly depend on ethnicity, can affect metabolic or transport pathways in the GI tract. CYP3A4 expression and activity is well known for its inter-individual variability. However, little is known on its variability in different ethnic groups and on clinically significant CYP3A4 polymorphisms [5]. Genetic polymorphisms in the CYP3A5 gene have shown variability between different ethnic groups. The reader is signposted to the following review on CYP450 enzymes and the impact of ethnicity in the variation of gene expression and enzyme activities [30]. Intestinal, instead of hepatic metabolism, is believed to be responsible for the ethnic differences in the PK of the immunosuppressant tacrolimus. The CYPA5 genotype can influence the oral bioavailability [31]. In African-American transplant patients, there is a low prevalence of the polymorphism of CYP3A5*3 compared to their Caucasian counterparts, where those homozygous for this gene are non-expressers [31, 32]. The FDA recommends higher doses of tacrolimus so patients achieve a comparable C_{max} to Caucasian patients [33].

Several polymorphisms of the ABCB1 gene have been described, with altered expression of the efflux transporter P-gp, resulting in potential changes in drug substrate PK [34]. Whilst polymorphisms of ABCB1 have been associated with changes in drug response, there are many interethnic differences in the distribution of ABCB1 polymorphisms and the findings are conflicting with limited clinical relevance [34].

13.4 Impact of Diet on Gastrointestinal Physiology

Modern diets are diverse and energy dense. In bioavailability and bioequivalence studies, the FDA and EMA advise the use of the FDA standard meal in the fed-state studies, which is a high-fat and high-calorie meal given designed to maximise the potential for a change in the extent or rate of oral drug bioavailability to occur [35, 36]. The EMA also encourages investigations into the effect of a 'moderate meal' and in certain cases, different food compositions such as a carbohydrate-rich meal [36]. Several review papers detail the interaction between food intake and drug absorption [37–39].

The intake of food induces complex and dynamic physiological changes in the gastrointestinal tract [38, 40]. Figure 13.2 shows the main gastrointestinal changes that occur after meal consumption, which could affect drug PK profiles of orally administered drugs. Postprandial changes to luminal fluid composition, properties and volume and patterns of motility are extensively reviewed by Pentafragka et al. [41] and in Chapter 11. The structural and physicochemical properties of food will influence the change in transit time, GI secretions and luminal fluid properties.

Various studies have shown that nutrient-sensing, by the entero-endrocrine system, causes the release of GI hormones, such as cholecystokinin (CCK), glucagon-like peptide 1 (GLP1), peptide YY (PYY) and gastric inhibitory polypeptide (GIP) amongst others, which in return influences patterns of GI motility and GI secretions, affecting drug absorption processes [42]. High-fat and to a lesser extent high-carbohydrate meals can delay gastric emptying, increasing the gastric residence in the fed acidic stomach for dissolution of drug products but prolonging the time for the drug to reach the small intestine, the main site of absorption [43]. Delayed gastric emptying is associated with an overall delay in T_{max} and reduced C_{max} for drugs with high oral bioavailability [44]. Vegetarians, compared with non-vegetarians, are believed to have a more rapid GI transit time, which could affect the residence time of drug

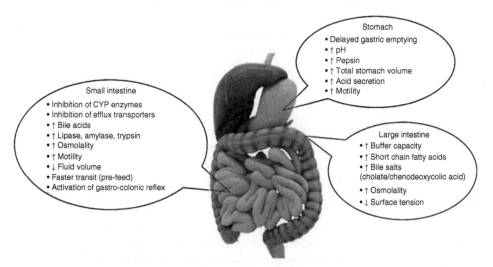

Figure 13.2 *Illustration of how food intake effects key physiological parameters involved in drug absorption and bioavailability Source: From Varum et al. [38] / With permission of Elsevier.*

product in the intestine [31]. Overall in a meal, the consistency of the food matrix or microstructure, texture, fat and calorie content can influence the rate of gastric emptying [45].

Fatty food will induce the secretions of bile salts and phospholipids, which together with the products of lipolysis can form micelles with drugs, improving the drug solubility of lipophilic drugs. The solubilisation in micellar structures is most significant for drugs with log $P > 3$ [44]. However, the incorporation of drugs into micelles can lead to a decrease in absorption of hydrophilic drugs. The different types of food ingested will affect the bile salt concentration to different extents. Bile acids can affect the membrane fluidity and enhance the paracellular drug transport [46]. Dietary lipids are assembled into lipoproteins that are absorbed by lymphatic vessels. The co-administration of highly lipophilic drug with dietary lipids tends to increase intestinal lymphatic transport, depending on the type and quantity of lipid in ingested food. Lymphatic transport prevents hepatic first-pass metabolism, increasing drug bioavailability [47]. Furthermore, fatty foods may irritate the GI tract, inducing diarrhoea, which may reduce drug absorption.

Food components can also influence drug absorption by interacting with metabolising enzymes and transporters. A well-known example is grapefruit juice, which interacts with numerous orally administered drugs by inhibiting intestinal proteins, such as the metabolising enzyme, CYP450 or membrane transporters such as OATP [48].

Lipids may inhibit epithelial efflux transporters such as P-gp, increasing the bioavailability of their substrates [43]. Protein in the diet is digested into peptides and amino acids, and may inhibit intestinal amino/peptide transporter, such as PEPT1, responsible for the absorption of peptidomimetic drugs such as levodopa [49]. Although clinically relevant examples where peptides in the diet affect absorption of peptidomimetic drugs have not been found in human subjects [50]. In addition, protein from precipitated food components has been shown to form a film around tablets, slowing the rate of disintegration [51].

Studies into food–drug interactions, the influence of diet and the food effect are mostly based on mechanistic *in vitro* and preclinical animal models, which examine specific food components or the high-fat FDA standard meal on drug delivery. The reader is signposted to an excellent review by O'Shea et al. where enabling formulations to overcome drug–food effects are discussed [52].

13.5   Pregnancy and Its Effect on Gastrointestinal Physiology

A number of reviews have previously reported that gastric pH is increased during pregnancy [53, 54] which has the capacity to alter drug bioavailability. In terms of gastric motility, ultrasound studies have revealed that gastric emptying time for fluids is not affected by pregnancy [55] and showed no differences in the performance of paracetamol over trimesters when compared with non-pregnant women with respect to C_{max} and T_{max} [56]. The authors reasoned that due to the very rapid absorption of paracetamol from the small intestine, gastric emptying will be absorption-limiting. Nevertheless, overall GI transit time was found to be longer in the third trimester, indicative of an overall lower intestinal motility [55]. Consequently, intestinal absorption may be delayed during pregnancy possibly resulting in slower T_{max} and lower C_{max} values for drugs in which the intestinal transit time is the rate-limiting step. Table 13.1 shows some examples of physiological changes during pregnancy and their potential effects on drug performance.

As discussed earlier in this chapter, the differential expression of drug metabolising enzymes and transporters can affect the ultimate bioavailability of drugs. In terms of pregnancy, such changes, however, are often secondary to hormonal changes in progesterone, oestrogens or cortisol. For example, drug substrates of CYP3A4 (e.g. midazolam) [57], CYP2D6 (e.g. fluoxetine) [58], CYP2C9 (e.g. glibenclamide), UCT1A1 (e.g. labetalol) [59] and UGT1A4 (e.g. lamotrigine) [60] are subject to differing drug performance from the increase of enzyme expression and activity during pregnancy which can lower drug bioavailability. The activity of CYP1A2 and CYP2C19, which caffeine and proguanil are substrates of respectively, declines in pregnancy which may increase drug bioavailability.

Table 13.1   *Physiological changes during pregnancy and its effects on drug performance.*

Parameter	Consequence on drug disposition
Delayed gastric emptying and increased gastric pH	Altered drug bioavailability and delayed time to peak levels following oral administration
Increased cardiac output	Increased hepatic blood flow; increased elimination for some drugs
Increased total body water, extracellular fluid	Altered drug disposition, increased V_d for hydrophilic drugs
Increased fat compartment	Decreased elimination of lipid-soluble drugs; increased V_d for hydrophobic drugs
Increased renal blood flow and glomerular filtration rate	Increased renal clearance
Decreased plasma albumin concentration	Increased free fraction of drugs
Altered CYP450 and UGT activity	Altered oral bioavailability and hepatic elimination

Source: Adapted from Pariente et al. [54].

However, as overall oral bioavailability is also determined by hepatic first-pass metabolism, it remains difficult to delineate to which extent the changes in drug PK are the result of changes in enzyme expression occurring in the gut. Although limited direct data are available on the expression of drug-metabolising enzymes and transporters in the intestine during pregnancy, changes in activity may be derived from clinical PK studies that have investigated drug disposition after oral administration during pregnancy and the non-pregnant population [61–63].

13.6 The Implication of Disease States on Gastrointestinal Physiology and Its Effect on Oral Drug Performance

There has been much research to increase the understanding of GI physiology and functional changes as a result of specific GI diseases [64–68], and the impact of systemic diseases on the GI tract [69]. The presence and severity of disease can be a significant factor towards inter-individual and intra-individual variability with respect to oral drug delivery and the behaviour of the formulation in the GI tract. In this section, we aim to focus on how GI diseased implicate 'normal' GI physiology and function, and consider drug pharmacokinetic and pharmacodynamic alterations in specific disease states.

13.7 Diseases that Affect the Gastrointestinal Tract

'Local' GI diseases include irritable bowel syndrome, inflammatory bowel disease, malabsorptive syndromes, neurogastroenterological diseases, infectious disorders and disorders of transit and GI morphology. The commonality between all these conditions is the predominant manifestation of symptoms appears to be of GI origin [68].

13.7.1 Irritable Bowel Syndrome

Irritable bowel syndrome (IBS) is the most extensively studied GI disorder and is a common reason for clinical referral. Exhibited symptoms range from intermittent and/or alternating episodes of alternating stool patterns (IBS-A), constipation (IBS-C), diarrhoea (IBS-D), abdominal pain and distension [70], and reduced left colonic transit. Gastric emptying time and small intestinal transit time were not significantly different in IBS patients when compared to controls measured with a SmartPill GI-monitoring system [71]. Differences, however, are exhibited between the IBS subtypes; IBS-D patients have a shorter small bowel transit time (3.3 ± 0.3 h) and total GI transit time (35 ± 5.0 h) than IBS-C patients (5.4 ± 0.3 and 87 ± 13 h, respectively) [72]. The implications for drug delivery to the gut in IBS are, therefore, differentially affected due to variations in transit time. Indeed, drug bioavailability may be reduced by accelerated intestinal transit where the time for dosage form disintegration, dissolution and drug absorption is also reduced by disease activity. Previous studies have demonstrated that food consumption correlates with the onset of ileal pain and that food and fibre intake may also accelerate colonic transit in IBS patients thus, potentially reducing drug absorption and bioavailability further [73].

The role of bile acids (BA) has also been underlined in IBS pathophysiology. Primary BA undergo dehydroxylation by bacteria in the small intestine which facilitate lipid digestion and absorption [74] forming secondary BA. A significant increase in primary BA was found in IBS-D and -C subtypes when compared with healthy subjects. Furthermore, a significant decrease in secondary BA was demonstrated in IBS-D patients against healthy participants. The role of BA in solubilising poorly soluble drugs and the impact on their absorption was noted previously.

Intestinal barrier dysfunction has been found to play a pathogenic role in IBS [75]. There is increasing evidence that heightened intestinal permeability is related to low-grade inflammation, visceral hypersensitivity and pain in IBS. With the use of oral probe excretion assays, increased small bowel and colonic permeability was demonstrated in both paediatric and adult IBS patients regardless of IBS-subtypes when compared with healthy controls [76].

13.7.2   Inflammatory Bowel Disease

Inflammatory bowel disease (IBD) is broadly recognised as a chronic, relapsing idiopathic inflammatory condition of the GI tract leading to long-term damage to the gut structure and function [77].

The two manifestations of IBD – ulcerative colitis (UC) and Crohn's disease (CD) – are distinct in their sites of inflammation. Numerous review and research articles on the pathology, clinical presentation and treatment of CD and UC can be found in the literature. The major treatment goals of current therapies are to decrease the inflammatory reaction, minimise symptoms, improve quality of life and minimize progression and complications of disease [5].

Patients with UC typically display a markedly lower colonic pH when compared with healthy individuals which has potential implications for the use of pH-sensitive polymer formulations namely, that of degradation and delayed opening of pH-dependent coatings [78–80]. Results obtained for pH values in the GI tract of UC patients are also highly variable ranging from pH 2.4–3.4 in the proximal colon [81] to pH 4.7–11.0 in the right colon [82, 83].

Prior studies revealed that GI transit times for patients with active UC show high variability [84–86] although it is widely known that high GI variability is evident even in healthy subjects [87–90]. Gastric emptying time is prolonged in UC patients during both active and inactive disease phases although again, the involvement of individual variability in both healthy and diseased subjects in these investigations is complicated [91]. The effect of the active and quiescent status of IBD on small intestinal transit time (SITT) was assessed by [92] without the interference from gastric residence times. Using a small-bowel video capsule endoscopy, the median SITT in non-UC participants was 216 minutes when investigated in 125 non-UC participants, although inter-subject variability ranged from 39 to 512 minutes. UC patients, however, were found to have a longer SITT at an average of 264 minutes (ranging from 216 to 326 minutes) when compared with healthy subjects [92] suggesting a delay in the onset of action of orally administered drugs targeted to the distal gut.

Studies have further identified physiological differences in the diseased UC gut when compared to healthy subjects in terms of the mucus barrier, expression of membrane transporters, metabolism and the composition of luminal contents. In terms of normal mucus barrier function, the protective phospholipid component of the GI mucosa, phosphatidylcholine, was strongly decreased in the colonic UC mucus barrier by approximately 70% [93, 94]. The ascending UC colon fluid composition in the fasted state was shown to be significantly different when compared with healthy subjects where increased concentrations of soluble proteins were observed in the remission and relapse state (19.0 ± 10.8 mg/mL; 18.9 ± 8.1 mg/mL) than in the healthy colon (9.8 ± 4.6 mg/mL). In addition, the buffer capacity of the ascending colon fluid in both remission (hydrochloric acid: 37.7 mmol/L/ΔpH; sodium hydroxide solution: 16.7 mmol/L/ΔpH) and relapse (hydrochloric acid: 32.0 mmol/L/ΔpH; sodium hydroxide solution: 18.3 mmol/L/ΔpH) states was found to be higher when compared with the control (hydrochloric acid: 21.4 mmol/L/ΔpH; sodium hydroxide solution: 10.3 mmol/L/ΔpH), although volume and surface tension were found to be similar regardless of the healthy and diseased state [95]. In terms of membrane transporters, the expression of peptide transporter 1 (PepT1) is upregulated in the colon during chronic inflammation in subjects with UC and therefore, can negatively modulate the influx of peptidomimetics [96].

The development of Crohn's disease (CD) is linked to a disordered inflammatory and immune response within the GI tract [97]. A population-based study carried out in Montreal revealed that the most common site of CD is in the terminal ileum (45%) followed by the colon (32%), ileo-colon (19%) and upper GI tract (4%) [98, 99].

CD is known to affect GI parameters ranging from delayed oro-caecal transit time (OCTT) to luminal pH changes, enzyme activity, bile acid malabsorption and enterohepatic recirculation and mucus levels. A study by Nishida et al. investigated the steady-state kinetics of enterohepatic circulation of bile acids in patients with CD using ^{2}H-labelled bile acid [100]. The mean biological half-life of the ^{2}H-labelled bile acid was significantly ($p < 0.01$) shorter (1.15 ± 0.08 days) than healthy subjects (1.95 ± 0.25 days). The mean values for the total bile acid pool were exhibited to be 1323.0 ± 173.6 mg in CD which were statistically significant ($p < 0.02$) when compared with healthy subjects (2290.9 ± 327.3 mg). It is known that conjugated bile acids are absorbed mainly at the ileum. Therefore, the diminished bile acid pool size in CD is attributed to the impaired absorption at the ileum [100].

Shaffer et al. investigated the absorption of prednisolone in both healthy subjects and CD patients [101]. It was identified that plasma concentrations of prednisolone in CD patients exhibited a higher degree of inter-subject variability following oral ingestion suggesting different severities and hence, different implications on absorption. The reason for apparent malabsorption in Crohn's patients is unclear. The reduced surface area of the diseased intestine, rapid transit, dispersion in unabsorbed luminal contents or changes in mucosal drug metabolising enzymes may all influence malabsorption [102]. Variable absorption mechanisms, therefore, should be taken into account when assessing drug responses in patients with CD.

Physiologically, it has been shown that the colonic pH of CD patients during both the inactive and active disease phases is much more acidic than that of healthy individuals – including those with resections [103] – measured as 5.3 ± 0.3 (CD) versus 6.8 ± 0.2 (control) in the right colon and 5.3 ± 0.7 (CD) versus 7.2 ± 0.3 (control) in the left colon by radiotelemetry [104].

Blokzijl et al. aimed to identify the intestinal MDR1 mRNA and protein expression in the uninflamed and inflamed intestinal epithelial of patients with IBD when compared with healthy controls. A slight decrease in the ileum and slight increase of MDR1 mRNA in the uninflamed colon of CD patients was identified, whereas a significant decrease of MDR1 expression was found in the inflamed ileum epithelia of the same patients [105]. The observed increase in MDR1 expression in the uninflamed colon could be an adaptive mechanism to consolidate for the decreased expression in inflamed tissue. The distinct significant effects were demonstrated in MDR1 expression in inflamed versus uninflamed intestinal tissue from individual patients where P-gp was strongly decreased in inflamed tissue of patients with CD and UC. As the main function of MDR1 is to export drugs out of cells, reduced P-gp expression in the diseased state can therefore increase the bioavailability of drugs used to treat IBD patients such as glucocorticosteroids and immunosuppressants.

13.7.3   Celiac Disease

Celiac disease is an autoimmune disorder occurring in genetically predisposed individuals in whom the ingestion of gluten, a protein from wheat, barley and rye, leads to an immune reaction in the small intestine. Celiac disease is manifested in the small intestine as an increased number of intraepithelial lymphocytes, crypt hyperplasia and villous atrophy [106]. Most often, villous atrophy is found in the duodenum, but it can also affect more distal parts of the small intestine. The areas of villous atrophy tend to be patchy and are dispersed between regions of normal mucosa [107]. Villous atrophy results in a flattening of the microvilli, which in turn gives a decreased absorptive surface area in the intestine. This can impair the ability to absorb nutrients [107, 108]. It is therefore possible that intestinal drug absorption will also be negatively affected. However, it is also well accepted that the intestinal permeability is increased, in part due to opening of tight junctions, which can potentially increase drug absorption [109].

The changes that occur along the length of the GI tract in celiac disease are also seen to correlate with disease severity and can impair normal patterns of drug absorption [110]. Disease presentation at the proximal jejunum is typically characterised by seemingly mild symptoms such as weight loss and diarrhoea, whereas that which manifests in the ileum is otherwise characterised by morphological changes in the GI mucosa [111].

Details on other changes in the small intestine in untreated celiac disease are scarce. A single study has examined the pH on the jejunal surface in untreated patients and compared it with the pH of patients on a gluten-free diet. The study showed a significantly higher pH in untreated celiac disease patients, which also might impact absorption of ionisable drugs [112]. Another study [113] found that the activity of CYP3A4 was decreased in adult untreated patients, compared to healthy subjects and Johnson et al. found the same for paediatric celiac disease patients [114]. This is likely to be related to the villus atrophy, as the CYP enzymes are primarily located in the apex of the enterocytes, in a band just below the microvillar border [113]. This decrease in CYP3A4 is normalised when patients were on a gluten-free diet [108, 113].

Celiac disease is associated with delayed gall bladder emptying and decreased pancreatic secretions [115]. The fasting gall bladder volume and the post-prandial residual volume are significantly higher in celiac disease patients compared to controls. These symptoms are reversible and disappear when celiac disease patients get on a gluten-free diet. Unfortunately, the intestinal bile salt concentrations have not been determined in any of these patient groups [115, 116].

The gastric emptying time is prolonged in untreated celiac disease patients but is normalised upon intake of a gluten-free diet. In addition, the transit time of a meal through the small intestine is delayed in celiac disease patients, probably due to functional alteration of small intestinal motility, whereas there are no changes in the colonic transit time. An abnormal gastrointestinal motor activity, indicating a neuropathic disorder, in untreated celiac disease patients has been observed and this is persisting in most adult celiac disease patients also during a gluten-free diet.

13.8   Infections in the Gastrointestinal Tract

13.8.1   *Helicobacter pylori* Infection

Helicobacter pylori (*H. pylori*) infection is known to induce alterations in some physiological functions of the stomach, particularly affecting gastric acid secretion and GI motility. The impact on gastric acidity depends on the severity of the infection and distribution of the affected tissue in the stomach. Inflammation in the antral region is associated with an increased production of gastrin, which in turn increases acid secretion in the gastric corpus. However, if the corpus is affected by severe inflammation, the response to gastrin is greatly reduced, resulting in reduced capacity to secrete acid [117]. Thus, *H. pylori* positive subjects can exhibit either increased or decreased stomach acidity. Between 25 and 40% of patients also exhibit antral hypomotility and a delayed gastric emptying [118].

The physiological changes induced by *H. pylori* infection can have a significant impact on the drug absorption process and thus important clinical implications as highlighted in several systematic reviews [11, 119, 120]. Particularly, oral compounds and formulations with pH-dependent solubility need careful evaluation when administered to *H. pylori* positive patients, as observed with levodopa [121], thyroxine [122] and delaviridin [123] where the main mechanism underlying the lack of absorption appears to be reduced gastric acid secretion [11]. However, more research is needed to capture the pathophysiology of *H. pylori* infections, its impact on the function of the entire GI tract and thus, the clinical implications on oral drug absorption.

Knowledge of bacterium-induced alterations in gastric pH during and following infection by *H. pylori* in the context of drug delivery may therefore necessitate the use of pH-sensitive formulations to achieve optimal clinical responsiveness, overcoming possible residual alterations in gastric acidity and secretion following apparent curation of the infection.

13.9   Systemic Diseases that Alter GI Physiology and Function

Cystic fibrosis, Parkinson's disease, diabetes, human immunodeficiency virus (HIV) and pain were selected here owing to available data. These disorders were chosen herein to raise awareness that diseases with no immediate association to the GI environment can, however, clinically manifest GI alterations with respect to physiology and function and, thus, influence formulation behaviour. In addition, although it is expected that patients will take medicines specifically for the aforementioned diseases, little is understood on the outcome of drug bioavailability in patients with multiple and diverse disease backgrounds. This section, therefore, addresses the implications of the altered GI tract as a consequence of systemic disease and its impact on oral drug bioavailability.

13.9.1 Cystic Fibrosis

Cystic fibrosis (CF) is a prominent example of a hereditary, severe, progressive and multi-systemic disorder with varying worldwide prevalence. People diagnosed with CF experience a build-up of thick sticky mucus in the lungs and the digestive system, manifesting symptoms at multiple bodily sites. The most common of these is the respiratory tract and distinctive – and often detrimental – changes to GI physiology and function; the latter is characterised by gastro-oesophageal reflux disease, malabsorption and pancreatic insufficiency [124]. Fat and nutrient malabsorption in CF patients is also common and is further affected by altered intestinal pH, motility and mucosal abnormalities [125].

Intestinal transit changes in CF patients are incompletely understood although factors such as mucosal inflammation [126] and small intestinal bacterial overgrowth (SIBO) [127] have been identified as contributing factors in animal models. Moreover, the inconsistencies in data of changes to GI tract transit times have been widely reported. For instance, oral-colon transit time (OCTT) has been shown to be unchanged in terms of the migrating motor complex [128] or prolonged following the administration of a single-unit dosage form [129, 130] at different regions of the GI tract in fasted CF patients, potentially contributing to wide variability in oral drug absorption and bioavailability. A study by Collins et al. demonstrated that gastric emptying (GE) is up to 30% faster in fed-state CF patients (average 53.0 minutes) when compared with healthy subjects (average 72.2 minutes), although marked discrepancies in gastric emptying time (GET) have also been observed including 58 minutes (range 6–107 minutes) and 41 minutes (range 4–125 minutes) noted within the same study [131].

In the small intestine, transit is significantly delayed in CF which is thought to be a consequence of abnormal intestinal mucous slowing normal transit. A study by Bali et al. [129] examined the small intestinal transit times (SITT) of 10 CF patients (seven males and three females between 17 and 24 years of age) against 15 control subjects (nine males and six females between 18 and 26 years of age). SITT in CF patients ranged from 160 to 390 minutes, whereas the control group demonstrated a regular SITT of 50 to 150 minutes. However, it has also been observed that while SITT is delayed in CF patients, duodenal and jejunal transit can be abnormally fast in some subjects until the point of pH neutralisation in the intestine and activation of the 'ileal brake' [132].

In terms of pH, the dysfunction of CFTR gene has been associated with reductions in pancreatic and duodenal – but not gastric – bicarbonate secretions [133], thereby lowering pre- and post-prandial duodenal pH by 1–2 units when compared with healthy controls (pH 5–7 versus pH 6, respectively). Low duodenal pH specifically in CF has been cited as the combined result of gastric hyperacidity and reduced pancreatic bicarbonate secretions [134]. Indeed, the efficacy of enteric-coated pancreatic enzyme preparations designed to dissolve above pH 5 for this purpose has been brought into question. Discrepancies in ranges observed for gastric pH have been noted according to different sources in fasted subjects with Youngberg et al. [135] citing between pH 0.9 and pH 1.8 in subjects with use of the Heidelberg Capsule and pH 1.3 according to Barraclough and Taylor [134]. However, sharp changes in pH from the terminal ileum (7.5) to the caecum (6.4) in CF patients when compared with the gradual increase in pH along the same length in healthy controls have been observed and, thus, could be responsible for aberrations in the delivery of pH-sensitive formulations [132].

The colonic microbiome of those with CF was found to be significantly less diverse when compared with non-CF controls. The slower SITT in CF patients can further lead to SIBO, characterised by the excessive concentrations of bacteria in the proximal small intestine. It is suggested that, owing to SIBO, the synthesis of enterotoxic and unabsorbable metabolites could result in mucosal damage and interfere with digestion and absorption. In addition, the type of microbial flora present contributes to the manifestation of signs and symptoms or overgrowth. For example, a predominance of bacteria that metabolise bile salts to unconjugated or insoluble compounds could lead to fat malabsorption or bile acid diarrhoea. The administration of the antibiotic ciprofloxacin, however, was shown to improve the digestion and absorption of fat CF patients with SIBO [136].

13.9.2 Parkinson's Disease

Parkinson's disease (PD) is the second-most common progressive and irreversible neuro-degenerative disease with evolving layers of complexity. In many cases, motor symptoms of PD are preceded by a series of milder prodromal symptoms such as fatigue and GI symptoms including constipation, dysphagia and defecatory dysfunction [137].

The appearance of early GI symptoms could also indicate that part of the pathological process originates in the gut given the abundance of enteric dopaminergic neurons collectively using ~50% of bodily dopamine [138], although GI symptoms are otherwise typically evident at all stages of PD. The gut–brain axis appears to dominate in PD because symptoms ranging from abdominal bloating, dyspepsia, nausea, vomiting and pain can be exacerbated by emotion [139]. Constipation is otherwise the most common GI symptom in PD reported by ~90% of patients and with the majority of cases associated with advanced disease progression [140].

Jost and Schimrigk [141] suggested that, in addition to impaired colonic motility, reduced tension of abdominal muscles and the diaphragm could contribute to the prolonged colon transit time in PD. Davies et al. (1996) administered oral mannitol to 15 PD patients as part of a study investigating intestinal permeability; in normal subjects, monosaccharides such as mannitol are absorbed by non-mediated diffusion through small channels in the enterocyte brush border membrane. Disaccharides like lactulose, however, are absorbed across tight junctions or between enterocytes. In PD, the percentage of mannitol absorption was reduced to 11.7% when compared with the healthy control group achieving 16.2% urinal mannitol recovery [142]. This suggested a reduction in the absorptive surface area of the small intestine owing to specific alterations in the enterocyte brush border membrane in PD patients.

Abnormal GET has been described in 43–83% of PD patients [143] and is currently the most investigated parameter in relation to GI symptoms. Most antiparkinsonian drugs are delivered orally; therefore, altered gastric motility and GE rate may affect the bioavailability of medications since the rate of absorption from the GI tract is fundamental in achieving a beneficial therapeutic effect. In a study by Edwards et al. [144], 28 untreated patients with PD were assessed and found to have an average time to empty half of the gastric contents

(GET½) of 59 minutes when compared with 44 minutes from a group of slightly younger healthy control individuals. An interesting association has also been made between delayed GET and the response fluctuations that develop after long-term levodopa therapy. Studies measuring gastric retention after one hour showed GET to be increased but GET½ significantly delayed in patients with fluctuations when compared with those without fluctuations. This demonstrates that GET is more significant in those PD patients displaying response fluctuations [145]. Another possible outcome of GET is the prolonged exposure of drugs to gastric acid and digestive enzymes, as well as the binding of food components such as proteins in the stomach. In this instance, dopa-decarboxylase normally present in the gastric mucosa would be unable to metabolise protein-bound levodopa to the active dopamine, leading to absence of any therapeutic effect.

Gastroparesis, the reduced motility of the stomach, can be present in early and advanced PD. Few attempts have been achieved to overcome gastroparesis in PD patients by either adjusting dietary requirements when administrating drugs or with the use of prokinetic agents such as cisapride, domperidone and metoclopramide to normalise GE [146]. Different formulation approaches have also been employed to prolong drug transit in the upper GI tract to maximise the opportunity for absorption. For example, in one study, the difference between solid and liquid emptying time in PD patients was exploited as a method of optimising drug delivery with results showing that dispersible or liquid formulations of levodopa can, in fact, decrease T_{max} [147]. However, levodopa absorption takes place primarily in the intestine and not the colon, rendering targeting difficult. In addition, it appears that these formulations are still dependent on erratic GET, to which end studies have suggested administering a levodopa gel solution directly into the duodenum or jejunum [148]. This gel-formulation approach was shown to achieve consistent plasma levodopa levels in treated patients with the effective management of motor complications.

Like CF patients, a significant proportion of PD patients demonstrates SIBO or the colonisation of *H. pylori* [149] where the eradication of these results in improvements of GI manifestations and motor fluctuations. In particular, the antibiotic treatment for *H. pylori* has been shown to improve levodopa absorption and bioavailability [150].

13.9.3 Diabetes

There are two manifestations of diabetes mellitus (DM) including type 1 DM (T1DM) and type 2 DM (T2DM). Both DM manifestations, however, feature significant links to impaired GI tract function [151].

There is ample evidence to suggest that diabetes and the GI tract are inextricably linked. Some studies have shown that pronounced hyperglycaemia (>250 mg/dL) affects the motility of the oesophagus, stomach, small intestine and colon in T1DM and T2DM individuals [152]. Diabetes is further believed to affect the morphology and function of the GI tract (Table 13.2), which can consequently influence oral drug performance [153].

Diabetic patients exhibit significantly reduced gastric acid secretion [154], which is more pronounced in diabetic gastroparesis. Furthermore, gastric pH is also increased in the fasted state of diabetic patients when compared with healthy individuals, leading to an impaired absorption of basic drugs in the stomach [155]. Changes in aboral pH can also impair the disintegration and dissolution of coated dosage forms, particularly those incorporating pH-sensitive materials. Indeed, modern enteric coatings such as cellulose acetate

Table 13.2   *Diabetes-mellitus-induced physiological, motor and sensory alterations in the intestinal tract.*

Alterations	Small intestine	Colon
Mucosa	• Increased mucosal thickness • Damaged tight junctions • Decreased membrane fluidity • Enhanced transport of glucose, amino acids, bile salts, phosphate, fatty acids, fatty alcohols and cholesterol • Decreased protein synthesis • Increased expression of monosaccharide transporters	• Increased thickness of the subepithelial collagen layer • Increased expression of AGE and RAGE •
Submucosa	• Increased thickness	• Increased thickness • Increased expression of AGE and RAGE
Wall	• Increased thickness • Increased expression of AGE and RAGE	
Motor	• Highly variable transit time • Decreased muscle tone • Increased jejunal and ileal contractility in response to distension • Dysmotility	• Increased transit time • Highly variable contractility • Highly variable spontaneous contractility • Impaired contraction and relaxation of circular muscle strips
Sensory	• Decreased duodenal sensitivity in response to mechanical, thermal and electrical stimulations • Increased jejunal sensitivity in response to mechanical stimulation	• Increased colonic sensitivity to mechanical stimulation

AGE, advanced glycation end-product; RAGE, advanced glycation of end-product receptor. Source: Adapted from Zhao et al. [153].

phthalate, polyvinylacetate phthalate and the polymethacrylates are almost entirely insoluble at normal gastric pH but begin to dissolve rapidly above pH 5 [156–159].

Dysbiosis can also indirectly affect the action of drugs. The discrete shifts in gut microbiota composition following sustained GLP administration may allow correction of dysbiosis, and thus improvement in glucose tolerance and insulin sensitivity [160]. For a fraction of patients with T2DM, oral medications such as metformin do not sufficiently control hyperglycaemia, and they must begin to regularly inject subcutaneous insulin. Though the prescription of insulin is often reserved for the most severe cases of T2DM, alteration of the gut microbiome can allow patients to reduce their insulin dose or even cease use altogether. It is entirely possible that patients with T2DM who adopt lifestyle changes that ameliorate gut dysbiosis can reduce the dose of exogenous insulin they need to inject [161].

13.9.4 HIV Infection

In vitro studies have postulated that there is a reduction in the number of intestinal entero-cytes differentiating in HIV/AIDS patients, ultimately leading to morphological changes in the intestinal lining and compromising intestinal epithelial barrier function [162], thus not only predisposing to opportunistic infections and more-extensive cellular damage but also impairing the absorption of orally delivered drugs. Gastric hypochlorhydria and hypoacid-ity have also been observed in AIDS patients up to a pH value ~1.5–3-times higher than that of healthy individuals [163], which can lead to reduction in the absorption and subse-quent bioavailability of basic drugs such as ketoconazole [164] and itraconazole [165]. An altered gastric pH owing to HIV/AIDS infection can also feature self-implicating knock-on effects; for example, the drugs – indinavir and delavirdine – used in infection treatment have been shown to be absorbed approximately 50% less at higher pH values [166]. Equally, up to 20% of patients with AIDS feature abnormal intestinal permeability [167]. As disease progresses, the functional and selective absorptive surface area of the intestine decreases, as evidenced by a study of mannitol permeability in HIV/AIDS patients with and without diarrhoea [168]. These changes can concomitantly influence the oral therapeu-tic efficacy of drugs for infections such as tuberculosis, which is seen to be the most com-mon opportunistic infection in HIV/AIDS patients worldwide. For instance, Gurumurthy et al. [169] demonstrated that peak concentrations for rifampicin and isoniazid were reduced in HIV patients which followed an earlier study that revealed that rifampicin and ethambutol plasma concentrations were significantly low in HIV/AIDS patients [170].

Reduced drug bioavailability could also be caused by fat malabsorption syndromes prev-alent in the majority of HIV patients [171]. Hyperlipidemia can have significant implica-tions, therefore, on the PKs of lipid-soluble drugs such as the antiretroviral drug zidovudine. Zidovudine PKs were highly altered when administered to HIV-infected patients; C_{max} was reduced to 6.39 ± 3.39 versus 11.51 ± 5.01 mmol/L and T_{max} was prolonged to 0.81 ± 0.51 versus 0.40 ± 0.14 h in HIV-infected patients when compared with healthy subjects. These data suggest a delayed absorption rather than an altered metabolism of zidovudine in AIDS-related small intestinal defect as a result of fat malabsorption [172].

A reduced microbial diversity has also been reported for HIV-infected patients. Specifically, a shift to Bacteroides and Prevotella predominance with a significant reduc-tion in the Firmucutes phyla, when compared to uninfected individuals, contributes to a loss in immune regulatory and probiotic activity [173]. The effect of HIV on the gut mucosal barrier might be a direct result from microbial dysbiosis, which could induce a leak in gut mucosa and, thus, trigger systemic inflammation from the circulation of micro-bial elements in the blood [174]. Combination therapy is often used to control the rapid replication rate of HIV in plasma [175]. For example, BILR355, a non-nucleotide reverse transcriptase inhibitor, is co-administered with ritonavir. The drug combination acts to interfere with the reproductive cycle of HIV and boosts BILR 355 efficacy by inhibiting its metabolism by CYP3A4-mediated catabolism [176]. A metabolite of BILR 355, however, is unusually increased to toxic levels when the drug is concomitantly administered. It is now recognised that ritonavir triggers BILR 344 metabolism which is reduced to an inter-mediate by the gut microbiota in patients with HIV, followed by further oxidation to the metabolite BILR 516 which exceeds the concentration of the administered drug [94]. This demonstrates how the gut microbiome can induce the biotransformation of drugs to create an alternative metabolic pathway that could result in adverse effects [177].

13.10   Age-related Influences on Gastrointestinal Tract Physiology and Function

13.10.1   Gastrointestinal Physiology and Function in Paediatrics

As the absorption, distribution, metabolism and excretion (ADME) can all be affected by the transformations that occur throughout childhood, a robust understanding of the physiological and PK differences between adults and children is required to design better and more appropriate paediatric medicines.

Traditionally, paediatric dosing (also referred to as allometric scaling) has been calculated by scaling adult values by comparing factors, such as body surface area, body weight and age. However, importantly, these approaches do not account for maturation changes, such as ontogeny of enzymes and transporters. In recent years, the development of more complex and representative mathematical models, such as physiologically-based pharmacokinetic (PBPK) modelling, has been pursed to try to deliver a better prediction of the appropriate paediatric doses [178]. In the following sections, the main changes in the GI tract that may influence the absorption and pharmacokinetics following oral drug administration in paediatric populations will be discussed.

In neonates, fasted gastric pH has been widely reported as being neutral shortly after birth (pH values of 6–8) [179]. Fasted gastric acidic pH values of 1.5–3 have been found to be reached between hours after birth up to the first two weeks of life and may display inter-individual variability [180, 181].

Available data have reported similar intestinal pH values in both adult and paediatric age groups, with a high pH variability observed in the fasted and fed state for both groups. Children and adolescents ($n = 12$, 8–14 years) were found to have similar fasted intestinal pH (6.4–7.4) [182], and similar mean fed intestinal pH values of 6.3 ($n = 16$, 7–16 years) [183]. More recent data have characterised pH osmolality and buffer capacity of paediatric GI fluids [184].

The composition of GI fluids is key to enable appropriate digestion and absorption of key nutrients and drugs into the systemic circulation. The importance and role of digestive enzymes in neonates and infants has been well summarised in the literature [185]. Pepsin, which is a protease secreted by the stomach, is not fully expressed at birth whereby lower pepsin secretions have been reported in younger cohorts such as neonates and infants <one year of age compared to older children and adults [186]. In a similar manner, pancreatic enzyme concentrations are lower at birth and appear to reach mature levels by about one year of age [187]. Additionally, bile secretion is widely known to be poor in the first 2–3 weeks of life is known to be poor with luminal concentrations somewhat lower than in adult intestines (2–4 vs. 3–5 mM respectively) [188, 189].

The volume of gastric fluids in children is not widely reported although a value of approximately 0.4 mL/kg has been reported in fasted children [190, 191]; this equates to an adult value of 28 mL (assuming a 70 kg male) which is lower than the typically reported 40 mL value in fasted adults [192]. The volumes of small intestinal fluids are less well characterised.

In general, it has been demonstrated that neonates and young infants have slower gastric emptying times compared with older children and adults [193]. The gastric emptying half-life ($GEt_{1/2}$) has been found to be 6.9 minutes for a liquid non-caloric meal (5 mL/kg) in neonates, measured by epigastric impedance using four electrodes [194]. However, other techniques for the measurement of gastric emptying of liquids have shown differing values, highlighting the need for a standardised method of analysis.

In the fed state, a recent review of mean gastric residence time showed that gastric emptying rates were not affected by age and confirmed the importance of food [195]. Indeed, the gastric emptying has been widely reported to be influenced by meal volume, meal type and composition and osmotic pressure [195–198]. Due to the differences in meal types, and also due to the higher frequency of feedings, neonate and infant subpopulations are most likely to show differences in the fed state compared to adults. For example, one study showed that in neonates and infants (four weeks to six months), $GEt_{1/2}$ was affected by administration of equal volumes of breast milk compared to infant formula where $GEt_{1/2}$ was 48 ± 15 and 78 ± 14 min, respectively, indicating that infant formula empties at slower rates than breast milk [199].

As both the intestinal length and radius increase with paediatric development, the functional surface area can increase significantly. For both neonates and adults, the central estimates of the physiological length (as measured in a living person) of the small intestine are approximately 1.6× body height. The mean anatomical length of the intestine changes with growth, with values ranging from ~275 cm at birth, 380 cm at 1 year, 450 cm at 5 years, 500 cm at 10 years and 575 cm at 20 years [200].

Furthermore, specific morphological features on the luminal surface, such as folds, villi and microvilli, naturally increase the surface area available for absorption [64]. Small intestinal villous patterns start developing at an early stage of gestation and mature and develop with age. In general, neonates and infants show a lower intestinal surface area compared with older children and adults, due to differences in both the structure and the quantity of the villi [197].

Intestinal permeability is high at birth for preterm neonates, with a decrease to adult values over the first week of postnatal life [201, 202]. Children over two years of age present similar permeability values to adults [203]. It has been shown that processes involved in active and passive transport are fully developed in infants by ~four months old, with the underpinning factors for development being growth factors, hormones, breast milk and changes in the thickness and viscosity of the intestinal mucus [203, 204]. Contradictory literature can be found on the ontogeny of the efflux transporter P-glycoprotein (Pgp), also referred to as multidrug resistance protein-1 (MDR1) [205]. Intestinal mRNA expression of MDR1, MRP2, PEPT1 and OATP2B1 was determined in surgical small bowel samples, with expression values for both MDR1 and MRP2 being similar to the adults [206]. Intestinal OATP2B1 expression in neonates was significantly higher than in adults. Furthermore, lower mRNA expression levels of PEPT1 have been observed in neonates/ infants compared with older children [207].

In general, the presence and activity of metabolising enzymes is low at birth and reaches adult levels and function by early childhood [198, 202]. Furthermore, anatomical changes occur along childhood, whereby due to a larger liver size and higher hepatic blood flow in older children, increased hepatic clearance is observed (when normalised with body weight).

Intestinal microflora is responsible for drug metabolism in the gut lumen, with differences in bacterial colonisation having a big impact on drug absorption [187, 202]. Studies have found that microbiota is present right after birth; however, the extent of microbial colonisation changes over time as a function of a variety of different factors, including method of delivery, food ingestion and gestational and postnatal age, amongst others [208]. Typically, an infant's intestinal microbiota will begin to resemble an adult's around one year [204], with full maturation being achieved at two to four years.

The ontogeny of intestinal wall metabolism requires further investigation, with infants and children being unrepresented in research [187]. To give the field a better understating on how gut wall enzymes changes with age, and hence how intestinal gut metabolism change, more research should be conducted including infant and children subpopulations.

13.10.2 Gastrointestinal Physiology and Function in Geriatrics

As age increases, physiological and morphological changes take place in the GI tract that result in a decline in molecular and cellular function [209]. These age-mediated changes can significantly affect the PK profile of certain orally administered drugs [5]. To add further complications, elderly patients are more commonly prescribed five or more medicines, named polypharmacy, and the incidence of disease is higher when compared to the younger population [210]. Certain comorbidities associated with ageing can impair the gut function, in particular degenerative disorders affecting the central nervous system, such as Parkinson's disease [69]. Chronological age is a poor indicator of physiological condition. Frailty can be a more important consideration than age.

Advancing age and age-related diseases are associated with a decline in oropharyngeal and oesophageal motility and the lower oesophageal sphincter pressure is found to be reduced [211]. Dysphagia and reflux are key barriers to effective oral drug administration in elderly patients [208] and saliva production is reduced [212]. Gastric emptying time was found to slow slightly with age [213, 214] and little to no differences in small intestinal transit time were observed [215, 216]. Colonic motility is reported to slow; in subjects from 74 to 85 years, the mean colonic transit time was 66 hours, compared with 39 hours in 20–53 year olds [215].

Decreased gastric acid production is found and elevated pH levels were reported in the ageing stomach, although the differences are not expected to influence drug absorption for the majority of elderly patients [217]. Postprandially, the rate of return of the gastric pH to the pre-prandial level was observed to be considerably slower than in young healthy subjects [217]. There is a 10% prevalence of achlorhydria in the elderly in comparison to <1% in the general population [218] and s Studies have found impaired bicarbonate secretion by the gastric mucosa [219] and the duodenum [220] and the rate of bile acid synthesis is impaired in elderly patients [221]. No differences were found in gall bladder contractibility and emptying kinetics [222]. There is limited information on the effect of ageing on the distal gut pH, buffer capacity, surface tension, fluid volumes, ionic composition and viscosity. Furthermore, little or no differences were found in the osmolality and bile salt concentration [223].

A modest change in the appearance of the villi was observed in comparison to younger subjects, with smaller leaf-shaped villi found in the older population as well as degeneration of villi, leading to decreased surface area [224]. GI protective mechanisms are believed to decline with age, with reduced mucus thickness in the upper GI tract and a decrease in the number of Goblet cells [208, 225]. Consequently, there is a higher incidence of GI conditions ranging from peptic ulceration, gastric cancer and inflammatory bowel diseases [225, 226]. The effect of age on intestinal CYP3A4 and P-gp activity is unknown, although age-mediated changes are seen in the activity and expression of P-gp in lymphocytes [208].

In older age, the mean total anaerobes are similar to younger subjects; however, differences are found in the dominant bacteria species [227] and the inter-individual variation. A decline in the numbers and species diversity of bifidobacteria and bacteriodes is reported. A consequence of the decrease in the colonic bacteriodes is a reduction in amylolytic activity and faecal concentration of short-chain fatty acids [227]. In the ageing gut, higher concentrations of facultative bacteria species are found, including enterobacteria, streptococci, staphylococci and yeasts [227]. External factors influence the microbiota composition in the elderly, including diet, fragility, nutritional status and inflammatory markers [228]. A summary of the key factors that affect biopharmaceutics in an older population is shown in Figure 13.3.

Older patients who are free of disease are often underrepresented in clinical trials and therefore there is a lack of information on the impact of ageing on oral drug absorption. Extrapolation from younger patients, however, may not be appropriate. The FDA published a guidance document in 1989 entitled 'Study of Drugs Likely to be used in the Elderly' for patients over 75 years encouraging evaluation of the effects on drugs in older patients [229]. The ICH E7 'Studies in support of special populations: geriatrics' published in 1993 for patients over 65 years, called for appropriate numbers of geriatric patients to be included in Phase 2 and Phase 3 studies [230]. Clinical studies with healthy elderly subjects should be

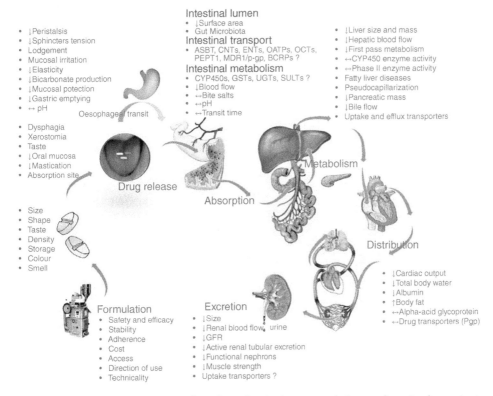

Figure 13.3 *Parameters to consider when developing an oral dosage form in the geriatric population and the impact of altered gastrointestinal physiology on drug pharmacokinetics. Source: From Khan et al. [231] / With permission of Elsevier.*

conducted and methods and tools harmonized to understand the effects of ageing on GI physiology and drug absorption, without the influence of polypharmacy and disease.

13.11   Conclusion

Drugs are extensively investigated regarding their PK pharmacodynamics and toxicity in healthy subjects by virtue of first-in-human studies and for its particular indication. Drug development, however, sometimes falls short in considering normal circumstances such as sex differences, the state of pregnancy, ethnic differences, the impact of ageing and co-morbidities on oral drug performance. Indeed, there remain strikingly large gaps in our knowledge of the GI tract physiology with respect to special populations; however, it is clear that one size does not fit all. Given that drug PK already demonstrate high inter-individual variability in healthy adults, the impact of special populations adds further complication to optimising drug response. The overarching theme is that although such populations are coined as 'special', such groups make up a large majority of the recipients of pharmaceutical products. As such, drug development should consider the potential differences in drug absorption between healthy and special population states. This should be the vision of improving models of the gut to accelerate the translation of important new drugs to patients.

References

[1] GAO (2001). Drug safety: most drugs withdrawn in recent years had greater health risk for women (cited 14 December 2017). https://www.gao.gov/products/gao-01-286r (accessed March 2021).

[2] Yang, Y., Carlin, A.S., Faustino, P.J. et al. (2009). Participation of women in clinical trials for new drugs approved by the food and drug administration in 2000–2002. *J. Women's Health (Larchmt)* **18** (3): 303–310.

[3] Yoon, D.Y., Mansukhani, N.A., Stubbs, V.C. et al. (2014). Sex bias exists in basic science and translational surgical research. *Surgery* **156** (3): 508–516.

[4] Liu, K.A. and Mager, N.A. (2016). Women's involvement in clinical trials: historical perspective and future implications. *Pharm. Pract (Granada)* **14** (1): 708.

[5] Stillhart, C., Vucicevic, K., Augustijns, P. et al. (2020). Impact of gastrointestinal physiology on drug absorption in special populations – an UNGAP review. *Eur. J. Pharm. Sci.* **147**: 105280.

[6] Madla, C.M., Gavins, F.K.H., Merchant, H., Orlu, M., Murdan, S., Basit, A.W. (2021). Let's Talk About Sex: Differences in Drug Therapy in Males and Females. *Adv. Drug Deliv. Rev.* DOI: 10.1016/j.addr.2021.05.014.

[7] Freire, A.C., Basit, A.W., Choudhary, R. et al. (2011). Does sex matter? The influence of gender on gastrointestinal physiology and drug delivery. *Int. J. Pharm.* **415** (1): 15–28.

[8] Gotch, F., Nadell, J., and Edelman, I.S. (1957). Gastrointestinal water and electrolytes. IV. The equilibration of deuterium oxide (D2O) in gastrointestinal contents and the proportion of total body water (T.B.W.) in the gastrointestinal tract. *J. Clin. Invest.* **36** (2): 289–296.

[9] Phan, J., Benhammou, J.N., and Pisegna, J.R. (2015). Gastric hypersecretory states: investigation and management. *Curr. Treat Opt. Gastroenterol.* **13** (4): 386–397.

[10] Feldman, M. and Barnett, C. (1991). Fasting gastric pH and its relationship to true hypochlorhydria in humans. *Dig. Dis. Sci.* **36** (7): 866–869.

[11] Lahner, E., Virili, C., Santaguida, M.G. et al. (2014). *Helicobacter pylori* infection and drugs malabsorption. *World J. Gastroenterol.* **20** (30): 10331–10337.

[12] Hutson, W.R., Roehrkasse, R.L., and Wald, A. (1989). Influence of gender and menopause on gastric emptying and motility. *Gastroenterology* **96** (1): 11–17.

[13] Senek, M., Nyholm, D., and Nielsen, E.I. (2018). Population pharmacokinetics of levodopa/carbidopa microtablets in healthy subjects and Parkinson's disease patients. *Eur. J. Clin. Pharmacol.* **74** (10): 1299–1307.

[14] Magallanes, L., Lorier, M., Ibarra, M. et al. (2016). Sex and food influence on intestinal absorption of ketoprofen gastroresistant formulation. *Clin. Pharmacol. Drug. Dev.* **5** (3): 196–200.

[15] Nandhra, G.K., Mark, E.B., Di Tanna, G.L. et al. (2020). Normative values for region-specific colonic and gastrointestinal transit times in 111 healthy volunteers using the 3D-Transit electromagnet tracking system: Influence of age, gender, and body mass index. *Neurogastroenterol. Motil.* **32** (2): e13734.

[16] Zimmermann, T., Laufen, H., Yeates, R. et al. (1999). The pharmacokinetics of extended-release formulations of calcium antagonists and of amlodipine in subjects with different gastrointestinal transit times. *J. Clin. Pharmacol.* **39** (10): 1021–1031.

[17] Krecic-Shepard, M.E., Barnas, C.R., Slimko, J. et al. (2000). Gender-specific effects on verapamil pharmacokinetics and pharmacodynamics in humans. *J. Clin. Pharmacol.* **40** (3): 219–230.

[18] Tamargo, J., Rosano, G., Walther, T. et al. (2017). Gender differences in the effects of cardiovascular drugs. *Eur. Heart J. Cardiovasc. Pharmacother.* **3** (3): 163–182.

[19] Waxman, D.J. and Holloway, M.G. (2009). Sex differences in the expression of hepatic drug metabolizing enzymes. *Mol. Pharmacol.* **76** (2): 215–228.

[20] Zhou, Q., Chen, Q.X., Ruan, Z.R. et al. (2013). CYP2C9*3(1075A > C), ABCB1 and SLCO1B1 genetic polymorphisms and gender are determinants of inter-subject variability in pitavastatin pharmacokinetics. *Pharmazie* **68** (3): 187–194.

[21] Smirnova, O.V. (2012). Sex differences in drug action: the role of multidrug-resistance proteins (MRPs). *Fiziol. Cheloveka* **38** (3): 124–136.

[22] Mai, Y., Ashiru-Oredope, D.A.I., Yao, Z. et al. (2020). Boosting drug bioavailability in men but not women through the action of an excipient. *Int. J. Pharm.* **587**: 119678.

[23] Mai, Y., Dou, L., Yao, Z., Madla, C.M., Gavins, F.K., Taherali, F., Yin, H., Orlu, M., Murdan, S., and Basit, A.W. (2021). Quantification of P-Glycoprotein in the Gastrointestinal Tract of Humans and Rodents: Methodology, Gut Region, Sex and Species Matter. *Mol Pharm.* **18** (5), 1895–1904.

[24] Ashiru, D.A., Patel, R., and Basit, A.W. (2008). Polyethylene glycol 400 enhances the bioavailability of a BCS class III drug (ranitidine) in male subjects but not females. *Pharm. Res.* **25** (10): 2327–2333.

[25] Mai, Y., Dou, L., Madla, C.M. et al. (2019). Sex-dependence in the effect of pharmaceutical excipients: polyoxyethylated solubilising excipients increase oral drug bioavailability in male but not female rats. *Pharmaceutics* **11** (5): 228.

[26] EMA (1998). ICH E5 (R1) ethnic factors in the acceptability of foreign clinical data https://www.ema.europa.eu/en/documents/scientific-guideline/ich-e-5-r1-ethnic-factors-acceptability-foreign-clinical-data-step-5_en.pdf (accessed March 2021).

[27] Chen, M.L. (2006). Ethnic or racial differences revisited: impact of dosage regimen and dosage form on pharmacokinetics and pharmacodynamics. *Clin. Pharmacokinet.* **45** (10): 957–964.

[28] Kaniwa, N. (2002). Japanese perspectives on pharmaceutical product release rate testing. *Drug Inf. J. DIJ/Drug Inf. Assoc.* **36** (2): 407–415.

[29] Johnson, J.A. (1997). Influence of race or ethnicity on pharmacokinetics of drugs. *J. Pharm. Sci.* **86** (12): 1328–1333.

[30] Zanger, U.M. and Schwab, M. (2013). Cytochrome P450 enzymes in drug metabolism: regulation of gene expression, enzyme activities, and impact of genetic variation. *Pharmacol. Ther.* **138** (1): 103–141.

[31] Chen, L. and Prasad, G.V.R. (2018). CYP3A5 polymorphisms in renal transplant recipients: influence on tacrolimus treatment. *Pharmgenom. Pers. Med.* **11**: 23–33.

[32] Ramamoorthy, A., Pacanowski, M.A., Bull, J., and Zhang, L. (2015). Racial/ethnic differences in drug disposition and response: review of recently approved drugs. *Clin. Pharmacol. Ther.* **97** (3): 263–273.

[33] FDA (1994). Prograf® (tacrolimus) capsules. FDA prescribing information. https://www.accessdata.fda.gov/drugsatfda_docs/label/2012/050709s031lbl.pdf (accessed February 2021).

[34] Wolking, S., Schaeffeler, E., Lerche, H. et al. (2015). Impact of genetic polymorphisms of ABCB1 (MDR1, P-Glycoprotein) on drug disposition and potential clinical implications: update of the literature. *Clin. Pharmacokinet.* **54** (7): 709–735.

[35] FDA 2002 Guidance for Industry (2002). Food-effect bioavailability and fed bioequivalence studies. https://www.fda.gov/files/drugs/published/Food-Effect-Bioavailability-and-Fed-Bioequivalence-Studies.pdf (accessed March 2021)

[36] EMA (2012). Guideline on the investigation of drug interactions. https://www.ema.europa.eu/en/documents/scientific-guideline/guideline-investigation-drug-interactions-revision-1_en.pdf (accessed March 2021)

[37] Fleisher, D., Li, C., Zhou, Y. et al. (1999). Drug, meal and formulation interactions influencing drug absorption after oral administration. *Clin. Pharmacokinet.* **36** (3): 233–254.

[38] Varum, F.J.O., Hatton, G.B., and Basit, A.W. (2013). Food, physiology and drug delivery. *Int. J. Pharm.* **457** (2): 446–460.

[39] Abuhelwa, A.Y., Williams, D.B., Upton, R.N., and Foster, D.J.R. (2017). Food, gastrointestinal pH, and models of oral drug absorption. *Eur. J. Pharm. Biopharm.* **112**: 234–248.

[40] Camilleri, M. (2006). Integrated upper gastrointestinal response to food intake. *Gastroenterology* **131** (2): 640–658.

[41] Pentafragka, C., Symillides, M., McAllister, M. et al. (2019). The impact of food intake on the luminal environment and performance of oral drug products with a view to *in vitro* and *in silico* simulations: a PEARRL review. *J. Pharm. Pharmacol.* **71** (4): 557–580.

[42] Karhunen, L.J., Juvonen, K.R., Huotari, A. et al. (2008). Effect of protein, fat, carbohydrate and fibre on gastrointestinal peptide release in humans. *Regul. Pept.* **149** (1): 70–78.

[43] Deng, J., Zhu, X., Chen, Z. et al. (2017). A review of food–drug interactions on oral drug absorption. *Drugs* **77** (17): 1833–1855.

[44] Koziolek, M., Carrière, F., and Porter, C.J.H. (2018). Lipids in the stomach – implications for the evaluation of food effects on oral drug absorption. *Pharm. Res.* **35** (3): 55.

[45] Kong, F. and Singh, R.P. (2008). Disintegration of solid foods in human stomach. *J. Food Sci.* **73** (5): R67–R80.

[46] Moghimipour, E., Ameri, A., and Handali, S. (2015). Absorption-enhancing effects of bile salts. *Molecules* **20** (8): 14451–14473.

[47] Porter, C.J.H., Trevaskis, N.L., and Charman, W.N. (2007). Lipids and lipid-based formulations: optimizing the oral delivery of lipophilic drugs. *Nat. Rev. Drug Discov.* **6** (3): 231–248.

[48] Grimm, M., Koziolek, M., Saleh, M. et al. (2018). Gastric emptying and small bowel water content after administration of grapefruit juice compared to water and isocaloric solutions of glucose and fructose: a four-way crossover mri pilot study in healthy subjects. *Mol. Pharm.* **15** (2): 548–559.

[49] Robertson, D., Higginson, I., Macklin, B. et al. (1991). The influence of protein containing meals on the pharmacokinetics of levodopa in healthy volunteers. *Br. J. Clin. Pharmacol.* **31** (4): 413–417.

[50] Takeo, N. and Ikumi, T. (2015). Interaction of drug or food with drug transporters in intestine and liver. *Curr. Drug Metab.* **16** (9): 753–764.

[51] Abrahamsson, B., Albery, T., Eriksson, A. et al. (2004). Food effects on tablet disintegration. *Eur. J. Pharm. Sci.* **22** (2): 165–172.

[52] O'Shea, J.P., Holm, R., O'Driscoll, C.M., and Griffin, B.T. (2019). Food for thought: formulating away the food effect – a PEARRL review. *J. Pharm. Pharmacol.* **71** (4): 510–535.

[53] Costantine, M.M. (2014). Physiologic and pharmacokinetic changes in pregnancy. *Front. Pharmacol.* **5**: 65.

[54] Pariente, G., Leibson, T., Carls, A. et al. (2016). Pregnancy-associated changes in pharmacokinetics: a systematic review. *PLoS Med.* **13** (11): e1002160.

[55] Chiloiro, M., Darconza, G., Piccioli, E. et al. (2001). Gastric emptying and orocecal transit time in pregnancy. *J. Gastroenterol.* **36** (8): 538–543.

[56] Whitehead, E.M., Smith, M., Dean, Y., and O'Sullivan, G. (1993). An evaluation of gastric emptying times in pregnancy and the puerperium. *Anaesthesia* **48** (1): 53–57.

[57] Hebert, M.F., Easterling, T.R., Kirby, B. et al. (2008). Effects of pregnancy on CYP3A and P-glycoprotein activities as measured by disposition of midazolam and digoxin: a University of Washington specialized center of research study. *Clin. Pharmacol. Ther.* **84** (2): 248–253.

[58] Heikkinen, T., Ekblad, U., Palo, P., and Laine, K. (2003). Pharmacokinetics of fluoxetine and norfluoxetine in pregnancy and lactation. *Clin. Pharmacol. Ther.* **73** (4): 330–337.

[59] Fischer, J.H., Sarto, G.E., Hardman, J. et al. (2014). Influence of gestational age and body weight on the pharmacokinetics of labetalol in pregnancy. *Clin. Pharmacokinet.* **53** (4): 373–383.

[60] Ohman, I., Beck, O., Vitols, S., and Tomson, T. (2008). Plasma concentrations of lamotrigine and its 2-N-glucuronide metabolite during pregnancy in women with epilepsy. *Epilepsia* **49** (6): 1075–1080.

[61] Broe, A., Pottegard, A., Lamont, R.F. et al. (2014). Increasing use of antibiotics in pregnancy during the period 2000–2010: prevalence, timing, category, and demographics. *BJOG* **121** (8): 988–996.

[62] Deligiannidis, K.M., Byatt, N., and Freeman, M.P. (2014). Pharmacotherapy for mood disorders in pregnancy: a review of pharmacokinetic changes and clinical recommendations for therapeutic drug monitoring. *J. Clin. Psychopharmacol.* **34** (2): 244–255.

[63] van der Galien, R., Ter Heine, R., Greupink, R. et al. (2019). Pharmacokinetics of HIV-integrase inhibitors during pregnancy: mechanisms, clinical implications and knowledge gaps. *Clin. Pharmacokinet.* **58** (3): 309–323.

[64] Bai, J.P.F., Burckart, G.J., and Mulberg, A.E. (2016). Literature review of gastrointestinal physiology in the elderly, in pediatric patients, and in patients with gastrointestinal diseases. *J. Pharm. Sci.* **105** (2): 476–483.

[65] Milovic, V. and Stein, J. (2010). Gastrointestinal disease and dosage form performance. In: *Oral Drug Absorption* (eds. J.D. Dressman and H. Lennernas), 127–135. Florisa: CRC Press, Taylor and Francis Group.

[66] Effinger, A., O'Driscoll, C.M., McAllister, M., and Fotaki, N. (2019). Impact of gastrointestinal disease states on oral drug absorption – implications for formulation design - a PEARRL review. *J. Pharm. Pharmacol.* **71** (4): 674–698.

[67] Grassi, R., Rambaldi, P.F., Di Grezia, G. et al. (2011). Inflammatory bowel disease: value in diagnosis and management of MDCT-enteroclysis and 99mTc-HMPAO labeled leukocyte scintigraphy. *Abdom. Imaging* **36** (4): 372–381.

[68] Hatton, G.B., Madla, C.M., Rabbie, S.C., and Basit, A.W. (2018). All disease begins in the gut: influence of gastrointestinal disorders and surgery on oral drug performance. *Int. J. Pharm.* **548** (1): 408–422.

[69] Hatton, G.B., Madla, C.M., Rabbie, S.C., and Basit, A.W. (2019). Gut reaction: impact of systemic diseases on gastrointestinal physiology and drug absorption. *Drug Discov. Today* **24** (2): 417–427.

[70] Malagelada, J.R. (2006). A symptom-based approach to making a positive diagnosis of irritable bowel syndrome with constipation. *Int. J. Clin. Pract.* **60** (1): 57–63.

[71] Lalezari, D. (2012). Gastrointestinal pH profile in subjects with irritable bowel syndrome. *Ann. Gastroenterol.* **25** (4): 333–337.

[72] Cann, P.A., Read, N.W., Brown, C. et al. (1983). Irritable bowel syndrome: relationship of disorders in the transit of a single solid meal to symptom patterns. *Gut* **24** (5): 405–411.

[73] Algera, J., Colomier, E., and Simren, M. (2019). The dietary management of patients with irritable bowel syndrome: a narrative review of the existing and emerging evidence. *Nutrients* **11** (9): 2162.

[74] Hundt, M. and John, S. (2018). Physiology, bile secretion (cited 16 April 2017). https://www.ncbi.nlm.nih.gov/books/NBK470209/ (accessed 16 February 2018).

[75] Brandtzaeg, P. (2011). The gut as communicator between environment and host: immunological consequences. *Eur. J. Pharmacol.* **668** (Suppl 1): S16–S32.

[76] Camilleri, M. (2012). Irritable bowel syndrome: how useful is the term and the 'diagnosis'? *Ther. Adv. Gastroenterol.* **5** (6): 381–386.

[77] Alimohammadi, N., Koosha, F., and Rafeian-Kopaei, M. (2020). Current, new and future therapeutic targets in inflammatory bowel disease: a systematic review. *Curr. Pharm. Des.* **26** (22): 2668–2675.

[78] Varum, F., Freire, A.C., Fadda, H.M., Bravo, R., and Basit, A.W. (2020). A dual pH and microbiota-triggered coating (Phloral™) for fail-safe colonic drug release. *Int. J. Pharm.* **583**: 119379.

[79] Ibekwe, V.C., Khela, M.K., Evans, D.F., and Basit, A.W. (2008). A new concept in colonic drug targeting: a combined pH-responsive and bacterially-triggered drug delivery technology. *Aliment. Pharmacol. Ther.* **28** (7): 911–916.

[80] McConnell, E.L., Liu, F., and Basit, A.W. (2009). Colonic treatments and targets: issues and opportunities. *J. Drug Target.* **17** (5): 335–363.

[81] Fallingborg, J., Christensen, L.A., Jacobsen, B.A., and Rasmussen, S.N. (1993). Very low intraluminal colonic pH in patients with active ulcerative colitis. *Dig. Dis. Sci.* **38** (11): 1989–1993.

[82] Nugent, S.G., Kumar, D., Rampton, D.S., and Evans, D.F. (2001). Intestinal luminal pH in inflammatory bowel disease: possible determinants and implications for therapy with amino-salicylates and other drugs. *Gut* **48** (4): 571–577.

[83] Press, A.G., Hauptmann, I.A., Hauptmann, L. et al. (1998). Gastrointestinal pH profiles in patients with inflammatory bowel disease. *Aliment. Pharmacol. Ther.* **12** (7): 673–678.

[84] Davis, S.S., Hardy, J.G., and Fara, J.W. (1986). Transit of pharmaceutical dosage forms through the small intestine. *Gut* **27** (8): 886–892.

[85] Hardy, J.G., Healey, J.N., Lee, S.W., and Reynolds, J.R. (1987). Gastrointestinal transit of an enteric-coated delayed-release 5-aminosalicylic acid tablet. *Aliment. Pharmacol. Ther.* **1** (3): 209–216.

[86] Hardy, J.G., Healey, J.N., and Reynolds, J.R. (1987). Evaluation of an enteric-coated delayed-release 5-aminosalicylic acid tablet in patients with inflammatory bowel disease. *Aliment. Pharmacol. Ther.* **1** (4): 273–280.

[87] Coupe, A.J., Davis, S.S., and Wilding, I.R. (1991). Variation in gastrointestinal transit of pharmaceutical dosage forms in healthy subjects. *Pharm. Res.* **8** (3): 360–364.

[88] Degen, L.P. and Phillips, S.F. (1996). Variability of gastrointestinal transit in healthy women and men. *Gut* **39** (2): 299–305.

[89] Fadda, H.M., McConnell, E.L., Short, M.D., and Basit, A.W. (2009). Meal-induced acceleration of tablet transit through the human small intestine. *Pharm. Res.* **26** (2): 356–360.

[90] Rabbie, S.C., Flanagan, T., Martin, P.D., and Basit, A.W. (2015). Inter-subject variability in intestinal drug solubility. *Int. J. Pharm.* **485** (1–2): 229–234.

[91] Bassotti, G., Antonelli, E., Villanacci, V. et al. (2014). Colonic motility in ulcerative colitis. *United Eur. Gastroenterol. J.* **2** (6): 457–462.

[92] Fischer, M., Siva, S., Wo, J.M., and Fadda, H.M. (2017). Assessment of small intestinal transit times in ulcerative colitis and Crohn's disease patients with different disease activity using video capsule endoscopy. *AAPS PharmSciTech* **18** (2): 404–409.

[93] Ehehalt, R., Wagenblast, J., Erben, G. et al. (2004). Phosphatidylcholine and lysophosphatidylcholine in intestinal mucus of ulcerative colitis patients. A quantitative approach by nanoelectrospray-tandem mass spectrometry. *Scand. J. Gastroenterol.* **39** (8): 737–742.

[94] Effinger, A., O'Driscoll, M., C, D. et al. (2020). Gastrointestinal diseases and their impact on drug solubility: ulcerative Colitis. *Eur. J. Pharm. Sci.* **152**: 105458.

[95] Vertzoni, M., Goumas, K., Soderlind, E. et al. (2010). Characterization of the ascending colon fluids in ulcerative colitis. *Pharm. Res.* **27** (8): 1620–1626.

[96] Estudante, M., Morais, J.G., Soveral, G., and Benet, L.Z. (2013). Intestinal drug transporters: an overview. *Adv. Drug Deliv. Rev.* **65** (10): 1340–1356.

[97] Effinger, A., O'Driscoll, C.M., McAllister, M., and Fotaki, N. (2020). Gastrointestinal diseases and their impact on drug solubility: Crohn's disease. *Eur. J. Pharm. Sci.* **152**: 105459.

[98] Baumgart, D.C. and Sandborn, W.J. (2012). Crohn's disease. *Lancet* **380** (9853): 1590–1605.

[99] Peyrin-Biroulet, L., Loftus, E.V. Jr., Colombel, J.F., and Sandborn, W.J. (2010). The natural history of adult Crohn's disease in population-based cohorts. *Am. J. Gastroenterol.* **105** (2): 289–297.

[100] Nishida, T., Miwa, H., Yamamoto, M. et al. (1982). Bile acid absorption kinetics in Crohn's disease on elemental diet after oral administration of a stable-isotope tracer with chenodeoxy-cholic-11, 12-d2 acid. *Gut* **23** (9): 751–757.

[101] Shaffer, J.A., Williams, S.E., Turnberg, L.A. et al. (1983). Absorption of prednisolone in patients with Crohn's disease. *Gut* **24** (3): 182–186.

[102] Matsuoka, K., Kobayashi, T., Ueno, F. et al. (2018). Evidence-based clinical practice guidelines for inflammatory bowel disease. *J. Gastroenterol.* **53** (3): 305–353.

[103] Fallingborg, J., Pedersen, P., and Jacobsen, B.A. (1998). Small intestinal transit time and intra-luminal pH in ileocecal resected patients with Crohn's disease. *Dig. Dis. Sci.* **43** (4): 702–705.

[104] Sasaki, Y., Hada, R., Nakajima, H. et al. (1997). Improved localizing method of radiopill in measurement of entire gastrointestinal pH profiles: colonic luminal pH in normal subjects and patients with Crohn's disease. *Am. J. Gastroenterol.* **92** (1): 114–118.

[105] Blokzijl, H., Vander Borght, S., Bok, L.I. et al. (2007). Decreased P-glycoprotein (P-gp/ MDR1) expression in inflamed human intestinal epithelium is independent of PXR protein levels. *Inflamm. Bowel Dis.* **13** (6): 710–720.

[106] Lebwohl, B., Sanders, D.S., and Green, P.H.R. (2018). Coeliac disease. *Lancet* **391** (10115): 70–81.

[107] Ciaccio, E.J., Bhagat, G., Lewis, S.K., and Green, P.H. (2016). Recommendations to quantify villous atrophy in video capsule endoscopy images of celiac disease patients. *World J. Gastrointest. Endosc.* **8** (18): 653–662.

[108] Samsel, A. and Seneff, S. (2013). Glyphosate, pathways to modern diseases II: celiac sprue and gluten intolerance. *Interdiscip. Toxicol.* **6** (4): 159–184.

[109] Heyman, M., Abed, J., Lebreton, C., and Cerf-Bensussan, N. (2012). Intestinal permeability in coeliac disease: insight into mechanisms and relevance to pathogenesis. *Gut* **61** (9): 1355–1364.

[110] Effinger, A., O'Driscoll, C.M., McAllister, M., and Fotaki, N. (2020). Gastrointestinal dis-eases and their impact on drug solubility: celiac disease. *Eur. J. Pharm. Sci.* **152**: 105460.

[111] Sergi, C.M., Caluseriu, O., McColl, H., and Eisenstat, D.D. (2017). Hirschsprung's disease: clinical dysmorphology, genes, micro-RNAs, and future perspectives. *Pediatr. Res.* **81** (1): 177–191.

[112] Kitis, G., Lucas, M.L., Bishop, H. et al. (1982). Altered jejunal surface pH in coeliac disease: its effect on propranolol and folic acid absorption. *Clin. Sci. (Lond.)* **63** (4): 373–380.

[113] Lang, C.C., Brown, R.M., Kinirons, M.T. et al. (1996). Decreased intestinal CYP3A in celiac disease: reversal after successful gluten-free diet: a potential source of interindividual variabil-ity in first-pass drug metabolism. *Clin. Pharmacol. Ther.* **59** (1): 41–46.

[114] Johnson, T.N., Tanner, M.S., Taylor, C.J., and Tucker, G.T. (2001). Enterocytic CYP3A4 in a paediatric population: developmental changes and the effect of coeliac disease and cystic fibrosis. *Br. J. Clin. Pharmacol.* **51** (5): 451–460.

[115] Benini, F., Mora, A., Turini, D. et al. (2012). Slow gallbladder emptying reverts to normal but small intestinal transit of a physiological meal remains slow in celiac patients during gluten-free diet. *Neurogastroenterol. Motil.* **24** (2): 100-7, e79–100-7, e80.

[116] Usai-Satta, P., Oppia, F., Lai, M., and Cabras, F. (2018). Motility disorders in celiac disease and non-celiac gluten sensitivity: the impact of a gluten-free diet. *Nutrients* **10** (11): 1705.

[117] Dacha, S., Razvi, M., Massaad, J. et al. (2015). Hypergastrinemia. *Gastroenterol. Rep. (Oxf.)* **3** (3): 201–208.

[118] Talley, N.J., Locke, G.R. 3rd, Lahr, B.D. et al. (2006). Functional dyspepsia, delayed gastric emptying, and impaired quality of life. *Gut* **55** (7): 933–939.

[119] Fiorini, G., Bland, J.M., Hughes, E. et al. (2015). A systematic review on drugs absorption modifications after eradication in *Helicobacter pylori* positive patients undergoing replacement therapy. *J. Gastrointestin. Liver Dis.* **24** (1): 95–100. 1 p following 100.

[120] Tshibangu-Kabamba, E., Yamaoka, Y. (2021) *Helicobacter pylori* infection and antibiotic resistance — from biology to clinical implications. *Nat. Rev. Gastroenterol Hepatol.* https:// doi.org/10.1038/s41575-021-00449-x

[121] Pierantozzi, M., Pietroiusti, A., Brusa, L. et al. (2006). *Helicobacter pylori* eradication and l-dopa absorption in patients with PD and motor fluctuations. *Neurology* **66** (12): 1824–1829.

[122] Centanni, M., Gargano, L., Canettieri, G. et al. (2006). Thyroxine in goiter, *Helicobacter pylori* infection, and chronic gastritis. *N. Engl. J. Med.* **354** (17): 1787–1795.

[123] Shelton, M.J., Akbari, B., Hewitt, R.G. et al. (2000). Eradication of *Helicobacter pylori* is associated with increased exposure to delavirdine in hypochlorhydric HIV-positive patients. *J. Acquir. Immune Defic. Syndr.* **24** (1): 79–82.

[124] Proesmans, M. and De Boeck, K. (2003). Omeprazole, a proton pump inhibitor, improves residual steatorrhoea in cystic fibrosis patients treated with high dose pancreatic enzymes. *Eur. J. Pediatr.* **162** (11): 760–763.

[125] Duffield, R.A. (1996). Cystic fibrosis and the gastrointestinal tract. *J. Pediatr. Health Care* **10** (2): 51–57.

[126] De Lisle, R.C., Meldi, L., Roach, E. et al. (2009). Mast cells and gastrointestinal dysmotility in the cystic fibrosis mouse. *PLoS One* **4** (1): e4283.

[127] de Lisle, R.C., Sewell, R., and Meldi, L. (2010). Enteric circular muscle dysfunction in the cystic fibrosis mouse small intestine. *Neurogastroenterol. Motil.* **22** (3): 341–e87.

[128] Hallberg, K., Abrahamsson, H., Dalenback, J. et al. (2001). Gastric secretion in cystic fibrosis in relation to the migrating motor complex. *Scand. J. Gastroenterol.* **36** (2): 121–127.

[129] Bali, A., Stableforth, D.E., and Asquith, P. (1983). Prolonged small-intestinal transit time in cystic fibrosis. *Br. Med. J. (Clin. Res. Ed.)* **287** (6398): 1011–1013.

[130] Hedsund, C., Gregersen, T., Joensson, I.M. et al. (2012). Gastrointestinal transit times and motility in patients with cystic fibrosis. *Scand. J. Gastroenterol.* **47** (8–9): 920–926.

[131] Collins, C.E., Francis, J.L., Thomas, P. et al. (1997). Gastric emptying time is faster in cystic fibrosis. *J. Pediatr. Gastroenterol. Nutr.* **25** (5): 492–498.

[132] Gilbert, J., Kelleher, J., Littlewood, J.M., and Evans, D.F. (1988). Ileal pH in cystic fibrosis. *Scand. J. Gastroenterol. Suppl.* **143**: 132–134.

[133] Tang, L., Fatehi, M., and Linsdell, P. (2009). Mechanism of direct bicarbonate transport by the CFTR anion channel. *J. Cyst. Fibros.* **8** (2): 115–121.

[134] Barraclough, M. and Taylor, C.J. (1996). Twenty-four hour ambulatory gastric and duodenal pH profiles in cystic fibrosis: effect of duodenal hyperacidity on pancreatic enzyme function and fat absorption. *J. Pediatr. Gastroenterol. Nutr.* **23** (1): 45–50.

[135] Youngberg, C.A., Berardi, R.R., Howatt, W.F. et al. (1987). Comparison of gastrointestinal pH in cystic fibrosis and healthy subjects. *Dig. Dis. Sci.* **32** (5): 472–480.

[136] Lisowska, A., Pogorzelski, A., Oracz, G. et al. (2011). Oral antibiotic therapy improves fat absorption in cystic fibrosis patients with small intestine bacterial overgrowth. *J. Cyst. Fibros.* **10** (6): 418–421.

[137] Cersosimo, M.G., Raina, G.B., Pecci, C. et al. (2013). Gastrointestinal manifestations in Parkinson's disease: prevalence and occurrence before motor symptoms. *J. Neurol.* **260** (5): 1332–1338.

[138] Abbott, R.D., Ross, G.W., Petrovitch, H. et al. (2007). Bowel movement frequency in late-life and incidental Lewy bodies. *Mov. Disord.* **22** (11): 1581–1586.

[139] Woitalla, D. and Goetze, O. (2011). Treatment approaches of gastrointestinal dysfunction in Parkinson's disease, therapeutical options and future perspectives. *J. Neurol. Sci.* **310** (1–2): 152–158.

[140] Pfeiffer, R.F. (2011). Gastrointestinal dysfunction in Parkinson's disease. *Parkinsonism Relat. Disord.* **17** (1): 10–15.

[141] Jost, W.H. and Schimrigk, K. (1991). Constipation in Parkinson's disease. *Klin. Wochenschr.* **69** (20): 906–909.

[142] Davies, K.N., King, D., Billington, D., and Barrett, J.A. (1996). Intestinal permeability and orocaecal transit time in elderly patients with Parkinson's disease. *Postgrad. Med. J.* **72** (845): 164–167.

[143] Tanaka, Y., Kato, T., Nishida, H. et al. (2011). Is there a delayed gastric emptying of patients with early-stage, untreated Parkinson's disease? An analysis using the 13C-acetate breath test. *J. Neurol.* **258** (3): 421–426.

[144] Edwards, L.L., Pfeiffer, R.F., Quigley, E.M. et al. (1991). Gastrointestinal symptoms in Parkinson's disease. *Mov. Disord.* **6** (2): 151–156.

[145] Muller, T., Erdmann, C., Bremen, D. et al. (2006). Impact of gastric emptying on levodopa pharmacokinetics in Parkinson disease patients. *Clin. Neuropharmacol.* **29** (2): 61–67.

[146] Abrahamsson, H. (2007). Treatment options for patients with severe gastroparesis. *Gut* **56** (6): 877–883.

[147] Contin, M., Riva, R., Martinelli, P. et al. (1999). Concentration-effect relationship of levodopa-benserazide dispersible formulation versus standard form in the treatment of complicated motor response fluctuations in Parkinson's disease. *Clin. Neuropharmacol.* **22** (6): 351–355.

[148] Nyholm, D. (2006). Enteral levodopa/carbidopa gel infusion for the treatment of motor fluctuations and dyskinesias in advanced Parkinson's disease. *Expert. Rev. Neurother.* **6** (10): 1403–1411.

[149] Bjarnason, I.T., Charlett, A., Dobbs, R.J. et al. (2005). Role of chronic infection and inflammation in the gastrointestinal tract in the etiology and pathogenesis of idiopathic parkinsonism. Part 2: response of facets of clinical idiopathic parkinsonism to *Helicobacter pylori* eradication. A randomized, double-blind, placebo-controlled efficacy study. *Helicobacter* **10** (4): 276–287.

[150] Pierantozzi, M., Pietroiusti, A., Sancesario, G. et al. (2001). Reduced l-dopa absorption and increased clinical fluctuations in *Helicobacter pylori*-infected Parkinson's disease patients. *Neurol. Sci.* **22** (1): 89–91.

[151] Deshpande, A.D., Harris-Hayes, M., and Schootman, M. (2008). Epidemiology of diabetes and diabetes-related complications. *Phys. Ther.* **88** (11): 1254–1264.

[152] Tambascia, M.A., Malerbi, D.A., and Eliaschewitz, F.G. (2014). Influence of gastric emptying on the control of postprandial glycemia: physiology and therapeutic implications. *Einstein (Sao Paulo)* **12** (2): 251–253.

[153] Zhao, M., Liao, D., and Zhao, J. (2017). Diabetes-induced mechanophysiological changes in the small intestine and colon. *World J. Diabet.* **8** (6): 249–269.

[154] Dotevall, G. (1961). Gastric emptying in diabetes mellitus. *Acta Med. Scand.* **170**: 423–429.

[155] Nowak, T.V., Johnson, C.P., Kalbfleisch, J.H. et al. (1995). Highly variable gastric emptying in patients with insulin dependent diabetes mellitus. *Gut* **37** (1): 23–29.

[156] Charman, W.N., Porter, C.J., Mithani, S., and Dressman, J.B. (1997). Physiochemical and physiological mechanisms for the effects of food on drug absorption: the role of lipids and pH. *J. Pharm. Sci.* **86** (3): 269–282.

[157] Alhnan, M.A., Kidia, E., and Basit, A.W. (2011). Spray-drying enteric polymers from aqueous solutions: a novel, economic, and environmentally friendly approach to produce pH-responsive microparticles. *Eur. J. Pharm. Biopharm.* **79** (2): 432–439.

[158] Liu, F. and Basit, A.W. (2010). A paradigm shift in enteric coating: achieving rapid release in the proximal small intestine of man. *J. Control. Release* **147** (2): 242–245.

[159] Liu, F., Lizio, R., Meier, C. et al. (2009). A novel concept in enteric coating: a double-coating system providing rapid drug release in the proximal small intestine. *J. Control. Release* **133** (2): 119–124.

[160] Madsen, M.S.A., Holm, J.B., Pallejà, A. et al. (2019). Metabolic and gut microbiome changes following GLP-1 or dual GLP-1/GLP-2 receptor agonist treatment in diet-induced obese mice. *Sci. Rep.* **9** (1): 15582.

[161] Ney, D., Hollingsworth, D.R., and Cousins, L. (1982). Decreased insulin requirement and improved control of diabetes in pregnant women given a high-carbohydrate, high-fiber, low-fat diet. *Diabet. Care* **5** (5): 529–533.

[162] Delezay, O., Yahi, N., Tamalet, C. et al. (1997). Direct effect of type 1 human immunodeficiency virus (HIV-1) on intestinal epithelial cell differentiation: relationship to HIV-1 enteropathy. *Virology* **238** (2): 231–242.

[163] Belitsos, P.C., Greenson, J.K., Yardley, J.H. et al. (1992). Association of gastric hypoacidity with opportunistic enteric infections in patients with AIDS. *J. Infect. Dis.* **166** (2): 277–284.

[164] Blum, R.A., D'Andrea, D.T., Florentino, B.M. et al. (1991). Increased gastric pH and the bioavailability of fluconazole and ketoconazole. *Ann. Intern. Med.* **114** (9): 755–757.

[165] Zimmermann, T., Yeates, R.A., Riedel, K.D. et al. (1994). The influence of gastric pH on the pharmacokinetics of fluconazole: the effect of omeprazole. *Int. J. Clin. Pharmacol. Ther.* **32** (9): 491–496.

[166] Tseng, A.L. and Foisy, M.M. (1997). Management of drug interactions in patients with HIV. *Ann. Pharmacother.* **31** (9): 1040–1058.

[167] Lim, S.G., Menzies, I.S., Lee, C.A. et al. (1993). Intestinal permeability and function in patients infected with human immunodeficiency virus. A comparison with coeliac disease. *Scand. J. Gastroenterol.* **28** (7): 573–580.

[168] Tepper, R.E., Simon, D., Brandt, L.J. et al. (1994). Intestinal permeability in patients infected with the human immunodeficiency virus. *Am. J. Gastroenterol.* **89** (6): 878–882.

[169] Gurumurthy, P., Ramachandran, G., Hemanth Kumar, A.K. et al. (2004). Malabsorption of rifampin and isoniazid in HIV-infected patients with and without tuberculosis. *Clin. Infect. Dis.* **38** (2): 280–283.

[170] Peloquin, C.A., Nitta, A.T., Burman, W.J. et al. (1996). Low antituberculosis drug concentrations in patients with AIDS. *Ann. Pharmacother.* **30** (9): 919–925.

[171] Poles, M.A., Fuerst, M., McGowan, I. et al. (2001). HIV-related diarrhea is multifactorial and fat malabsorption is commonly present, independent of HAART. *Am. J. Gastroenterol.* **96** (6): 1831–1837.

[172] Kapembwa, M.S., Fleming, S.C., Orr, M. et al. (1996). Impaired absorption of zidovudine in patients with AIDS-related small intestinal disease. *AIDS* **10** (13): 1509–1514.

[173] Dinh, D.M., Volpe, G.E., Duffalo, C. et al. (2015). Intestinal microbiota, microbial translocation, and systemic inflammation in chronic HIV infection. *J. Infect. Dis.* **211** (1): 19–27.

[174] Bandera, A., De Benedetto, I., Bozzi, G., and Gori, A. (2018). Altered gut microbiome composition in HIV infection: causes, effects and potential intervention. *Curr. Opin. HIV AIDS* **13** (1): 73–80.

[175] Wei, X., Ghosh, S.K., Taylor, M.E. et al. (1995). Viral dynamics in human immunodeficiency virus type 1 infection. *Nature* **373** (6510): 117–122.

[176] Huang, F., Drda, K., MacGregor, T.R. et al. (2009). Pharmacokinetics of BILR 355 after multiple oral doses coadministered with a low dose of ritonavir. *Antimicrob. Agents Chemother.* **53** (1): 95–103.

[177] Wilkinson, E.M., Ilhan, Z.E., and Herbst-Kralovetz, M.M. (2018). Microbiota–drug interactions: Impact on metabolism and efficacy of therapeutics. *Maturitas* **112**: 53–63.

[178] Maharaj, A. and Edginton, A. (2014). Physiologically based pharmacokinetic modeling and simulation in pediatric drug development. *CPT Pharmacometr. Syst. Pharmacol.* **3** (11): 1–13.

[179] Batchelor, H.K., Kendall, R., Desset-Brethes, S. et al. (2013). Application of *in vitro* biopharmaceutical methods in development of immediate release oral dosage forms intended for paediatric patients. *Eur. J. Pharm. Biopharm.* **85** (3, Part B): 833–842.

[180] Strolin Benedetti, M., Whomsley, R., and Baltes, E.L. (2005). Differences in absorption, distribution, metabolism and excretion of xenobiotics between the paediatric and adult populations. *Expert Opin. Drug Metab. Toxicol.* **1** (3): 447–471.

[181] Bartelink, I.H., Rademaker, C.M., Schobben, A.F., and van den Anker, J.N. (2006). Guidelines on paediatric dosing on the basis of developmental physiology and pharmacokinetic considerations. *Clin. Pharmacokinet.* **45** (11): 1077–1097.

[182] Fallingborg, J., Christensen, L.A., Ingeman-Nielsen, M. et al. (1990). Measurement of gastrointestinal pH and regional transit times in normal children. *J. Pediatr. Gastroenterol. Nutr.* **11** (2): 211–214.

[183] Robinson, P., Smith, A., and Sly, P. (1990). Duodenal pH in cystic fibrosis and its relationship to fat malabsorption. *Dig. Dis. Sci.* **35** (10): 1299–1304.

[184] Pawar, G., Papadatou-Soulou, E., Mason, J. et al. (2020). Characterisation of fasted state gastric and intestinal fluids collected from children. *Eur. J. Pharm. Biopharm.* **158**: 156–165.

[185] Kamstrup, D., Berthelsen, R., Sassene, P.J. et al. (2017). *in vitro* model simulating gastro-intestinal digestion in the pediatric population (neonates and young infants). *AAPS PharmSciTech* **18** (2): 317–329.

[186] Gharpure, V., Meert, K.L., Sarnaik, A.P., and Metheny, N.A. (2000). Indicators of postpyloric feeding tube placement in children. *Crit. Care Med.* **28** (8): 2962–2966.

[187] Nicolas, J.M., Bouzom, F., Hugues, C., and Ungell, A.L. (2017). Oral drug absorption in pediatrics: the intestinal wall, its developmental changes and current tools for predictions. *Biopharm. Drug Dispos.* **38** (3): 209–230.

[188] Moreno, M.P.d.l.C., Oth, M., Deferme, S. et al. (2006). Characterization of fasted-state human intestinal fluids collected from duodenum and jejunum. *J. Pharm. Pharmacol.* **58** (8): 1079–1089.

[189] Navarro, J., Schmitz, J., and Barefoot, B. (1992). *Paediatric Gastroenterology*. USA: Oxford University Press.

[190] Meakin, G., Dingwall, A., and Addison, G. (1987). Effects of fasting and oral premedication on the pH and volume of gastric aspirate in children. *Br. J. Anaesth.* **59** (6): 678–682.

[191] Schwartz, D.A., Connelly, N.R., Theroux, C.A. et al. (1998). Gastric contents in children presenting for upper endoscopy. *Anesth. Analg.* **87** (4): 757–760.

[192] Goetze, O., Treier, R., Fox, M. et al. (2009). The effect of gastric secretion on gastric physiology and emptying in the fasted and fed state assessed by magnetic resonance imaging. *Neurogastroenterol. Motil.* **21** (7): 725–e42.

[193] Bar-Shalom, D. and Rose, K. (2014). *Pediatric Formulations: A Roadmap*, vol. **11**. Springer.

[194] Lange, A., Funch-Jensen, P., Thommesen, P., and Schiøtz, P.O. (1997). Gastric emptying patterns of a liquid meal in newborn infants measured by epigastric impedance. *Neurogastroenterol. Motil.* **9** (2): 55–62.

[195] Bonner, J.J., Vajjah, P., Abduljalil, K. et al. (2015). Does age affect gastric emptying time? A model-based meta-analysis of data from premature neonates through to adults. *Biopharm. Drug Dispos.* **36** (4): 245–257.

[196] Van Den Driessche, M. and Veereman-Wauters, G. (2003). Gastric emptying in infants and children. *Acta Gastro-enterologica Belgica (English Ed.)* **66** (4): 274–282.

[197] Edginton, A. and Fotaki, N. (2010). *Oral Drug Absorption: Prediction and Assessment*, vol. **193**, 108–126. New York: Informa Healthcare.

[198] Guimarães, M., Statelova, M., Holm, R. et al. (2019). Biopharmaceutical considerations in paediatrics with a view to the evaluation of orally administered drug products – a PEARRL review. *J. Pharm. Pharmacol.* **71** (4): 603–642.

[199] Cavkll, B. (1981). Gastric emptying in infants fed human milk or infant formula. *Acta Paediatr.* **70** (5): 639–641.

[200] Weaver, L., Austin, S., and Cole, T. (1991). Small intestinal length: a factor essential for gut adaptation. *Gut* **32** (11): 1321–1323.

[201] Van Elburg, R., Fetter, W., Bunkers, C., and Heymans, H. (2003). Intestinal permeability in relation to birth weight and gestational and postnatal age. *Arch. Dis. Child Fetal Neonatal Ed.* **88** (1): F52–F55.

[201] Batchelor, H.K., Fotaki, N., and Klein, S. (2014). Paediatric oral biopharmaceutics: key considerations and current challenges. *Adv. Drug Deliv. Rev.* **73**: 102–126.

[203] Batchelor, H. and Initiative, E.P.F. (2014). Paediatric biopharmaceutics classification system: current status and future decisions. *Int. J. Pharm.* **469** (2): 251–253.

[204] Kerr, C.A., Grice, D.M., Tran, C.D. et al. (2015). Early life events influence whole-of-life metabolic health via gut microflora and gut permeability. *Crit. Rev. Microbiol.* **41** (3): 326–340.

[205] Batchelor, H.K. and Marriott, J.F. (2015). Paediatric pharmacokinetics: key considerations. *Br. J. Clin. Pharmacol.* **79** (3): 395–404.

[206] Mooij, M.G., Schwarz, U.I., De Koning, B.A. et al. (2014). Ontogeny of human hepatic and intestinal transporter gene expression during childhood: age matters. *Drug Metab. Dispos.* **42** (8): 1268–1274.

[207] Mooij, M.G., de Koning, B.E., Lindenbergh-Kortleve, D.J. et al. (2016). Human intestinal PEPT1 transporter expression and localization in preterm and term infants. *Drug Metab. Dispos.* **44** (7): 1014–1019.

[208] Merchant, H.A., Liu, F., Orlu Gul, M., and Basit, A.W. (2016). Age-mediated changes in the gastrointestinal tract. *Int. J. Pharm.* **512** (2): 382–395.

[209] An, R., Wilms, E., Masclee, A.A.M. et al. (2018). Age-dependent changes in GI physiology and microbiota: time to reconsider? *Gut* **67** (12): 2213–2222.

[210] Petrini, E., Caviglia, G.P., Pellicano, R. et al. (2020). Risk of drug interactions and prescription appropriateness in elderly patients. *Irish J. Med. Sci. (1971)* **189** (3): 953–959.

[211] Liu, F., Ranmal, S., Batchelor, H.K. et al. (2014). Patient-centered pharmaceutical design to improve acceptability of medicines: similarities and differences in paediatric and geriatric populations. *Drugs* **74** (16): 1871–1889.

[212] Perrie, Y., Badhan, R.K., Kirby, D.J. et al. (2012). The impact of ageing on the barriers to drug delivery. *J. Control. Release* **161** (2): 389–398.

[213] Evans, M.A., Triggs, E.J., Cheung, M. et al. (1981). Gastric emptying rate in the elderly: implications for drug therapy. *J. Am. Geriatr. Soc.* **29** (5): 201–205.

[214] Brogna, A., Ferrara, R., Bucceri, A.M. et al. (1999). Influence of aging on gastrointestinal transit time: an ultrasonographic and radiologic study. *Investig. Radiol.* **34** (5): 357.

[215] Madsen, J.L. and Graff, J. (2004). Effects of ageing on gastrointestinal motor function. *Age Age.* **33** (2): 154–159.

[216] Fischer, M. and Fadda, H.M. (2016). The effect of sex and age on small intestinal transit times in humans. *J. Pharm. Sci.* **105** (2): 682–686.

[217] Russell, T.L., Berardi, R.R., Barnett, J.L. et al. (1993). Upper gastrointestinal ph in seventy-nine healthy, elderly, north american men and women. *Pharm. Res.* **10** (2): 187–196.

[218] Gidal, B.E. (2006). Drug absorption in the elderly: biopharmaceutical considerations for the antiepileptic drugs. *Epilepsy Res.* **68**: 65–69.

[219] Feldman, M. and Cryer, B. (1998). Effects of age on gastric alkaline and nonparietal fluid secretion in humans. *Gerontology* **44** (4): 222–227.

[220] Kim, S.W., Parekh, D., Townsend, C.M. Jr., and Thompson, J.C. (1990). Effects of aging on duodenal bicarbonate secretion. *Ann. Surg.* **212** (3): 332–338.

[221] Einarsson, K., Nilsell, K., Leijd, B., and Angelin, B. (1985). Influence of age on secretion of cholesterol and synthesis of bile acids by the liver. *N. Engl. J. Med.* **313** (5): 277–282.

[222] Khalil, T., Walker, J.P., Wiener, I. et al. (1985). Effect of aging on gallbladder contraction and release of cholecystokinin-33 in humans. *Surgery* **98** (3): 423–429.

[223] Annaert, P., Brouwers, J., Bijnens, A. et al. (2010). ex vivo permeability experiments in excised rat intestinal tissue and *in vitro* solubility measurements in aspirated human intestinal fluids support age-dependent oral drug absorption. *Eur. J. Pharm. Sci.* **39** (1): 15–22.

[224] Webster, S.G.P. and Leeming, J.T. (1975). The appearance of the small bowel mucosa in old age. *Age Ageing* **4** (3): 168–174.

[225] Newton, J.L. (2004). Changes in upper gastrointestinal physiology with age. *Mech. Ageing Dev.* **125** (12): 867–870.

[226] Pullan, R.D., Thomas, G.A., Rhodes, M. et al. (1994). Thickness of adherent mucus gel on colonic mucosa in humans and its relevance to colitis. *Gut* **35**: 353–359.

[227] Woodmansey, E.J., McMurdo, M.E., Macfarlane, G.T., and Macfarlane, S. (2004). Comparison of compositions and metabolic activities of fecal microbiotas in young adults and in antibiotic-treated and non-antibiotic-treated elderly subjects. *Appl. Environ. Microbiol.* **70** (10): 6113–6122.

[228] Claesson, M.J., Jeffery, I.B., Conde, S. et al. (2012). Gut microbiota composition correlates with diet and health in the elderly. *Nature* **488** (7410): 178–184.

[229] FDA (1989). Study of drugs likely to be used in the elderly. https://www.fda.gov/downloads/Drugs/GuidanceComplianceRegulatoryInformation/Guidances/UCM072048.pdf (accessed February 2021).

[230] ICH (1993). ICH harmonised tripartite guideline E7: Studies in support of special populations: geriatrics. https://www.ich.org/fileadmin/Public_Web_Site/ICH_Products/Guidelines/Efficacy/E7/Step4/E7_Guideline.pdf (accessed February 2021).

[231] Khan, M.S. and Roberts, M.S. (2018). Challenges and innovations of drug delivery in older age. *Adv. Drug Deliv. Rev.* **135**: 3–38.

14

Inhalation Biopharmaceutics

Precious Akhuemokhan*, Magda Swedrowska and Ben Forbes

Institute of Pharmaceutical Sciences, King's College London, London, United Kingdom

14.1   Introduction

The term 'inhalation biopharmaceutics' encompasses the features that determine regional drug deposition by affecting aerosol behaviour in the respiratory tract and, the processes that determine the post-deposition fate of aerosol medicines [1]. Factors affecting deposition may be patient-specific such as inspiratory profile, lung luminal conditions, and airway geometry or drug product dependent such as inhalation device or drug formulation. The interplay between these factors may be further complicated by physiological processes like mucociliary clearance, macrophage uptake, drug metabolism and absorptive clearance, all of which will contribute to the performance of the inhaled drug product.

Drugs delivered via inhalation include gaseous anaesthetics and aerosols for respiratory disease. This chapter focuses on the latter, which can be classified collectively in regulatory terms as orally inhaled drug products. Compared to systemic therapy, inhalation for local action in the lungs enables the use of smaller doses and avoids extra-pulmonary side effects. A complication of using inhalation to treat patients with respiratory disease is that lung exposure is affected by respiratory inflammation, immune responses or impaired lung diseases such as asthma, chronic obstructive pulmonary disease, infection and cystic fibrosis. Inhalation is used less commonly for systemic delivery of drugs in circumstances where

* Currently with Advanced Drug Delivery, Pharmaceutical Sciences, R&D, AstraZeneca, Gothenburg, Sweden

Biopharmaceutics: From Fundamentals to Industrial Practice, First Edition. Edited by Hannah Batchelor.
© 2022 John Wiley & Sons Ltd. Published 2022 by John Wiley & Sons Ltd.

inhalation can provide advantages over oral or parenteral therapy, e.g. for delivery of large molecules such as peptides and proteins, e.g. insulin, or to achieve rapid drug action in drugs that are subject to high first-pass metabolism, e.g. nicotine.

Inhalation medicines have been well described and comprehensive information can be found in recent reviews [2–4]. This chapter focuses specifically on inhalation biopharmaceutics by providing a synopsis of the factors determining respiratory and systemic exposure to drugs delivered to the lungs by oral inhalation.

In an early contribution to the field, key questions for biopharmaceutics were posited by Schanker in the 1970s [5] and are paraphrased below:

- What happens to an aerosol droplet or particle upon depositing on the surface of the alveolar or bronchiolar epithelium?
- What is the influence of the epithelial lining fluid on the fate of inhaled drugs?
- How rapidly do powder aerosols dissolve in the epithelial lining fluid and what factors determine dissolution rate?
- What is the permeability of the epithelium and what physiologic and pathological factors influence this throughout the respiratory tract?
- What types of transporters are present in the respiratory epithelium and what is their distribution?
- What are the effects of environmental, physiological and disease variables on the absorption rate of inhaled substances?
- Do inhaled or systemically administered drugs alter the permeability characteristics of the pulmonary epithelium?

Cross-sector working groups from the pharmaceutical industry and academia have come together to reconsider these questions [6, 7], extend them to include drug deposition in the lungs [6, 8] and review progress [9]. Clearly, the pharmacokinetics of inhaled formulations depends upon a complex combination of simultaneous dynamic processes. The aim is to understand these and describe full temporal relationships between pharmacokinetic (PK) and pharmacodynamic (PD) measurements to help mitigate the risk during drug development of not engaging successfully or persistently with the drug target, as well as identifying the potential for drug accumulation in the lung or excessive systemic exposure [6].

This chapter describes developments in modelling lung deposition, understanding the impact of mucociliary clearance, exploring the resurgent interest in aerosol particle dissolution, an update on lung permeability and drug absorption, the opportunities offered by pulmonary metabolism, the role of non-clinical studies in drug development, recent applications of physiologically-based pharmacokinetic (PBPK) modelling to inhaled drug delivery and the prospect of an inhaled version of the biopharmaceutics classification system for orally administered drugs.

14.2   Structure of the Lungs

The physiology and anatomy of the lungs are described in brief.

14.2.1   Basic Anatomy

The lungs are a gas-filled organ that serves the essential function of gas exchange, i.e. bringing oxygen into the body and expelling carbon dioxide. Air enters and leaves the lungs through a series of bifurcating tubes known as the conducting airways, which are generally idealised as beginning with the trachea followed by 16 generations of bronchial

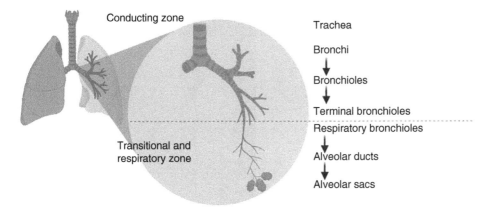

Figure 14.1   *Idealised structure of the respiratory tract. Source: Image made in © BioRender – biorender.com*

airways of diminishing length and diameter (Figure 14.1). The succeeding transitional and respiratory zone consists of the smaller respiratory bronchioles, alveolar ducts and sacs. The lungs receive the entire cardiac output of blood and perfusion is intimately matched with the air spaces over a large surface area, 70–100 m^2, and has a short, 1–2 μm, air-to-blood gas transfer (and drug absorption) pathway in the alveolar region.

14.2.2   Epithelial Lining Fluid

When a drug is inhaled, the aerosol particles deposit onto the walls of the airspaces. These mucosal surfaces are lined with lung fluid that has surfactants to reduce surface tension and keep the airspaces from collapse. The lining fluid also contains protective components such as antibacterial peptides and antioxidants, thus enabling the defence properties of the lungs. The conducting airways are additionally overlaid with a dynamic mucus layer that serves an additional protective function by trapping and removing inhaled material from the airways. The epithelial lining fluid volume of the lungs has been estimated to be 10–30 mL [7, 10].

14.2.3   Epithelium

The respiratory epithelium varies in cytology from the larger airways (trachea/bronchi) to the bronchioles and the alveoli (Figure 14.2). The epithelium of the conducting airways is pseudostratified and consists of ciliated columnar cells, mucus-secreting goblet cells and basal cells, which do not interface with the mucosal surface. The alveolar epithelium in contrast is a monolayer of type I and type II pneumocytes in a 1 : 2 ratio. Type I cells are thin, extended cells that present 95% of the surface area of the pulmonary region, while the type II cells are cuboidal and produce lung surfactant.

14.3   Molecules, Inhalation Devices, Formulations

14.3.1   Inhaled Molecules

The physicochemical properties of inhaled molecules show that most are small molecules (200–600 Da), with a wide range of log D values and with different charges [11, 12]. This includes a number of molecules that are prodrugs, e.g. beclomethasone and ciclesonide,

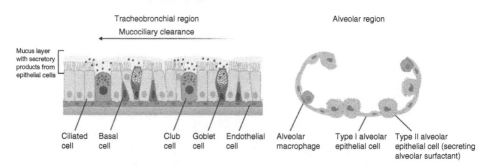

Figure 14.2   *Cellular composition of the epithelium in the conducting airways and alveoli. Source: Image made in © BioRender – biorender.com*

which require metabolic activation to their pharmacologically active form. Newer molecules and those in development present new larger and more complex modalities such as peptides and proteins, new agents such as nucleotide-based medicines (oligonucleotides, ribonucleic acids, genes), PROTACs and CRISPR, as well as more poorly soluble compounds.

Inhaled molecules require high potency as the amount of drug that can be delivered by aerosol in a single inhalation is limited to microgram or a few milligrams. Molecular properties can be optimised to promote lung retention via mechanisms such as tissue binding, cellular trapping, prolonged receptor occupancy. Lung retention can also be achieved by limiting the dissolution rate (e.g. corticosteroids [13]; Section 14.4.3) or the permeability (e.g. antibiotics [14]; Section 14.4.4).

14.3.2   Inhalation Devices

Inhalation devices have a long history but reached the modern era via the evolution of improved nebuliser technologies for producing aerosols from aqueous liquids, the development of pressurised metered-dose inhalers (pMDI) in the 1950s, the advent of devices for the inhalation of powders in the 1960s and 1970s and the more recent appearance of soft mist inhalers as a new category of metered, portable inhalation device.

Importantly, inhalation devices should be tailored to the patient with consideration of factors such as age (paediatric or elderly), cognitive ability and dexterity, disease-related changes in lung structure and function. Training in the correct use of the device is another important component of treatment adherence and a key factor in the success of inhalation therapy.

The main technologies currently in clinical use are summarised below.

14.3.2.1   Nebulisers

Nebulisers generate droplet aerosols from solutions or suspensions, which patients can inhale during tidal breathing. Jet nebulisers generate aerosols using the Venturi effect and are the oldest of the popular technologies in use today. They exist in a variety of forms, are widely used and are inexpensive. Ultrasonic nebulisers are the traditional rival format and produce droplets by applying high-frequency pulses to the solution from an oscillating piezo element. The heat generated with these devices makes them unsuitable for temperature-sensitive drug products. More recently, vibrating mesh systems have become increasingly popular as a more efficient option. The main challenges with nebulisers are to manage delivery times, increase efficiency and portability. Advantages include the ability to administer to unconscious patients, ready interfacing with electronics to improve the efficiency of delivery and the ability to deliver high doses.

14.3.2.2   Pressurised Metered-Dose Inhalers

Pressurised metered-dose inhalers (pMDI) were developed in the 1950s by formulating drugs in chlorofluorocarbon (CFC) propellants in inert canisters with metering valves that provide dose control. When actuated, pMDI emit a dynamic aerosol plume where the propellent flash evaporates forming respirable aerosol of drug particles together with non-volatile excipients. The CFCs were phased out in the 1990s in favour of hydrofluoroalkanes (HFA), which required reformulation of pMDI to adapt to the physicochemical properties of the new propellants.

14.3.2.3   Dry Powder Inhalers

Dry powder inhalers deliver drugs to the lungs by using the inspiration of the patient to disaggregate a fine particle powder into a respirable aerosol. The development of powder inhalers has evolved from original capsule-based inhalers, through reservoir devices to the development of efficient inhalers that deliver multiple individually packaged doses of drugs or drug combinations. New powder formulation technologies and more efficient devices for generating respirable aerosols have increased the dose that can be delivered to the lungs using dry powder inhalers.

14.3.2.4   'Soft Mist' Inhalers

Soft mist inhalers are the most recent inhaler development that incorporates a liquid formulation. Once actuated, the drug solution is mechanically forced through engineered orifices to generate respirable aerosol clouds that can be inhaled without the need for actuation–inhalation coordination. Soft mist inhalers produce a slow-moving mist that allows for relatively high lung deposition. These devices differ from the traditional nebulisers by being hand-held, portable devices that do not require an external power source.

14.3.3   Inhaled Medicine Formulation

Formulation requirements vary according to delivery devices. Nebulisers and soft mist inhalers may include surfactants, buffering agents or viscosity modifiers. For pMDI, the drug is dissolved or suspended in propellent, and excipients may be included to facilitate and control the dynamics of particle formation when actuated. Powders come in a variety of forms including micronised particles with carriers, engineered aggregates or low density spray dried formulations and require excipients to assist disaggregation when inhaled.

From a biopharmaceutics perspective, formulations in most licensed inhaled products have few excipients and are designed to help devices to emit drug efficiently in an aerosol form that favours good lung penetration and meets regulatory requirements. The engineering of these formulations is complex and important for the delivered lung dose and optimal regional distribution [9]. As few excipients are licensed for inhalation [15], the formulation options for pharmacokinetic-modification are limited and there is less research on the effects of formulation on the disposition of drugs in the lungs after deposition. The excipients used in pMDI can affect the dynamics of aerosol formation with results of biopharmaceutical relevance, e.g. effects on drug crystallisation and solid-state properties, formation of solvates, presence of residual non-volatile excipients [16]. For powders, particles size, solid-state chemistry and wettability may affect the rate at which particles dissolve [17].

More complex formulations are in development for improving the selectivity of current inhaled drugs or to enable effective or targeted delivery of more challenging molecules, such as the potential new classes of inhaled molecules discussed in Section 14.3.2. Many academic, industrial and clinical researchers worldwide are investigating the risk-benefits of nanoparticles, lipid formulations, surfactants for the next generation of inhaled medicines [18, 19].

14.4 Inhaled Drug Delivery and Models for Studying Inhalation Biopharmaceutics

The fate of inhaled medicines depends on a complex combination of simultaneous dynamic processes, which are considered individually below and illustrated in Figure 14.3.

14.4.1 Dosimetry and Deposition

Only the fraction of the aerosol emitted from an inhaler that passes the oropharynx is delivered to the lungs. This is often represented as the fine particle fraction or fine particle dose, which expressed as the percentage and mass of the emitted dose, respectively.

The deposition mechanisms by which particles deposit in the lungs depend on a balance between inertial, gravitational and diffusional forces, as well as the drag force of the moving air that counteracts deposition. For some aerosols, electrostatic forces or interception may also be important. The three major mechanisms are as follows:

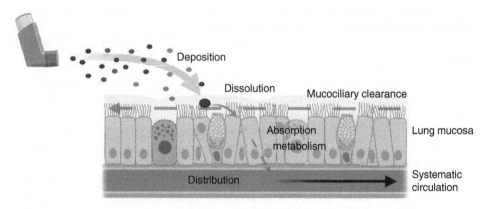

Figure 14.3 *The competing processes that determine the fate of aerosol medicines in the lungs. Source: Image made in © BioRender – biorender.com*

i) Inertial impaction, which is the predominant deposition mechanism for particles with an aerodynamic diameter larger than 5–10 μm in the upper airways, where the air velocity is high and the airflow turbulent. The probability of impaction increases with the square of the particle diameter, particle density and particle velocity.

ii) Sedimentation, which is the time-dependent settling of particles under the influence of gravity. The gravitational force increases with the mass and the stationary settling velocity (counteracted by the drag force) with the square of the particle diameter. This mechanism prevails when both the air velocity is low and has the same order of magnitude as the settling velocity, and the residence time is high, which is the case in the peripheral airways.

iii) Diffusion or Brownian motion, which is the random movement of particles in a gas. Diffusion is greatest for smaller particle diameters and only very fine particles deposit by this mechanism, i.e. particles < 0.5 μm. Deposition by diffusion is time-dependent and occurs mainly in the peripheral airways but is limited by the residence time of particles in the respiratory tract such that particles with diameters below 0.5 μm may be exhaled rather soon after being inhaled.

Thus, the extent and location of particle deposition in the airways depend on particle size, density, velocity, and the residence time in the lungs. The concept of aerodynamic diameter standardises the size parameter by incorporating particle density and shape. The aerodynamic diameter of a particle is defined as the diameter of a sphere of unit density that settles with the same velocity in still air (thus under the influence of gravity). In general, particles with an aerodynamic diameter of 1–5 μm are expected to reach the lungs. Particle size is commonly expressed as the mass median aerodynamic diameter, i.e. the aerodynamic diameter below which 50% of the emitted mass is contained, and the polydispersity is indicated by the geometric standard deviation.

Aerodynamic particle size is typically measured using methods for fractionating aerosols based on their aerodynamic diameter. Measurements are performed using pharmacopeial apparatus such as the twin-stage impinger (TSI), Andersen cascade impactor (ACI) or next-generation impactor (NGI). The ACI and NGI are used in marketing authorisation applications and for quality control purposes. However, adaptations such as using a realistic oropharyngeal geometry to introduce aerosols to the impactor, simulating environmental temperature and humidity and using age- and disease-relevant breathing profiles can make the measurements more realistic and provide good *in vitro–in vivo* correlations with lung deposition in humans measured by imaging [9].

There is increasing interest in modelling regional deposition and lung doses. Models include a one-dimensional algebraic whole lung model, originally introduced for environmental protection purposes such as ICRP 1994, or three-dimensional computational fluid-particle dynamics simulations. The former can be used to provide dose estimates for PBPK models, whereas three-dimensional modelling has the potential to evaluate deposition in individual patient geometries to evaluate the effect of disease or drug treatment on regional deposition [20].

14.4.2 Mucociliary Clearance

Particles depositing in the conducting airways are subject to mucociliary clearance. The mucus lining of the bronchi and bronchioles traps particles, which are transported with the mucus by the synchronised action of cilia towards the oropharynx and out of the lungs in a non-absorptive clearance mechanism [21].

The interaction with airway mucus affects particle and drug transport through mucus, thereby determining access to the epithelial cells and susceptibility to mucociliary clearance. The interactions vary according to factors such as particle size, surface chemistry and dissolution rate and drug characteristics such as size and charge [22, 23]. Mucociliary clearance reduces the amount of time a formulation or drug is in the lungs with important consequences for the bioavailability of the delivered drug. There are a variety of methods available to measure cilia beat frequency, mucociliary transport *in vitro*, mucociliary transport and clearance *in vivo* and relationship between ciliary beat frequency and mucociliary transport rate [24]. Mucociliary clearance rates are generally faster in larger airways, vary widely between individuals and may be impaired in patients with lung disease. Thus, the impact of mucociliary clearance for inhaled therapy is complex and depends on a combination of drug characteristics, formulation and patients factors.

14.4.3   Dissolution

Although methods for evaluating the dissolution of inhaled products were reported in the late 1990s and early 2000s, a working group report in 2012 reinforced the conclusion of the USP Inhalation Ad Hoc Advisory Panel in 2008, which 'could not find compelling evidence suggesting that such dissolution testing is kinetically and/or clinically crucial for currently approved inhalation drug products'. [25]. However, the Food and Drug Administration (FDA) in the United States of America sponsored research projects in this area in 2014, which stimulated a wave of method development by researchers in academia and industry [17, 26] and chapters on inhaled drug dissolution began to appear in textbooks on aerosol medicines [27, 28] and biopharmaceutics [29].

Of the methods have been developed, some are more suited to quality control purposes, and others attempt to be biorelevant. There has been little attempt at standardisation, but common features are (i) a means of collecting respirable fraction(s) of the aerosol; (ii) a dissolution apparatus in which the aerosol particles are introduced to the dissolution medium; (iii) data interpretation, including mathematical description, statistical assessment and comparison of dissolution profiles. The various apparatus for collection of particles and drug release are generally adaptations of pharmacopeial apparatus or, less commonly, bespoke designs. The mathematical treatment is typical of that used in biopharmaceutical assessment of other dosage forms. It is generally accepted that the dissolution methods may have value, even if not strictly *in vivo* predictive, by discriminating between formulations with similar aerodynamic properties, but different drug release profiles. This may be particularly the case for poorly soluble molecules.

An important consideration for biorelevant methods includes the collection of a relevant particle fraction and presentation of the particles to the dissolution medium in a disaggregated form, the use of a dissolution medium that mimics lung fluid, simulation of the fluid restriction in the lungs (drug mass:fluid volume ratio) and considerations around the use of stirring and sink conditions. The deposition of particles to the surface of air-interfaced epithelial cell layers to study dissolution permeability in an integrated *in vitro* system has also been explored by some researchers.

To date, there has been limited use of dissolution measurements for *in vitro–in vivo* correlations. Relationships have been observed between *in vitro* dissolution and lung residence time *in vivo* for corticosteroids. Mechanistic pharmacokinetic models are beginning to incorporate experimental dissolution data and can be used to explore the impact of dissolution on PK. Drug dissolution, which is consequent to drug solubility, may also have a role in the inhalation biopharmaceutics classification system (Section 14.5).

14.4.4   Lung Permeability, Absorption and Retention

The major barrier to lung permeability is the epithelium, which has been described in Figure 14.2. Mechanisms of transport across the epithelium are similar to those observed in other organs, for example the intestine, although the presence and significance of transporters in the lungs are not as well understood [30]. Notably, luminal conditions in the lungs are very different to the intestine. In the lungs, drug deposits at the absorptive mucosal surface in a small volume of epithelial lining fluid that is not subject to the variations of fed and fasted state and has less enzymatic activity than the gastrointestinal tract. Generally favourable luminal conditions, together with the large surface area of the lungs, the short absorption pathway and rich perfusion with blood, make absorption from the lungs rapid for most small molecules and retention in the lungs for local activity a challenge.

The physicochemical properties of inhaled molecules determine their pulmonary absorption rate. Pulmonary absorption of a wide range of compounds in rats is positively correlated with compound lipophilicity and negatively correlated with molecular weight [31]. Similar findings have been reported for inhaled compounds in rats [12], demonstrated using isolated perfused rat lungs [32–35] and explored *in vitro* using a variety of cell culture models [32, 35–37].

Lung retention can be achieved if molecules have features that delay their clearance, e.g. if the absorption of molecules is dissolution rate limited because of their solubility or permeability rate limited because of large molecular weight. There are also a variety of mechanisms by which molecules may interact with the lungs slowing or reducing their absorption, including extracellular interactions with mucus or surfactant phospholipids, cell sequestration by ion trapping, e.g. formoterol, or trapping of lipid esters, e.g. budesonide. Molecular modifications, such as PEGylation, and a wide variety of formulations [19] have been studied as potential controlled release or lung-retentive strategies, without progressing to clinical use. Because it is technically difficult to determine free drug concentrations in the lungs, plasma concentration-time profiles are commonly used as a measure to estimate lung exposure [6]. However, the latter approach is of limited value where retention of the drug at the receptor provides prolonged duration of action but is driven by a small fraction of the dose that engages with the target following inhalation.

Several biopharmaceutical tools have been used to study permeability barriers in the lungs. These include models of lung mucus, a wide variety of cell models of the airway or alveolar epithelium based on immortalised epithelial cells or primary cell cultures, which are amenable to mechanistic drug transport studies. The isolated perfused lung is a model that preserves the architecture of the lungs and allows absorption to be studied in the intact organ under controlled conditions. The majority of *in vivo* experiments to study drug absorption or retention have been performed in rats.

In vitro–in vivo relationships between physicochemical properties or *in vitro* permeability data and absorption in rats have been described. More realistic *in vitro* models of the lung absorption barriers are being developed but have yet to show advantages over using simpler and more readily available Caco-2 cell permeability data. As mechanistic modelling of pulmonary absorption to predict the pharmacokinetics of inhaled molecules develops, experimental estimates of pulmonary permeability and lung absorption will be required as inputs for the *in silico* modelling or validation of simulations.

14.4.5   Metabolism

The lungs possess many xenobiotic metabolising enzymes, which influence the pharmacokinetics and safety of inhaled medicines. Current knowledge regarding Phase I and II metabolism in the lungs and the impact of metabolism on inhaled medicines including small molecules, biologics and macromolecular formulation excipients has recently been reviewed [38].

Anticipating metabolism in the lungs provides an opportunity to optimise new inhaled medicines, although this is not without challenges as lung enzyme activity varies with age, race, smoking status, diet, drug exposure and pre-existing lung disease. As well as utilising metabolism for advantage in the design of inhaled medicines, e.g. as with prodrugs, there is potential for metabolism to lead to inhalation toxicity.

Although there are currently limited options for measuring *in situ* metabolism of inhaled drugs within the human lungs, there are several methods that rely on either *ex vivo* animal models or *in vitro* human/animal models. These include isolated perfused lungs, precision-cut lung slices, lung explants, primary cells and cell lines, lung cell homogenates and S9 fractions. Considerations in model selection include the ability to work with human material, the degree of disruption to the lung sample and complexity, availability and cost of the method. Developments in experimental technologies to study lung metabolism include tissue-engineered models, improved analytical capability and *in silico* models to predict lung metabolism.

Metabolism offers the opportunity to optimise drugs by molecular modification to reduce systemic exposure using a 'soft drug' approach, improve bioavailability by resisting metabolism or through the use of a prodrug approach to overcome pharmacokinetic limitations. Drugs that are very labile in the lungs for example nucleotide therapies may require a protective formulation to avoid premature degradation. In contrast, some of the drug carriers that are being investigated as PK-modifying formulations rely on lung enzymes to trigger drug release or for biodegradation of the carrier matrix.

14.4.6   Non-Clinical Inhalation Studies

Non-clinical studies are an essential part of inhaled drug development. Shortcomings in models of human lung disease in which the efficacy of new drugs are evaluated [39] and complications in inhalation toxicology, such as the observation of macrophage responses at doses required to provide safety margins for non-clinical safety testing [40], can hinder the translation of research between laboratory and clinic and are challenges for the successful development of new inhaled medicines [6].

In early non-clinical studies, drugs are often administered by intratracheal instillation, but it is important that realistic aerosol exposures are used early in the process to support transitioning to first in man studies. The regulatory toxicology pathway for inhaled products is well established [41]. Improved methods to study efficacy are in development, e.g. new challenge agents to study airway hyperresponsiveness and systems that can deliver aerosols with greater accuracy to individual animals [42].

14.4.7   Mechanistic Computer Modelling

Following the successful application of physiologically based pharmacokinetic modelling to oral drug delivery, mechanistic computer modelling is becoming increasingly popular for predicting lung exposure following inhalation of novel locally acting pulmonary drug molecules and generic inhaled products [43]. Estimating local exposure has the potential to speed up and improve the chances of successful inhaled API and product development. Wider application of mechanistic models to understand local drug exposure after inhalation and support product development requires reliable, biorelevant means to acquire experimental data, and proven models that combine aerosol deposition and post-deposition processes in physiologically based pharmacokinetic models that predict free local tissue concentrations.

14.5   Bioequivalence and an Inhalation Bioclassification System

International regulatory agencies, including the European Medicines Association and the FDA in the United States of America, have clear guidelines for the development of orally inhaled drug products. For generic locally acting OIDP, similarity is required in formulation and inhaler device compared to reference, as well as equivalent performance in *in vitro* tests, *in vivo* pharmacokinetics and pharmacodynamics and clinical studies. Some differences exist in the requirements of regulatory agencies regarding whether these similarities can be shown in a stepwise manner or must be demonstrated more holistically using a weight of evidence approach [44, 45].

The feasibility of developing a systematic framework to classify pulmonary drugs in a manner analogous to the biopharmaceutics classification system for immediate release orally administered compounds were evaluated at a cross-sector workshop in the United States in 2015. Factors influencing drug delivery and action in the lungs were considered along with the concepts of dose, dissolution and absorption numbers as they would apply to pulmonary drug delivery. The workshop concluded that an inhalation biopharmaceutics classification system would be a useful tool for formulators and discovery chemists [10] and proposals for a framework for such a system are under development with anticipated publication in 2021.

14.6   Conclusion

The discipline of inhalation biopharmaceutics has been established, and there has been much progress over the last 50 years. Current areas of unmet need include establishing clinically relevant assays and specifications to assure product quality, biorelevant methods to acquire experimental data, and development and validation of mechanistic models of aerosol deposition and post-deposition pharmacokinetics. The increasing use of the inhaled route for more diverse chemical and pharmacological classes of molecules will bring new challenges and stimulate fresh progress in inhalation biopharmaceutics.

References

[1] Ehrhardt, C. (2017). Inhalation biopharmaceutics: progress towards comprehending the fate of inhaled medicines. *Pharm. Res.* **34** (12): 2451–2453.

[2] Hickey, A.J. and Mansour, H.M. (2020). *Inhalation Aerosols. Physical and Biological Basis for Therapy*. CRC Press.

[3] Kassinos, S., Bäckman, P.E.R., and Hickey, A.J. (2021). Introduction. In: *Inhaled Medicines* (eds. S. Kassinos, P. Bäckman, J. Conway and A.J. Hickey), xvii–xviii. Academic Press.

[4] Newman, S. and Anderson, P. (2009). *Respiratory Drug Delivery: Essential Theory and Practice*. Richmond, VA: Respiratory Drug Delivery Online.

[5] Schanker, L.S. (1978). Drug absorption from the lung. *Biochem. Pharmacol.* **27** (4): 381–385.

[6] Forbes, B., Asgharian, B., Dailey, L.A. et al. (2011). Challenges in inhaled product development and opportunities for open innovation. *Adv. Drug Deliv. Rev.* **63** (1–2): 69–87.

[7] Patton, J.S., Brain, J.D., Davies, L.A. et al. (2010). The particle has landed--characterizing the fate of inhaled pharmaceuticals. *J. Aerosol. Med. Pulm. Drug Deliv.* **23** (Suppl 2): S71–S87.

[8] Borgström, L., Clark, A., and Olsson, B. (2010). Introduction. 1000 years of pharmaceutical aerosols: what remains to be done? *J. Aerosol. Med. Pulm. Drug Deliv.* **23** (Suppl 2): S1–S4.

[9] Forbes, B., Bäckman, P., Christopher, D. et al. (2015). *in vitro* testing for orally inhaled products: developments in science-based regulatory approaches. *AAPS J.* **17** (4): 837–852.

[10] Hastedt, J.E., Bäckman, P., Clark, A.R. et al. (2016). Scope and relevance of a pulmonary biopharmaceutical classification system AAPS/FDA/USP Workshop March 16-17th, 2015 in Baltimore, MD. *AAPS Open* **2** (1): 1.

[11] Bonn, B. and Perry, M. (2021). Chapter 2 - The API. In: *Inhaled Medicines* (eds. S. Kassinos et al.), 13–34. Academic Press.

[12] Tronde, A., Nordén, B., Jeppsson, A.-B. et al. (2003). Drug Absorption from the Isolated perfused rat lung–correlations with drug physicochemical properties and epithelial permeability. *J. Drug Target.* **11** (1): 61–74.

[13] Bhagwat, S., Schilling, U., Chen, M.J. et al. (2017). Predicting pulmonary pharmacokinetics from *in vitro* properties of dry powder inhalers. *Pharm. Res.* **34** (12): 2541–2556.

14] Marchand, S., Boisson, M., Mehta, S. et al. (2018). Biopharmaceutical characterization of nebulized antimicrobial agents in rats: 6. Aminoglycosides. *Antimicrob. Agents Chemother.* **62** (10): 01261-18.

[15] Pilcer, G. and Amighi, K. (2010). Formulation strategy and use of excipients in pulmonary drug delivery. *Int. J. Pharm.* **392** (1): 1–19.

[16] Grainger, C.I., Saunders, M., Buttini, F. et al. (2012). Critical characteristics for corticosteroid solution metered dose inhaler bioequivalence. *Mol. Pharm.* **9** (3): 563–569.

[17] Velaga, S.P., Djuris, J., Cvijic, S. et al. (2018). Dry powder inhalers: an overview of the *in vitro* dissolution methodologies and their correlation with the biopharmaceutical aspects of the drug products. *Eur. J. Pharm. Sci.* **113**: 18–28.

[18] Loira-Pastoriza, C., Todoroff, J., and Vanbever, R. (2014). Delivery strategies for sustained drug release in the lungs. *Adv. Drug Deliv. Rev.* **75**: 81–91.

[19] Smyth, H.D.C. and Hickey, A. (2011). *Controlled Pulmonary Drug Delivery*. Springer.

[20] Koullapis, P., Stylianou, F., Lin, C.-L. et al. (2021). Chapter 7 - *In silico* methods to model dose deposition. In: *Inhaled Medicines* (eds. S. Kassinos et al.), 167–195. Academic Press.

[21] Donnelley, M., Gardner, M., Morgan, K., and Parsons, D. (2021). Chapter 8 - Non-absorptive clearance from airways. In: *Inhaled Medicines* (eds. S. Kassinos et al.), 197–223. Academic Press.

[22] Cingolani, E., Alqahtani, S., Sadler, R.C. et al. (2019). *in vitro* investigation on the impact of airway mucus on drug dissolution and absorption at the air-epithelium interface in the lungs. *Eur. J. Pharm. Biopharm.* **141**: 210–220.

[23] Murgia, X., Loretz, B., Hartwig, O. et al. (2018). The role of mucus on drug transport and its potential to affect therapeutic outcomes. *Adv. Drug Deliv. Rev.* **124**: 82–97.

[24] Lansley, A.B. (1993). Mucociliary clearance and drug delivery via the respiratory tract. *Adv. Drug Deliv. Rev.* **11** (3): 299–327.

[25] Riley, T., Christopher, D., Arp, J. et al. (2012). Challenges with developing *in vitro* dissolution tests for orally inhaled products (OIPs). *AAPS PharmSciTech* **13** (3): 978–989.

[26] Radivojev, S., Zellnitz, S., Paudel, A., and Fröhlich, E. (2019). Searching for physiologically relevant *in vitro* dissolution techniques for orally inhaled drugs. *Int. J. Pharm.* **556**: 45–56.

[27] Amini, E. and Hochhaus, G. (2021). Chapter 9 - Dissolution and drug release. In: *Inhaled Medicines* (eds. S. Kassinos et al.), 225–266. Academic Press.

[28] Forbes, B., Richer, N.H., and Buttini, F. (2015). Dissolution: a critical performance characteristic of inhaled products? In: *Pulmonary Drug Delivery: Advances and Challenges* (eds. A. Nokhodchi and G.P. Martin). Wiley.

[29] Mercuri, A. and Fotaki, N. (2019). *in vitro* dissolution for inhalation products. In: *in vitro Drug Release Testing of Special Dosage Forms* (eds. N. Fotaki and S. Klein), 119–153. John Wiley & Sons Ltd.

[30] Ehrhardt, C., Bäckman, P., Couet, W. et al. (2017). Current progress toward a better understanding of drug disposition within the lungs: summary proceedings of the first workshop on drug transporters in the lungs. *J. Pharm. Sci.* **106** (9): 2234–2244.

[31] Taylor, G. (1990). The absorption and metabolism of xenobiotics in the lung. *Adv. Drug Deliv. Rev.* **5** (1): 37–61.

[32] Bosquillon, C., Madlova, M., Patel, N. et al. (2017). A comparison of drug transport in pulmonary absorption models: isolated perfused rat lungs, respiratory epithelial cell lines and primary cell culture. *Pharm. Res.* **34** (12): 2532–2540.

[33] Edwards, C.D., Luscombe, C., Eddershaw, P., and Hessel, E.M. (2016). Development of a novel quantitative structure-activity relationship model to accurately predict pulmonary absorption and replace routine use of the isolated perfused respiring rat lung model. *Pharm. Res.* **33** (11): 2604–2616.

[34] Eriksson, J., Sjögren, E., Thörn, H. et al. (2018). Pulmonary absorption - estimation of effective pulmonary permeability and tissue retention of ten drugs using an ex vivo rat model and computational analysis. *Eur. J. Pharm. Biopharm.* **124**: 1–12.

[35] Tronde, A., Nordén, B., Marchner, H. et al. (2003). Pulmonary absorption rate and bioavailability of drugs *in vivo* in rats: structure-absorption relationships and physicochemical profiling of inhaled drugs. *J. Pharm. Sci.* **92** (6): 1216–1233.

[36] Grainger, C.I., Greenwell, L.L., Martin, G.P., and Forbes, B. (2009). The permeability of large molecular weight solutes following particle delivery to air-interfaced cells that model the respiratory mucosa. *Eur. J. Pharm. Biopharm.* **71** (2): 318–324.

[37] Mathia, N.R., Timoszyk, J., Stetsko, P.I. et al. (2002). Permeability characteristics of calu-3 human bronchial epithelial cells: *in vitro-in vivo* correlation to predict lung absorption in rats. *J. Drug Target.* **10** (1): 31–40.

[38] Enlo-Scott, Z., Bäckström, E., Mudway, I., and Forbes, B. (2021). Drug metabolism in the lungs: opportunities for optimising inhaled medicines. *Expert Opin. Drug Metab. Toxicol.* **17**: 611–625.

[39] Sécher, T., Bodier-Montagutelli, E., Guillon, A., and Heuzé-Vourc'h, N. (2020). Correlation and clinical relevance of animal models for inhaled pharmaceuticals and biopharmaceuticals. *Adv. Drug Deliv. Rev.* **167**: 148–169.

[40] Forbes, B., O'Lone, R., Allen, P.P. et al. (2014). Challenges for inhaled drug discovery and development: induced alveolar macrophage responses. *Adv. Drug Deliv. Rev.* **71**: 15–33.

[41] Owen, K. (2013). Regulatory toxicology considerations for the development of inhaled pharmaceuticals. *Drug Chem. Toxicol.* **36** (1): 109–118.

[42] Lexmond, A.J., Keir, S., Terakosolphan, W. et al. (2018). A novel method for studying airway hyperresponsiveness in allergic guinea pigs *in vivo* using the PreciseInhale system for delivery of dry powder aerosols. *Drug Deliv. Transl. Res.* **8** (3): 760–769.

[43] Bäckman, P., Arora, S., Couet, W. et al. (2018). Advances in experimental and mechanistic computational models to understand pulmonary exposure to inhaled drugs. *Eur. J. Pharm. Sci.* **113**: 41–52.

[44] Evans, C., Cipolla, D., Chesworth, T. et al. (2012). Equivalence considerations for orally inhaled products for local action-ISAM/IPAC-RS European workshop report. *J. Aerosol. Med. Pulm. Drug Deliv.* **25** (3): 117–139.

[45] Lu, D., Lee, S.L., Lionberger, R.A. et al. (2015). International guidelines for bioequivalence of locally acting orally inhaled drug products: similarities and differences. *AAPS J.* **17** (3): 546–557.

15

Biopharmaceutics of Injectable Formulations

Wang Wang Lee and Claire M. Patterson

Seda Pharmaceutical Development Services, Alderley Edge, Alderley Park, Cheshire, United Kingdom

15.1 Introduction

Most patients tend to prefer to take their medication orally. However, in some circumstances, it may be advantageous to administer medication via injection (parenterally). For example, drugs that are not orally bioavailable – perhaps due to poor solubility, poor permeability, instability in the GI tract or are subject to extensive first-pass metabolism, may be more effectively administered via injection. Insulin, for example, is digested by proteases within the GI tract and is subsequently very challenging to administer orally.

The most commonly used routes of injection for systemic exposure of pharmaceutical formulations are intravenous (IV), subcutaneous (SC) and intramuscular (IM). IV injections, which are administered directly into either a central or peripheral vein as either bolus or infusion, are considered 100% bioavailable. With no absorption phase, maximum plasma concentration (C_{max}) is achieved almost instantaneously following bolus administration or at steady state during an infusion or at the end of an infusion where steady state is not reached (see Chapter 2). Due to achieving 100% bioavailability, IV administered formulations are used as the reference against which absolute bioavailability of medicines administered via other routes is determined. Care must be taken to understand the potential impact of any excipients used on clearance which could complicate data interpretation [1].

Biopharmaceutics: From Fundamentals to Industrial Practice, First Edition. Edited by Hannah Batchelor.
© 2022 John Wiley & Sons Ltd. Published 2022 by John Wiley & Sons Ltd.

IV solution formulations do not usually require a great deal of input from a biopharmaceutical scientist as they are simple and have no rate-limited steps to availability. More complex IV formulations such as nanomedicines or micellar solutions have quite different considerations and are reviewed extensively elsewhere e.g. [2, 3].

Following extravascular injection (injection into anything other than a vein) such as SC or IM administration, drug must first be absorbed before reaching the systemic circulation. Absorption may be rapid and complete, reaching C_{max} within one hour or less, or may be slow and sometimes incomplete. The absorption rate can be deliberately slowed down to achieve sustained release. Given that the duration of absorption is not limited by GI residence time as it is with oral formulations, the delivery system can remain in place for weeks/months acting as a slow releasing depot. This can help patient concordance and can improve safety profile by minimising peak:trough ratios. Examples of PK profiles that can be obtained using the different parenteral routes of administration and different formulations compared with oral dosing are illustrated in Figure 15.1. The drug, formulation and physiological factors affecting absorption, plus the biorelevant models available for studying them, will be the main subject of this chapter.

Subcutaneous (SC) injections are administered into the adipose layer of the skin. The main advantage is that they can be self-administered by the patient, making SC a preferred route of administration for the management of chronic conditions, such as insulin-dependent

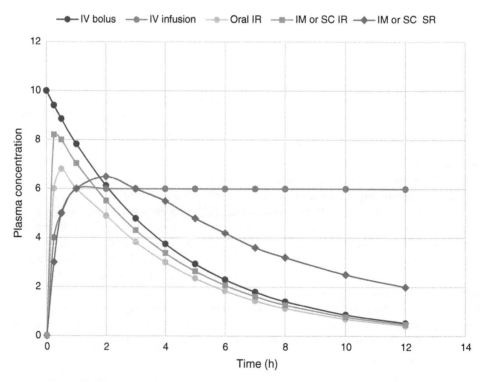

Figure 15.1 *Typical PK profiles following IV bolus, IV infusion, oral IR, IM/SC IR and IM/SC SR formulation dosing (IR = immediate release; SR = sustained release).*

diabetes mellitus. Formulation types are similar to those administered intramuscularly, and also include polymeric implants. The subcutaneous route may also be of interest in immunology and for targeting the lymphatic system e.g. in metastatic disease. Needles are typically 12–16 mm (1/2–5/8″) long, typically shorter than used for IM. Using too long a needle can result in inadvertent intramuscular injection of a SC formulation which may alter absorption kinetics and bioavailability. Doses administrable by SC or IM route are typically limited by the volume of injection and the ability to maintain injectability of the formulation. Highly concentrated solutions can be limited by API solubility, or can quickly become highly viscous, and while suspensions are allowed, the particle size and suspension concentration must be such that it can still pass readily through a needle gauge which is acceptable to the patient population e.g. 23G for a slow release suspension administered less frequently, but smaller e.g. 27G is preferred for daily injections. The addition of recombinant human hyaluronidase PH20 (rHuPH20) as a permeation enhancer has enabled SC dose volumes of 5 mL or larger in several commercial products [4, 5].

Intramuscular (IM) injections are administered into muscles around the body, usually the deltoid (upper arm), quadricep (thigh) or gluteal (buttock) muscles. They generally require administration by a trained healthcare professional. Injection volume varies according to the size of the muscle into which it is injected, with the gluteal muscle being the largest and accommodating up to 5 mL. Formulations are generally aqueous or oily solutions, suspensions, emulsions or depot systems. Needles for IM injections are typically 20–25 gauge with a needle length of 16–38 mm (1–2 in.) depending on site of injection and patient characteristics such as body mass index. As well as immediate release, the IM route also provides a means for prolonged release of drugs formulated as aqueous or oily solutions or suspensions. Where compound properties dictate, the IM route may be preferable to achieve the desired PK profile for certain drugs. For instance, administration of epinephrine via the IM route results in faster and higher peak plasma concentrations compared with SC administration which is essential in the treatment of life-threatening indications such as anaphylaxis. Drugs commonly injected by IM administration include lidocaine, cephalosporins, aminoglycosides, diazepam, phenytoin, insoluble salts of penicillin G (procaine penicillin G), corticosteroids, narcotics, narcotic antagonists and contraceptive steroids.

Key characteristics of the various administration routes are summarised in Table 15.1.

Understanding the physiological, and drug/formulation/device, factors affecting the rate and extent of absorption from these injection sites can help design formulations to achieve the desired therapeutic profile for maximum patient benefit.

Preclinically, SC or IM injections can be useful in exploring PK/PD for compounds which perhaps lack appropriate pharmaceutical properties for oral absorption (poor solubility or permeability, GI instability or extensive first pass metabolism). They can also be used to achieve slow release, thus facilitating early efficacy studies for compounds with undesirable PK properties e.g. rapid clearance that would result in multiple daily administrations to maintain plasma concentrations above the target level for the required duration. Occasionally, SC or IM injections can be used in preclinical toxicology studies as a 'top up' to the oral dose to enable higher systemic concentrations to be reached. Maximum injection volumes for each route in common laboratory species are presented in Table 15.2. Caution is required when translating rate and extent of absorption of SC or IM administered formulations from animals to humans – this is discussed further in the biorelevant models section.

Table 15.1 *Typical characteristics of common parenteral injection types.*

Injection route	Max injection volume per site (mL)	Formulation options	Suitable for self-administration?	Slow release achievable?
Intravenous bolus	5	Solution	No	Indirectly – by prodrug, complexation, nanomedicine
Intravenous Infusion	200	Solution	No	Yes – by slow infusion (long infusions limit clinical acceptability) or prodrug/nanomedicine
Subcutaneous	1.5	Solution, suspension, implant	Yes	Yes – oily depots, polymeric microspheres, *in situ* forming depots, implants
Intramuscular	5	Solution, suspension	No	Yes – oily depots, polymeric microspheres

Table 15.2 *IM and SC injection volumes in preclinical species.*

Species	Dose volume	IM (mL/kg)	SC (mL/kg)
Mouse	Ideal	0.05 total mL/site (2 sites/day)	5
	Maximum	0.1 total mL/site (2 sites/day)	20 (divided in 2–3 sites)
Rat	Ideal	0.1 total mL/site (2 sites/day)	5
	Maximum	0.2 total mL/site (2 sites/day)	10 (divided in 2–3 sites)
Dog	Ideal	0.25	1
	Maximum	0.5 (max 3 mL limit)	2 (divided in 2–3 sites)
Macaque	Ideal	0.25	1
	Maximum	0.5 (max 2 mL limit)	2 (divided in 2–3 sites)

Source: Adapted from Recommended Dose Volumes for Common Laboratory Animals, IQ 3Rs Leadership Group – Contract Research Organization Working Group.

15.2 Subcutaneous Physiology and Absorption Mechanisms

15.2.1 Physiology

Subcutaneous injection targets the fibrous hyaluronic acid (HA)/collagen matrix within the subcutaneous tissue, which is bathed in interstitial fluid (ISF). ISF is an ultrafiltrate of the plasma. It has the same pH (pH 7.4) and electrolyte composition. It is however acellular and has a lower protein concentration (albumin concentration is ~7.36 g/L, only ~15% of that in plasma). Bicarbonate concentration is typically maintained at 25 mM. Table 15.3 describes the typical electrolyte composition. As well as connective tissue, the subcutaneous layer also contains adipocytes (fat cells) organised as lobules, which can act as a reservoir for lipophilic compounds. These adipocytes are surrounded by the extracellular matrix

Table 15.3　*Composition of ISF.*

Cation	Concentration (Eq/L)	Anion	Concentration (Eq/L)
Na^+	0.137	Cl^-	0.111
K^+	0.003	HCO_3^-	0.031
Mg^{2+}	0.002	SO_4^{2-}	0.001
Ca^{2+}	0.001	CO_3^{2-}	0.000045
Total cations	**0.143**	**Total anions**	**0.143**

Source: From Kinnunen and Mrsny [6] / With permission of Elsevier.

(ECM). The temperature in subcutaneous tissue is slightly lower than core body temperature, at ~34 °C in humans. The subcutaneous tissue is perfused with a network of blood capillaries.

15.2.2　Absorption Mechanisms

Following SC administration of an aqueous solution of API, small molecules, peptides and small proteins primarily diffuse through the ECM and through pores in the blood capillary walls. Blood capillaries empty directly into the venous circulation; thus, subcutaneously administered drugs are not subject to first-pass clearance in the liver. There is also a network of lymphatic capillaries, which collect ISF and return it, via lymphatic ducts, through lymph nodes before emptying into the systemic circulation. The pores in the lymphatic capillaries are much larger (compared to blood capillary pores); hence, the lymphatic route is the predominant route of absorption for larger molecules [7]. Convective flow of ISF from capillaries to lymph vessels also helps transport API within the ISF.

Blood capillaries are non-fenestrated and non-sinusoidal but with loose junctions with a pore size of approximately 5 nm [8]. Rate of diffusion through the extracellular matrix is affected by API hydrodynamic radius (as in Ficks law of diffusion; see Chapter 6). pK_a (or isoelectric point for peptides) and the charge state of the molecule at physiological pH (pH 7.4 in ISF) affect potential interactions with charged components of the ECM (e.g. negatively charged hyaluronic acid and positively charged collagen). A molecule's lipophilicity will influence its propensity to partition into local adipocytes which can act as a reservoir and slow the absorption of such compounds. Compounds may precipitate or aggregate at the injection site as it transitions from the formulation to the aqueous environment of the extracellular matrix. Redissolution of the resulting form is then required before absorption can take place, and may retard absorption, or can be intentionally employed to achieve sustained release (e.g. insulin glargine).

Local proteases and other enzymes, macrophages and dendritic cells can contribute to degradation of API at the injection site and during passage through the lymphatic system, which can result in incomplete bioavailability. Differences in blood and capillary lymphatic density and fluid fluxes can contribute to differences in absorption rate/extent between injection sites and in different pathological states. Movement, or external factors such as heat and message, can increase blood and lymphatic flow rates at the injection site which may increase absorption rate.

Formulation approaches can be used to modify absorption rate to achieve the target PK profile. For example, polymeric microparticles may be injected subcutaneously, whereby

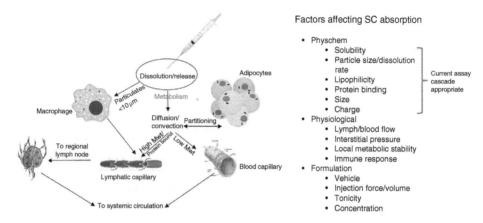

Figure 15.2 *Mechanisms and factors affecting SC absorption.*

drug release rate can be retarded over a period of days to months and can be controlled by diffusion through the polymer matrix, water-mediated transport processes and/or polymer hydrolysis and erosion. Processes involved in SC absorption are illustrated in Figure 15.2. A foreign body response to depot formulations injected subcutaneously may be observed – please see below in IM section.

Complete absorption is achievable for certain compounds administered subcutaneously, for example testosterone, which is predictable for a highly soluble steroid administered parenterally to reach the systemic circulation without first-pass hepatic clearance [9]. However, it is not uncommon to observe incomplete absorption following SC injection. The exact mechanism for such losses is not always understood or explained, and translation of SC bioavailability from animal to human is often poor. First-pass catabolism at the subcutaneous administration site or in the draining lymphatics may contribute to reduced bioavailability.

Additional processes are involved in the absorption of subcutaneously administered monoclonal antibodies (mAbs). In addition to factors common to other molecules, binding to FcRn is potentially important, as it can protect against degradation at the injection site and provide an additional uptake mechanism for mAbs in the form of FcRn-mediated transcytosis from the interstitium to the blood.

15.3 Intramuscular Physiology and Absorption Mechanisms

15.3.1 Physiology

An IM injection is an injection directly into the body of a relaxed skeletal muscle. It is highly compartmentalised and each component is considered as a separate entity (such as bicep muscle). These individual muscles are composed of single cells or fibres embedded in a matrix of collagen. Each fibre contains string-like structures called myofilaments that sit within the sarcoplasm, a fluid similar to cytoplasm which is held in by the fibre's membrane (sarcolemma). The tendon connects the muscle to the bone at either end of the muscle component. The pH of the normal resting muscle **cell is 5.99,** which was determined by direct measurements of intracellular pH [10].

Various muscle sites are available for delivery, including the gluteal, deltoid, triceps, pectoral and vastus lateralis muscles. For rapid absorption and small volumes (<2 mL), the deltoid muscle is preferred, as some studies suggest that blood flow in the deltoid muscle is 7% greater than that of the vastus lateralis and 17% greater than that of the gluteus maximus [11]. In infants and small children, the vastus lateralis of the thigh is often preferred because it is better developed than other muscle groups. The main precaution with IM is to avoid entering a blood vessel (especially an artery), which might lead to infusion of a toxic agent or a toxic vehicle directly to an organ or tissue. This can be prevented usually by pulling back on the plunger of the syringe; if blood does not appear, the needle is probably not in a vessel.

15.3.2 Absorption Mechanisms

Drug absorption from long-acting injectable suspensions of practically water-insoluble drugs is generally considered to be dissolution rate limited and, therefore, driven by the physical properties of the nano-/microparticles and the depot being formed upon IM or SC injection. Nevertheless, the drug release and absorption might be influenced by several other factors such as the agglomeration or the local spreading of the drug particles, the particle size and surface properties (e.g. the host protein adsorption) that dictate the phagocytosis efficiency, the susceptibility to the local enzymatic (pro)drug degradation and lymphatic uptake, all of which can be modulated by the local inflammatory response.

Darville and colleagues investigated these challenges associated with the parenteral administration of long-acting injectables (LAI), especially relating to the complex *in vivo* disposition and pharmacokinetics [12]. Although all LAI nano-/microsuspension constituents are generally well tolerated and biodegradable, the particles are recognized as nonself, leading to local inflammatory reactions.

It is important to have interim assessment ahead of significant degradation of injected material to evaluate host response. This is most relevant for slowly dissolving injectables since the local tissue response to IM injections is a possible determinant of the drug dissolution and pharmacokinetics [13]. These include granuloma formation, extensive intracellular accumulation of formulation by phagocytosis, the depot vascularization, the formation of dense fibrous capsule around bioerodible implants, the active processing by inflammatory components have all been hypothesized to be drug absorption rate limiting [14, 15]. Finally, the physical behaviour of the formulation (e.g. the formulation spreading within the connective tissues, particle agglomeration, etc.) also plays an important role in the *in vivo* drug release and absorption [16].

In summary, a detailed chronological characterization of the nature and dynamics of the host responses caused by the IM injection and the formulation characteristics potentially influence the drug absorption rate *in vivo*.

Future mechanistic *in vivo* studies, supported by *in silico* simulations, are necessary to elucidate the impact of the complex tissue dynamics on the formulation behaviour and drug release/absorption and vice versa.

15.4 *In Vitro* Performance and IVIVC

It is acknowledged that *in vitro* drug release testing for parenterals that are non-solutions could benefit from more regulatory guidance and compendial information [17]. Method selection can be challenging for many reasons, and each formulation carries its own unique considerations. The *in vitro* test conditions should mimic specific aspects of the intended physiological environment, such as the osmolarity, pH or buffer capacity. Although desirable, *in vitro* and *in vivo* correlations may not be possible for modified-release parenteral dosage forms due to the complexity of the release mechanisms and the lack of knowledge about *in vivo* release conditions. Lack of compendial *in vitro* release methods for complex parenterals has created a development challenge for *in vivo* performance [18]. Standard sample-and-separate [19–21], membrane dialysis [22, 23] and flow through [24, 25] have been utilized to determine *in vitro* drug release characteristics and to develop IVIVCs. Each method has their apparent benefits to detect different formulation profiles. For example, the sample-and-separate method captures drug release from PLGA microspheres in muscular tissues, thus giving a better correlation to IM route rather than the SC route [19]. Meanwhile, the membrane dialysis and flow through (USP apparatus 4) methods have more complex apparatus setup and are appear to better mimic *in vivo* drug release conditions of parenteral microspheres/implants where formulation is exposed to a limited volume of release media at a time. The USP apparatus 4 method has also demonstrated better discrimination against compositionally equivalent microsphere formulations with manufacturing differences while capturing the initial burst release phase.

Different mathematical models such as Higuchi, Loo-Riegelman and Weibull have often been used to help understand drug release mechanisms and guide the establishment of IVIVC for slow release formulations if required by introducing a parameter that represents the degree of dissolution [18, 26]. The challenge with complex microspheres or implant formulations is to capture the multi-phasic release characteristics with simple mathematical model. Another consideration is the relationship between accelerated and real-time *in vitro* release data which could follow the same release mechanism with a one-to-one correlation between the release profiles [18]. However, drug mechanism(s) may change since accelerated release tests are typically performed under extreme conditions (e.g. high temperatures, extreme pH conditions). Risperdal® Consta® achieved this by developing an IVIVC based on accelerated testing for PLGA microspheres [24]. Burgess' group has claimed achieving IVIVC using the USP apparatus 4 method to discriminate PLGA microspheres that are equivalent in formulation composition but with manufacturing differences and predicting their *in vivo* performance in the rabbit model with risperidone and naltrexone [27, 28].

The limitations of current IVIVC model which have been discussed include insufficient sampling time, overprediction of C_{max}, slower *in vitro* tests and differences between *in vivo* release across species. Hence, to achieve a successful IVIVC requires reiterative steps and multiple target optimizations.

More recently, an *in vitro* instrument for assessment of SC mAbs has been developed, termed the Scissor model, which claims to mimic solubility, supersaturation, aggregation, diffusion and pH within the SC space. Through the diffusion of mAb from the Scissor system injection cartridge into a large volume physiological buffer, the model emulates passive diffusion from the injection site into the systemic circulation [27–29]. In this model, parameters such as protein charge at neutral pH, pI, viscosity and mAb concentration were found to influence mAb movement. While the Scissor system is not intended to reproduce the entirety or complexity of events that occur at the SC injection site, the data generated provided a reasonable prediction of human percent bioavailability for a test set of eight mAbs [29].

15.4.1   *In Silico* Models

In silico models for prediction of rate and extent of absorption for subcutaneously injected formulations are still in their infancy relative to models for oral drug absorption. Bottom-up, mechanistic *in silico* physiologically based biopharmaceutics models (PBBMs), are in development by two leading commercial absorption software providers: Simulations Plus and Simcyp [30]. Models are being developed to simulate SC absorption of small molecules, peptides/proteins and mAbs. In the Simcyp model, bioavailability must be empirically derived (i.e., entered as user input), with the only drug-specific parameter being hydrodynamic radius. Other empirical and semi-mechanistic models have been developed with varying degrees of success [31].

15.4.2   Preclinical Models

Translation of SC absorption kinetics from preclinical species to human is acknowledged as challenging. Body temperatures are subtly different between species, while extracellular fluid composition is similar. Lymphatic and vascular uptake can proceed at markedly different rates between species and between injection sites [32]. Correlation between preclinical and human bioavailability for mAbs has also been shown to be poor [33]. Subcutaneous connective tissue in animals (except pigs) is less fibrous so can accommodate relatively larger injected volumes. The Panniculus carnosus muscle (subdermal striated muscle for twitching skin) affects studies of injection mechanics. For this reason, minipig is often regarded as the best preclinical model for injectability testing, but care should be taken after tissue is excised since its biomechanical properties change rapidly. Rats, cynomolgus monkey and minipig are most often used for preclinical assessment of absorption kinetics from subcutaneous and intramuscular formulations.

15.5   Bioequivalence of Injectable Formulations

In the case of other parenteral routes, e.g. intramuscular or subcutaneous, and when the test product is a solution that contains the same concentration of the same active substance and the same excipients in similar amounts as the medicinal product currently approved, bioequivalence studies are not required. Moreover, a bioequivalence study is not required for an aqueous parenteral solution with comparable excipients in similar amounts, if it can be demonstrated that the excipients have no impact on the viscosity.

   For suspensions or complexes or any kind of matrix intended to delay or prolong the release of the active substance for IM or SC administration, demonstration of bioequivalence is generally required. Currently, the most relevant guidance available relates to oral/transdermal formulations.

   The complex nature of parenteral nanomedicines means that minute variations in the manufacturing process can substantially change the composition and performance of final products, and this poses a challenge for the development of regulatory guidelines. The establishment of equivalence – pharmaceutical equivalence and/or bioequivalence – is recognised as a major challenge [34].

15.6 Summary

The subcutaneous and intramuscular routes of administration represent attractive routes for a number of reasons: improved bioavailability of non-orally available compounds, improved concordance by reducing dose frequency and improved tolerability through minimising peak:trough concentration ratios. The myriad factors influencing selection of preferred administration route have been recently reviewed [35]. Safety, efficacy, patient preference and pharmacoeconomics are four principles governing the choice of injection route.

Understanding the factors affecting absorption and formulation strategies to tailor the PK profile is developing but is acknowledged as challenging. The complex interplay of processes makes the development of biorelevant *in vitro* and *in silico* models difficult. This has been acknowledged, and concerted efforts between regulatory authorities and academia/industry to improve available models and guidance are being made [17, 33], to enable exploitation of these routes of administration for optimal delivery of new medicines to patients.

References

[1] Buggins, T.R., Dickinson, P.A., and Taylor, G. (2007). The effects of pharmaceutical excipients on drug disposition. *Adv. Drug Deliv. Rev.* **59**: 1482–1503.

[2] Crommelin, D.J.A. and de Vlieger, J.S.B. (2015). *Non-Biological Complex Drugs: The Science and the Regulatory Landscape*. Cham: Springer.

[3] Zheng, N., Sun, D.D., and Zou, P. (2017). Scientific and regulatory considerations for generic complex drug products containing nanomaterials. *AAPS J.* **19**: 619–631.

[4] Frost, G.I. (2007). Recombinant human hyaluronidase (rHuPH20): an enabling platform for subcutaneous drug and fluid administration. *Expert. Opin Drug Deliv.* **4** (4): 427–440.

[5] Locke, K.W., Maneval, D.C., and LaBarre, M.J. (2019). ENHANZE(®) drug delivery technology: a novel approach to subcutaneous administration using recombinant human hyaluronidase PH20. *Drug Deliv.* **26** (1): 98–106.

[6] Kinnunen, H.M. and Mrsny, R.J. (2014). Improving the outcomes of biopharmaceutical delivery via the subcutaneous route by understanding the chemical, physical and physiological properties of the subcutaneous injection site. *J. Control. Release* **182**: 22–32.

[7] Porter, C.J. and Charman, W.N. (1997). Uptake of drugs into the intestinal lymphatics after oral administration. *Adv. Drug Deliv. Rev.* **25** (1): 71–89.

[8] Sarin, H. (2010). Physiologic upper limits of pore size of different blood capillary types and another perspective on the dual pore theory of microvascular permeability. *J. Angiogenes. Res.* **2**: 14. https://doi.org/10.1186/2040-2384-2-14.

[9] Bhasin, S. (1996). *Pharmacology, Biology, and Clinical Applications of Androgens: Current Status and Future Prospects*, 462. Wiley.

[10] Carter, N.W., Rector, F.C. Jr., Campion, D.S., and Seldin, D.W. (1967). Measurement of intracellular pH of skeletal muscle with pH-sensitive glass microelectrodes. *J. Clin. Investig.* **46** (6): 920–933.

[11] Duma, R.A. and Akers, M.J. (1984). *Parenteral Drug Administration, Routes, Precaution, Problems, and Complications*, vol. **1**. New York and Basel: Marcel Dekker.

[12] Darville, N., van Heerden, M., Erkens, T. et al. (2016). Modelling the time course of the tissue responses to intramuscular long-acting. *Toxicol. Pathol.* **44** (2): 189–210.

[13] van't Klooster, G., Hoeben, E., Borghys, H. et al. (2010). Pharmacokinetics and disposition of rilpivirine (TMC278) nanosuspension as a long-acting injectable antiretroviral formulation. *Antimicrob. Agents Chemother.* **54**: 2042–2050.

[14] Darville, N., van Heerden, M., Vynckier, A. et al. (2014). Intramuscular administration of paliperidone palmitate extended-release injectable microsuspension induces a subclinical inflammatory reaction modulating the pharmacokinetics in rats. *J. Pharm. Sci.* **103**: 2072–2087.

[15] Anderson, F.D., Archer, D.F., Harman, S.M. et al. (1993). Tissue response to bioerodible, subcutaneous drug implants: a possible determinant of drug absorption kinetics. *Pharm. Res.* **10**: 369–380.

[16] Zuidema, J., Kadir, F., Titulaer, H.A.C., and Oussoren, C. (1994). Release and absorption rates of intramuscularly and subcutaneously injected pharmaceuticals (II). *Int. J. Pharm.* **105**: 189–207.

[17] Gray, V., Cady, S., Curran, D. et al. (2018). *in vitro* release test methods for drug formulations for parenteral applications. *Dissol. Technol.* **25**: 8–13.

[18] Shen, J. and Burgess, D.J. (2015). *in vitro-in vivo* correlation for complex non-oral drug products: where do we stand? *J. Control. Release* **219**: 644–651.

[19] Chu, D.F., Fu, X.Q., Liu, W.H. et al. (2006). Pharmacokinetics and *in vitro* and *in vivo* correlation of huperzine A loaded poly(lactic-co-glycolic acid) microspheres in dogs. *Int. J. Pharm.* **325**: 116–123.

[20] Blanco-Prieto, M.J., Campanero, M.A., Besseghir, K. et al. (2004). Importance of single or blended polymer types for controlled *in vitro* release and plasma levels of a somatostatin analogue entrapped in PLA/PLGA microspheres. *J. Control. Release* **96**: 437–448.

[21] Negrn, C.M., Delgado, A., Llabres, M., and Evora, C. (2001). In vivo-*in vitro* study of biodegradable methadone delivery systems. *Biomaterials* **22**: 563–570.

[22] D'Souza, S., Faraj, J.A., Giovagnoli, S., and Deluca, P.P. (2014). IVIVC from long acting olanzapine microspheres. *Int. J. Biomater.* **2014**: 1–11.

[23] D'Souza, S., Faraj, J.A., Giovagnoli, S., and Deluca, P.P. (2014). *in vitro-in vivo* correlation from lactide-co-glycolide polymeric dosage forms. *Prog. Biomater.* **3**: 131–142.

[24] Rawat, A., Bhardwaj, U., and Burgess, D.J. (2012). Comparison of *in vitro-in vivo* release of risperdal (R) consta(R) microspheres. *Int. J. Pharm.* **434**: 115–121.

[25] Zolnik, B.S. and Burgess, D.J. (2008). Evaluation of *in vivo-in vitro* release of dexamethasone from PLGA microspheres. *J. Control. Release* **127**: 137–145.

[26] Schliecker, G., Schmidt, C., Fuchs, S. et al. (2004). *J. Control. Release* **94**: 25–37.

[27] Shen, J., Choi, S., Qu, W. et al. (2015). *in vitro*-in vito correlation of parenteral risperidone polymeric microspheres. *J. Control. Release* **218**: 2–12.

[28]] Andhariya, J.V., Shen, J., Choi, S. et al. (2017). Development of in vitro-*in vivo* correlation of parenteral naltrexone loaded polymeric microspheres. *J. Control. Release* **255**: 27–35.

[29] Bown, H.K., Bonn, C., Yohe, S. et al. (2018). *in vitro* model for predicting bioavailability of subcutaneously injected monoclonal antibodies. *J. Control. Release* **273**: 13–20.

[30] Gill, K.L., Gardner, I., Li, L., and Jamei, M. (2016). A bottom-up whole-body physiologically based pharmacokinetic model to mechanistically predict tissue distribution and the rate of subcutaneous absorption of therapeutic proteins. *AAPS J.* **18**: 156–170.

[31] Kagan, L. (2014). Pharmacokinetic modeling of the subcutaneous absorption of therapeutic proteins. *Drug Metab. Dispos.* **42** (11): 1890.

[32] Martinez, M.N. (2011). Factors influencing the use and interpretation of animal models in the development of parenteral drug delivery systems. *AAPS J.* **13**: 632–649.

[33] Sánchez-Félix, M., Burke, M., Chen, H.H. et al. (2020). Predicting bioavailability of monoclonal antibodies after subcutaneous administration: open innovation challenge. *Adv. Drug Deliv. Rev.* **167**: 66–77. https://doi.org/10.1016/j.addr.2020.05.009.

[34] de Vlieger, J.S.B., Crommelin, D.J.A., Flühmann, B. et al. (2019). A progress report on the 3rd international symposium on scientific and regulatory advances in biological and non-biological complex drugs: A to Z in bioequivalence. *GaBi J.* **8**: 128–131.

[35] Jin, J.-F., Zhu, L.-L., Chen, M. et al. (2015). The optimal choice of medication administration route regarding intravenous, intramuscular, and subcutaneous injection. *Pat. Pref. Adher.* **9**: 923–942.

16

Biopharmaceutics of Topical and Transdermal Formulations

Hannah Batchelor

Strathclyde Institute of Pharmacy and Biomedical Sciences,
University of Strathclyde, Glasgow, United Kingdom

16.1 Introduction

Topical pharmaceutical products are usually liquids or semisolids, whereas transdermal products are often formulated as patches that deliver drugs to or through the skin to provide therapy. Topical products are often used for local treatment; there is also potential for systemic uptake from transdermal products where oral therapy may not be desired or where prolonged exposure is required. This chapter will outline the key factors that control exposure from topically and administered products where the skin is the barrier to uptake.

The pharmacokinetic response to a topical/transdermal formulation depends on the drug release, then penetration of the drug through the skin then distribution to the site of action. Topically applied formulations are retained in the upper layers of skin and offer localized therapy, whereas transdermal formulations access the deeper layers of skin (richly supplied with blood vessels) offer systemic therapeutics. The relative efficiency of these processes will determine the exposure of the drug at the site of action, hence the biopharmaceutics.

Key biopharmaceutics parameters associated with topical products include the following:

- The distribution of the drug in the dosage form
- Drug release from the dosage form

- Drug partitioning from the dosage form into the skin, also termed drug penetration
- Drug diffusion within the skin itself; termed drug permeation not the skin
- Extent of systemic exposure following permeation through the skin

16.2 Skin Structure

Human skin consists of three principal layers: epidermis, dermis and hypodermis, where the epidermis is the outermost layer (Figure 16.1). The exterior layer of the epidermis is the stratum corneum, which is 10–30 μm thick [1] and acts as the major barrier of the skin.

The stratum corneum is comprised of a nonviable layer of dead, flattened, keratin-rich cells, known as corneocytes. These corneocytes are surrounded by proteins and lipids where the lipids are ceramides, cholesterol and free fatty acids. This results in a two-component system of the corneocytes and an extracellular lipid matrix. It is often depicted as bricks and mortar reflecting the elongated corneocytes being submerged in a lipid matrix. However, this is a dynamic system where lamellar bodies secrete: lipid processing enzymes; proteases and anti-proteases, and antimicrobial peptides; all of which may have

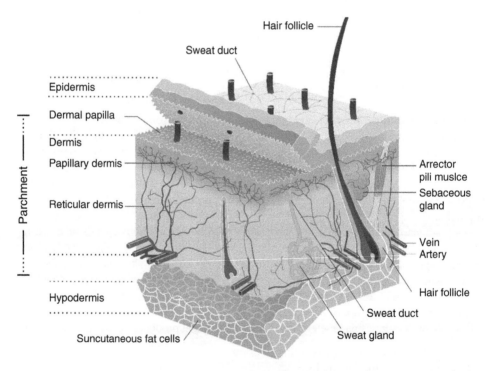

Figure 16.1 *Structure of skin and the layers typically present. Source: Sean P Doherty / Wikimedia Commons / CC BY-SA 4.0, https://commons.wikimedia.org/w/index.php?curid= 90870177.*

an effect on the effective absorption of drugs [2]. The stratum corneum is the major barrier-limiting percutaneous absorption.

The concentration gradient between the formulation and site of action provides the driving force for penetration of drug through the skin, hence a well-designed formulation can maximise this concentration gradient by maximising the solubility within the formulation. Flux measurement across human skin is the most relevant test to perform to understand permeation of a drug.

16.2.1 Transport of Drugs Through Skin

The major route of passive transport is via the intercellular pathway where the drug diffuses through the lipid domains within the stratum corneum to reach the epidermis and dermis. This is often depicted is a tortuous route resembling travelling along the mortar within a brick wall. Transcellular diffusion is also possible where drugs penetrate through the corneocytes within the stratum corneum.

It is also possible for drugs to permeate the skin via the sweat gland pores or hair follicle orifices, which are 40–80 and 50–150 μm in diameter, respectively [3, 4]. However, these pores are sparsely distributed in skin in terms of surface area and are often secondary to inter or transcellular pathways.

16.2.2 Skin Metabolism

The skin contains a wide range of enzymes that can metabolise drugs within the skin. The relative metabolic activity of the skin is much lower than the gut wall or the liver, yet esterase activity within the skin has been reported for a range of drugs [5]. The extent of metabolism within the skin has been reported to vary with anatomical site, sex and age [6].

16.3 Active Pharmaceutical Ingredient Properties

Candidate drugs that are likely to be successful for transdermal delivery should have moderate lipophilicity (log P 2–3), sufficient solubility both in aqueous and lipophilic phases, molecular weight <500 Da, high potency and a low melting point [7]. Formulation strategies can incorporate the use of penetration enhancers or agents that modify the structure of the stratum corneum, alternatively physical methods can also be applied, for example the use of microneedles [8].

16.4 Topical and Transdermal Dosage Forms

A range of products are used for topical administration due to the diverse range of conditions to be treated and the diversity of patient needs. Most products are semi-solid formulations where the drug is dissolved/dispersed wither in a single- or a two-phase system, for example an emulsion. The product will be developed to achieve the target therapy whether this is delivery to the upper surface of the skin or for penetration through the skin. The relative bioavailability of the drug will be influenced by the composition of the formulation where excipients to improve solubility within the formulation will drive drug uptake.

Topical application of a cream or ointment results in a product thickness of not greater than 10 μm; thus, the area to be covered will dictate the dose applied. Transdermal patches range in size with most commercial products being between 5 and 40 cm² [9].Transdermal patches are formulated as reservoir products where either the matrix or a membrane control the rate of release of the drug.

Standard analytical techniques are used to evaluate the distribution of drug in the dosage form. In cases where the drug is in solution this is a simple content test. In products where drug is suspended or in a two-phase system, it can be important to understand the homogeneity within the formulation as the fraction dissolved; size distribution of particles/globules; concentration of drug in the continuous phase; drug polymorphic state and the pH of the formulation. Often a formulation is optimized to maximise drug solubility whilst providing a suitable stability and shelf-life. The range of tests undertaken to evaluate topical dosage forms are reported in relevant pharmacopoeia, and the reader is directed to this paper for full details [10].

16.5   Measurement of *In Vitro* Drug Release

Topical semi-solid formulations can be more complex that solid oral dosage forms as they are often composed of two phases (oil and water). The active is dispersed in both phases and the extent of dispersion depends upon the solubility of the drug in each phase. The rate of drug release can depend upon several factors including active pharmaceutical ingredient (API) particle size; API partition coefficient between the two phases; product rheology and the interfacial tension between the two phases.

Drug release from topical and transdermal products is measured using a modified dissolution apparatus (see Chapter 6 for further details on dissolution).

16.5.1   Diffusion Cells

Diffusion cells, for example Franz cells are commonly used to evaluate drug release. These systems separate a donor compartment that contains the drug formulation from a receiver compartment containing the drug release media via a membrane, an image is shown in Figure 16.2. The vertical diffusion cell can often also be referred to as an *in vitro* release test (IVRT).

Diffusion cells can be vertical as in Figure 16.2; they can also be horizontal, which can reduce the likelihood of air bubbles at the membrane surface or flow through systems.

The diffusion cell is maintained at 32 °C to reflect the normal skin temperature, and this is often achieved using a circulating water jacket. The fluid in the receptor chamber is stirred at a constant rate (normally 600 rpm). Following application of the donor formulation, the fluid in the receptor chamber is collected (typically at one-hour intervals) and analysed to measure the cumulative drug release over time.

In this experiment, it is important to select the most appropriate membrane, stirrer speed and receptor medium. Whereas for oral drug delivery, there are compendial media that represent gastric and small intestinal fluids, and there is no similar fluid for topical

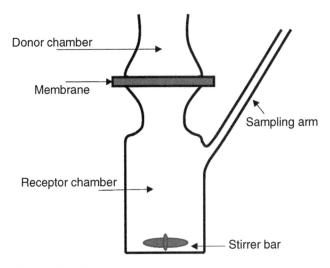

Figure 16.2 *Diagram of a vertical diffusion cell.*

products. The receptor fluid is usually an aqueous buffer at pH 7.4 with surfactant or co-solvents added to ensure adequate solubility of the drug of interest.

When conducting a release test, the membrane selected should offer the least possible resistance as the test is designed to measure the impact of the formulation on the rate of release rather than the permeation through a membrane. Appropriate inert and commercially available synthetic membranes include polysulfone, cellulose acetate/nitrate mixed ester or polytetrafluoroethylene membrane of appropriate size to fit the diffusion cell diameter [11]. It should be noted that the choice of membrane can affect the rate of release [12]. The solvent in the receptor chamber can also affect release rate; the more soluble the drug in the receptor fluid the faster it will traverse the membrane [13].

The rate of release is usually normalised to the mass released per unit surface area such that the size of the orifice between the donor and receptor is normalised across apparatus. Where the formulation controls the rate of drug release the release will be proportional to the square root of time ($\sqrt{t}$)-based Higuchi kinetics [14].

Efforts are usually focused on the development of a discriminatory test where the membrane and receptor fluid are selected to detect physiochemical differences in topical formulations; this enables optimisation of a given formulation [15].

16.5.2 Compendial Dissolution Apparatus

Compendial dissolution apparatus was described in Chapter 6 with specific sections on topical and transdermal tests in Sections 6.4.4–6.4.6.

16.6 Measurement of Skin Permeation

A recent review provides a detailed description of a range of methods to evaluate skin permeation, and the reader is directed to this review for further information [16].

16.6.1   Tape-Stripping 'Dermatopharmacokinetics' (DPK)

The 'cutaneous stress test' depletes the stratum corneum of its complement of lipids using any type of acute barrier disruption, for example application of an organic solvent or physical disruption using tape stripping [17]. This action removes the extracellular lipid matrix and reduces the barrier function of the skin.

Tape stripping is also used as a tool to measure the concentration of drug in the stratum corneum over time, often also termed dermatopharmacokinetic (DPK) methods. At a given time, tape is applied to the skin, then stripped away; the concentration of drug that is measured in the stripped stratum corneum can demonstrate the permeation over time [18, 19]. Tape stripping is associated with variability, and there is need to evaluate multiple skin sites; although this was previously a method of choice it has fallen from favour in recent times, being removed as a test in FDA guidance in 2002 [20].

16.6.2   Confocal Laser Scanning Microscopy (CLSM)

Confocal laser scanning microscopy can be used as a method for non-invasive imaging of skin superficial layers, which can provide insights into the dynamic processes associated with skin permeation. This can be used to assess drug or vehicle permeation in the skin that provides feedback to optimize delivery systems [21].

16.6.3   Diffusion Cells Using Biorelevant Membranes to Model Permeation

A diffusion cell can be used to measure permeation of skin or skin mimetics, where the simple support membrane used to separate the donor from the receptor chambers is replaced with a membrane that replicates the properties of skin, or uses skin itself. The apparatus is the same as that shown in Figure 16.2, this can also be termed an *in vitro* permeation test.

The use of human skin within a diffusion cell is most frequently limited by availability. There are also limitations associated with the high inter- and intra-individual variability associated with human skin [22], particularly in relation to gender, age, race, and anatomical site. Human skin can be obtained from cadavers or surgery; it may be further processed to use full thickness skin; dermatomed skin; or to separate the stratum corneum or epidermis [23].

Analysis of skin permeation follows the same procedure as for drug release in a diffusion cell; however, it is often of interest to quantify the drug concentration within the skin substrate. The entire skin can be homogenized to measure drug concentration or the skin can be separated into layers to better understand uptake of the drug [24].

16.6.3.1   *Alternative Skin Substrates Used for Permeability Studies*

Ex vivo animal skin has been used as an alternative to human skin. Due to the difference in hair coverage between animals and humans, porcine ear skin and hairless mouse skin are the most widely used [22].

The histology of porcine ear skin is similar to human skin. It has a similar stratum corneum thickness and hair-follicle density. However, the lipidic organisation within the stratum corneum differs with a hexagonal lattice in porcine tissue and an orthorhombic lattice in humans [22, 25]. Porcine skin is usually harvested from slaughterhouses where the material is a by-product of the food industry, ethical approval for use of tissue from a deceased animal is not required.

Hairless mice can be advantageous as in centres where animal research is ongoing, they can be more readily available; however, there is a need for ethical approval for their use. The permeation rate through mouse skin is slower than in humans, which is a further disadvantage [26].

Synthetic membranes have been developed that mimic skin yet avoid the inherent variability associated with human and animal skin. The most established synthetic membrane is Strat-M™ (Merck Millipore, Burlington, Massachusetts, USA). Strat-M™ membrane has multiple layers to mimic human skin. Several studies have explored correlations between permeability in human skin and Strat-M™ membranes [27, 28].

The PAMPA membrane for gastrointestinal permeability studies is well established, and the skin-PAMPA was developed by the addition of synthetic certramides, cholesterol and stearic acid to mimic the components of human skin [29]. Another synthetic option uses a layer of liposomes immobilized on a filter as an artificial skin membrane [30]. Artificial membranes have demonstrated discriminatory power [31] and offer an alternative to human skin for formulation development and optimisation.

Tissue-culture-derived skin equivalents are made from human cells grown from tissue culture, and they can mimic the substructures within the skin [32]. These systems are generally more permeable than skin due to their structure [33]. There are multiple commercially available three-dimensional skin models that are used in dermal toxicity assessment [34].

16.6.4   Dermal Microdialysis

Dermal microdialysis enables *in situ* measurement of drug concentration in the dermis using an extremely thin dialysis probe that is located within the dermal layer [35]. The probe is perfused with a compatible buffered fluid such that it provides a site for drug exchange within the dermal layer and thus to detect drug levels to demonstrate permeation [36]. This method offers significant benefits over *in vitro* methods as it provides real-time information on absorption, as well as assessment of the impact of disease on permeation [37]. However, there remains a need to use *in vitro* diffusion testing to better understand formulation attributes that may influence the absorption of the drug.

16.6.5   Skin Biopsy

A skin biopsy enables quantification of drug levels based on the depth of penetration into the skin; however, this is an invasive technique that may leave a permanent scar in human volunteer studies. There is also a risk of contamination within the biopsy as drug at surface layers is carried into deeper layers with the tool used to cut the skin.

16.6.6   *In Silico* Models of Dermal Absorption

There have been efforts to mathematically model skin absorption and permeation with a detailed review published in 2003 [38]. The model needs to account for the rate of release from a formulation, as well as diffusion through skin, which is an inhomogeneous substrate. Commercial software is available, which provides a mechanistic physiologically based model of dermal absorption [39].

The permeation coefficient (K_p) of a drug is related to the relative partitioning of the drug between the stratum corneum and the formulated product $(P_{sc/w})$, as well as the diffusivity of the drug into the stratum corneum (D_{sc}) and the thickness of the stratum corneum layer (h_{sc}). When developing a formulation, optimization of the partition to drive drug into the stratum corneum is critical. This is shown in Eq. (16.1).

$$K_p = \frac{P_{sc/w} * D_{sc}}{h_{sc}} \text{ (cm/h)} \tag{16.1}$$

The diffusivity into the stratum corneum (D_{sc}) is a measure of how quickly a drug diffuses through the stratum corneum and is related to the drug properties itself and the mechanism by which it penetrates the skin.

The overall uptake of a drug into the skin is a function of the permeation coefficient (K_p); the concentration of the drug at the absorption surface; the contact time and the contact area as shown in Eq. (16.2).

$$\text{Dermal uptake} = K_p \times [\text{drug}] \times \text{contact time} \times \text{contact area} \tag{16.2}$$

Thus, critical inputs to drive dermal absorption are the concentration of drug at the site of absorption and the permeation coefficient into the stratum corneum.

The MechDermA model is a commercially available *in silico* model that is multi-phase and multidimensional based on skin physiology. The model incorporates parameters associated with the API, formulation, physiology and environmental parameters, combining all these aspects to simulate diffusion through the stratum corneum. The model can also simulate drug partitioning and absorption through the hair follicular pathway [40]. A schematic of this model is shown in Figure 16.3. The multilayer structure of this model that replicates the structure of skin accounts for the longitudinal diffusion and distribution processes in relation to skin physiology (i.e. tortuosity of the diffusion pathway, keratin adsorption kinetics, stratum corneum (SC) hydration state, hair follicular transport, pH at the skin surface and within the SC layer) [41].

The MechDermA model incorporates aspects of skin physiology, including skin layer thickness, lipid contents, blood flow rates, hydration state and other parameters. This mechanistic information can enable predictions of absorption in populations where these parameters may change, for example the elderly [41].

16.6.7   Pre-Clinical Models

Animal models are used to evaluate the dermal absorption of drugs from topical products. The major difference between animals and humans is the hair density, and this can limit the choice of species used or the site of administration for a topical product. A detailed review on a range of animal species is available for further reading [42]. In general, absorption in more common laboratory animals (rat or rabbit) is often higher than in humans although hairless variants (hairless mice or hairless guinea pigs) provide more similar data [42].

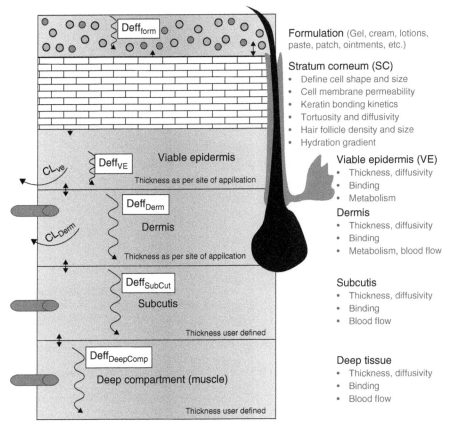

Figure 16.3 *Structure of the MechDermA mechanistic multi-layer dermal absorption model showing the variables that may impact on absorption through each layer of the skin. Source: From Chetty et al [41] / With permission of Elsevier.*

16.7   Bioequivalence Testing of Topical/Transdermal Products

Demonstration of bioequivalence between topical products is required for interchangeability of products or to bridge formulations during product development. There are regulatory guidelines that describe the testing procedures required to demonstrate equivalence between products [43]. There are a range of *in vitro* and *in vivo* tests that can support risk assessments associated with bioequivalence, and these can be useful in the design and optimization of a topical product.

Pharmacokinetic studies for topically applied drugs are limited to those where systemic exposure is quantifiable; in these cases, a maximal usage trial is conducted. A maximum usage study is designed to maximise the potential for drug absorption taking into account the frequency, duration and body surface area for the proposed clinical use [44]. It is recognised that pharmacokinetic analysis is complex due to the typically low concentrations achieved in the systemic circulation, as well as the slow absorption rates [45].

In silico tools that demonstrated virtual bioequivalence were recently used to leverage FDA approval of a generic version of a sodium diclofenac gel, this demonstrates the value of biopharmaceutical tools in product development and approval [46].

16.8 Conclusions

Topical and transdermal administration of medicines provides an alternative route of administration for products that act locally within the skin or are not suitable for oral administration. The extent of absorption is a function of the formulation applied and the barrier function of the skin at the site of administration. An understanding of factors that govern permeation is present, yet there are limited mechanistic *in vitro* methods that mimic the *in vivo* environment.

References

[1] Czekalla, C., Schönborn, K.H., Lademann, J., and Meinke, M.C. (2019). Noninvasive determination of epidermal and stratum corneum thickness *in vivo* using two-photon microscopy and optical coherence tomography: impact of body area, age, and gender. *Skin Pharmacology and Physiology* **32** (3): 142–150.

[2] Elias, P.M. (2005). Stratum corneum defensive functions: an integrated view. *Journal of Investigative Dermatology* **125** (2): 183–200.

[3] Flament, F., Francois, G., Qiu, H. et al. (2015). Facial skin pores: a multiethnic study. *Clinical, Cosmetic and Investigational Dermatology* **8**: 85.

[4] Otberg, N., Richter, H., Schaefer, H. et al. (2004). Variations of hair follicle size and distribution in different body sites. *Journal of Investigative Dermatology* **122** (1): 14–19.

[5] Qian, Z., Jeffrey, E.G., Guangji, W., and Michael, S.R. (2009). Cutaneous metabolism in transdermal drug delivery. *Current Drug Metabolism* **10** (3): 227–235.

[6] Pyo, S.M. and Maibach, H.I. (2019). Skin metabolism: relevance of skin enzymes for rational drug design. *Skin Pharmacology and Physiology* **32** (5): 283–294.

[7] Prausnitz, M.R., Mitragotri, S., and Langer, R. (2004). Current status and future potential of transdermal drug delivery. *Nature Reviews Drug Discovery* **3** (2): 115–124. doi: 10.1038/nrd1304. PMID: 15040576.

[8] Kirkby, M., Hutton, A.R.J., and Donnelly, R.F. (2020). Microneedle mediated transdermal delivery of protein, peptide and antibody based therapeutics: current status and future considerations. *Pharmaceutical Research* **37** (6): 117.

[9] Prausnitz, M.R. and Langer, R. (2008). Transdermal drug delivery. *Nature Biotechnology* **26** (11): 1261–1268.

[10] Ueda, C.T., Shah, V.P., Derdzinski, K. et al. (2009). Topical and transdermal drug products. *Pharmacopeial Forum* **35**: 750–764. https://doi.org/10.14227/DT170410P12.

[11] FDA (1997). Guidance for industry. Nonsterile semisolid dosage forms: scale-up and postapproval changes: chemistry, manufacturing, and controls; *in vitro* release testing and *in vivo* bioequivalence documentation. U.S. Department of Health and Human Services, Food and Drug Administration, Center for Drug Evaluation and Research (CDER). https://www.fda.gov/media/71141/download (accessed February 2021).

[12] Shah, V.P., Elkins, J., Lam, S.-Y., and Skelly, J.P. (1989). Determination of *in vitro* drug release from hydrocortisone creams. *International Journal of Pharmaceutics* **53** (1): 53–59.

[13] Shah, V.P., Elkins, J.S., and Williams, R.L. (1999). Evaluation of the test system used for *in vitro* release of drugs for topical dermatological drug products. *Pharmaceutical Development and Technology* **4** (3): 377–385.

[14] Higuchi, T. (1961). Rate of release of medicaments from ointment bases containing drugs in suspension. *Journal of Pharmaceutical Sciences* **50** (10): 874–875.

[15] Dandamudi, S. (2017). *in vitro* bioequivalence data for a topical product. In: *FDA Workshop on Bioequivalence Testing of Topical Drug Products. Maryland.* https://www.fda.gov/media/110389/ download (accessed May 2021).

[16] Chaturvedi, S. and Garg, A. (2021). An insight of techniques for the assessment of permeation flux across the skin for optimization of topical and transdermal drug delivery systems. *Journal of Drug Delivery Science and Technology* **62**: 102355.

[17] Monach, S. and Blank, H. (1958). Location and re-formation of the epithelial barrier to water vapor. *AMA Archives of Dermatology* **78** (6): 710–714.

[18] Benson, H.A. and Watkinson, A.C. (2012). *Topical and Transdermal Drug Delivery: Principles and Practice.* John Wiley & Sons.

[19] Russell, L.M. and Guy, R.H. (2012). Novel imaging method to quantify stratum corneum in dermatopharmacokinetic studies: proof-of-concept with acyclovir formulations. *Pharmaceutical Research* **29** (9): 3362–3372. https://doi.org/10.1007/s11095-012-0831-4.

[20] Yacobi, A., Shah, V.P., Bashaw, E.D. et al. (2014). Current challenges in bioequivalence, quality, and novel assessment technologies for topical products. *Pharmaceutical Research* **31** (4): 837–846.

[21] Rossetti, F.C., Depieri, L.V., Bentley, M. et al. (2013). Confocal laser scanning microscopy as a tool for the investigation of skin drug delivery systems and diagnosis of skin disorders. In: *Confocal Laser Microscopy – Principles and Applications in Medicine, Biology, and the Food Sciences* (ed. N. Lagali), 99–140. IntechOpen https://doi.org/10.5772/55995.

[22] Praça, F.S.G., Medina, W.S.G., Eloy, J.O. et al. (2018). Evaluation of critical parameters for *in vitro* skin permeation and penetration studies using animal skin models. *European Journal of Pharmaceutical Sciences* **111**: 121–132.

[23] Haigh, J.M. and Smith, E.W. (1994). The selection and use of natural and synthetic membranes for *in vitro* diffusion experiments. *European Journal of Pharmaceutical Sciences* **2** (5): 311–330.

[24] Praça, F.S.G., Bentley, M.V.L.B., Lara, M.G., and Pierre, M.B.R. (2011). Celecoxib determination in different layers of skin by a newly developed and validated HPLC-UV method. *Biomedical Chromatography* **25** (11): 1237–1244.

[25] Debeer, S., Le Luduec, J.-B., Kaiserlian, D. et al. (2013). Comparative histology and immunohistochemistry of porcine versus human skin. *European Journal of Dermatology* **23** (4): 456–466.

[26] Roy, S., Hou, S.Y., Witham, S., and Flynn, G. (1994). Transdermal delivery of narcotic analgesics: comparative metabolism and permeability of human cadaver skin and hairless mouse skin. *Journal of Pharmaceutical Sciences* **83** (12): 1723–1728.

[27] Uchida, T., Kadhum, W.R., Kanai, S. et al. (2015). Prediction of skin permeation by chemical compounds using the artificial membrane, Strat-M™. *European Journal of Pharmaceutical Sciences* **67**: 113–118.

[28] Neupane, R., Boddu, S.H.S., Renukuntla, J. et al. (2020). Alternatives to biological skin in permeation studies: current trends and possibilities. *Pharmaceutics* **12** (2): 152.

[29] Sinkó, B., Garrigues, T.M., Balogh, G.T. et al. (2012). Skin-PAMPA: a new method for fast prediction of skin penetration. *European Journal of Pharmaceutical Sciences* **45** (5): 698–707.

[30] Palac, Z., Engesland, A., Flaten, G.E. et al. (2014). Liposomes for (trans)dermal drug delivery: the skin-PVPA as a novel *in vitro* stratum corneum model in formulation development. *Journal of Liposome Research* **24** (4): 313–322.

[31] Luo, L., Patel, A., Sinko, B. et al. (2016). A comparative study of the *in vitro* permeation of ibuprofen in mammalian skin, the PAMPA model and silicone membrane. *International Journal of Pharmaceutics* **505** (1–2): 14–19.

[32] Netzlaff, F., Lehr, C.-M., Wertz, P., and Schaefer, U. (2005). The human epidermis models EpiSkin®, SkinEthic® and EpiDerm®: an evaluation of morphology and their suitability for testing phototoxicity, irritancy, corrosivity, and substance transport. *European Journal of Pharmaceutics and Biopharmaceutics* **60** (2): 167–178.

[33] Schmook, F.P., Meingassner, J.G., and Billich, A. (2001). Comparison of human skin or epidermis models with human and animal skin in in-vitro percutaneous absorption. *International Journal of Pharmaceutics* **215** (1–2): 51–56.

[34] De Wecer, B., Petersohn, D., and Mewes, K. (2013). Overview of human three-dimensional (3D) skin models used for dermal toxicity assessment. *HPC Today* **8**: 18–22.

[35] Anderson, C., Andersson, T., and Molander, M. (1991). Ethanol absorption across human skin measured by *in vivo* microdialysis technique. *Acta Dermato-Venereologica* **71** (5): 389–393.

[36] Walicka, A. and Iwanowska-Chomiak, B. (2018). Drug diffusion transport through human skin. *International Journal of Applied Mechanics Engineering* **23** (4): 977–988.

[37] Ortiz, P.G., Hansen, S.H., Shah, V.P. et al. (2008). The effect of irritant dermatitis on cutaneous bioavailability of a metronidazole formulation, investigated by microdialysis and dermatophar-macokinetic method. *Contact Dermatitis* **59** (1): 23–30.

[38] Yamashita, F. and Hashida, M. (2003). Mechanistic and empirical modeling of skin permeation of drugs. *Advanced Drug Delivery Reviews* **55** (9): 1185–1199.

[39] Polak, S., Ghobadi, C., Mishra, H. et al. (2012). Prediction of concentration–time profile and its inter-individual variability following the dermal drug absorption. *Journal of Pharmaceutical Sciences* **101** (7): 2584–2595.

[40] Puttrevu, S.K., Arora, S., Polak, S., and Patel, N.K. (2020). Physiologically based pharmacoki-netic modeling of transdermal selegiline and its metabolites for the evaluation of disposition differences between healthy and special populations. *Pharmaceutics* **12** (10): 942.

[41] Chetty, M., Johnson, T.N., Polak, S. et al. (2018). Physiologically based pharmacokinetic mod-elling to guide drug delivery in older people. *Advanced Drug Delivery Reviews* **135**: 85–96.

[42] Jung, E.C. and Maibach, H.I. (2015). Animal models for percutaneous absorption. *Journal of Applied Toxicology* **35** (1): 1–10.

[43] EMA (2018). Draft guideline on quality and equivalence of topical products. https://www.ema.europa.eu/en/documents/scientific-guideline/draft-guideline-quality-equivalence-topical-products_en.pdf (accessed February 2021).

[44] FDA (2019). Maximal usage trials for topically applied active ingredients being considered for inclusion in an over-the - counter monograph: study elements and considerations. https://www.fda.gov/media/125080/download (accessed February 2021).

[45] Bashaw, E.D., Tran, D.C., Shukla, C.G., and Liu, X. (2015). Maximal usage trial: an overview of the design of systemic bioavailability trial for topical dermatological products. *Therapeutic Innovation & Regulatory Science* **49** (1): 108–115.

[46] Certara (2019). Blog post: Certara's Simcyp MechDermA model achieves regulatory approval: demonstrates virtual bioequivalence in dermal drug development. https://www.certara.com/blog/certaras-simcyp-mechderma-model-achieves-regulatory-approval-demonstrates-virtual-bioequivalence-in-dermal-drug-development/ (accessed March 2021).

17

Impact of the Microbiome on Oral Biopharmaceutics

Laura E. McCoubrey[1], Hannah Batchelor[2], Abdul W. Basit[1], Simon Gaisford[1] and Mine Orlu[1]

[1] *Department of Pharmaceutics, UCL School of Pharmacy, University College London, London, United Kingdom*
[2] *Strathclyde Institute of Pharmacy and Biomedical Sciences, University of Strathclyde, Glasgow, United Kingdom*

17.1 Introduction

The human microbiome is composed of the genomes of trillions of microbes inhabiting the human body: namely, bacteria, viruses, fungi and protozoa. The identities and locations of these microbiota have only recently been characterised, with the culmination of the first Human Microbiome Project (HMP) in 2012 [1]. The HMP was a multi-site collaboration spanning several years, utilising genomic sequencing to map the human microbiome for the first time. At the conclusion of the project, the sheer scale of microbiota living on and within humans was elucidated; for example, bacteria alone are known to encode for 100-fold more unique genes than their human host [2]. Subsequently, the intrinsic role of microbiota in human health and metabolism has begun to emerge in a period of intense seminal research [3]. Microbiota are now recognised to colonise almost every surface of the human body, with species generally specialised to suit the niche in which they reside [4]. As such, species of bacteria living within arid crevices of exposed skin are markedly distinct to those colonising the smooth surface of teeth, and again different to those inhabiting the GI tract [5].

The GI tract contains the majority of microbiota, housing over 4500 distinct species of bacteria alone [6]. It is estimated that 93.5% of these intestinal bacteria belong to just four phyla: Proteobacteria, Firmicutes, Actinobacteria and Bacteroidetes [7]. A significant proportion of GI microbiota are anaerobes, dependent on the intestine' low oxygen concentrations for survival. As such, it has been historically challenging to investigate GI microbiota in laboratory settings, though recent advances in genomic sequencing have facilitated significant developments [8]. GI microbiota have now been implicated in a myriad of pathologies; from local diseases such as inflammatory bowel disease (IBD), to systemic illnesses including neurological disorders and cardiovascular disease [9–11]. Disease can result when populations of microbiota become unbalanced, or dysbiotic, leading to a resultant imbalance in microbial metabolites and functions.

The realisation that a dysbiotic microbiome can influence the onset of both local and systemic diseases has transformed the concept of the pharmaceutical–microbiome relationship. In addition to affecting the intestinal environment and the wider patient disease state, it is increasingly recognised that the gut microbiome may impact the biopharmaceutics of orally administered drugs. The microbiome now represents an intermediate, capable of metabolising and altering drug pharmacokinetics to enhance or inhibit clinical response. Microbiome composition is also highly individual and sensitive to external factors, making microbiome–biopharmaceutical implications complex to predict. GI microbiome composition can be affected by many factors, including diet [12], age [13] and drug use, especially antibiotics [14, 15]. Whilst microbiome composition does present substantial inter- and intra-individual variability, broad functional capacity is generally conserved owing to the microbiota's functionally diverse nature [16]. That said, there are circumstances in which the presence and concentration of certain microbial strains can have significant impacts on the pharmacokinetics of drugs [17, 18].

17.2   Microbiome Distribution in the GI Tract

As the conditions across the GI tract vary, so too does the distribution of resident microbiota; an overview is shown in Figure 17.1. Physiological parameters such as availability of nutrients, water and oxygen; luminal transit rate; and host immune activity all affect where microbiota can reside in the GI tract [19]. At the beginning of the GI tract the oral cavity provides several distinct microbial niches, resulting in a complex and physiologically unique ecosystem. Prevalent oral bacterial species include *Streptococcus, Prevotella* and *Fusobacterium,* whilst notable fungal representatives include *Candida* species [20, 21]. Oral microbiota typically function within symbiotic niches or biofilms, exerting substantial impacts on both dental and systemic health [22]. Within the acidic environment of the stomach microbial diversity declines, with just 10 bacteria per gram content [23]. A problematic bacterium found commonly in the stomach is *Helicobacter pylori*, whose urease activity affords it viability in acidic conditions [24]. Involved in gastric ulcer and cancer pathogenesis, *H. pylori* are typically eradicated upon detection with a potent combination of oral antibiotics. Moving beyond the pyloric sphincter and into the duodenum, bacterial concentration increases to 10^3 bacteria/gram content. Predominant bacteria within the duodenum are thought to be *Firmicutes* and *Actinobacteria,* with key roles including amino acid, carbohydrate and lipid metabolism [25]. In the jejunum and ileum bacterial concentration rises again to 10^4 and 10^7 bacteria/g content, respectively [23].

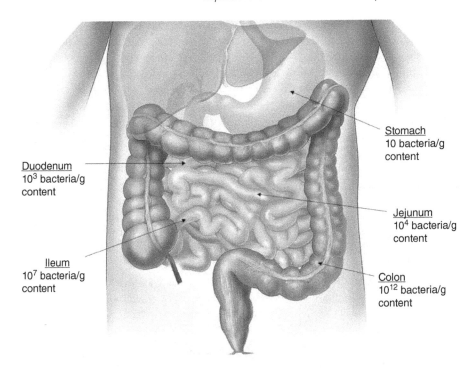

Stomach
10 bacteria/g
content

Duodenum
10^3 bacteria/g
content

Jejunum
10^4 bacteria/g
content

Ileum
10^7 bacteria/g
content

Colon
10^{12} bacteria/g
content

Figure 17.1   *Average bacterial concentration in key areas of the GI tract.*

A healthy small intestinal microbiome is compositionally distinct from the oral cavity, stomach and colon [26, 27]. Microbiota in the small intestine are less concentrated and diverse compared to the colon yet more dynamic due to the need to respond to influx of food, digestive enzymes and bile which alter the luminal conditions radically throughout the day [28]. Characterisation of bacteria within the small intestine has revealed inconsistencies between studies, likely to be related to both sample collection techniques, analytical method differences and the impact of host diet. The temporal and inter-individual variability has been noted to be greater in the small intestine compared with the colon [29]. Commonly reported small intestinal bacteria genera include *Lactobacillus, Clostridium, Staphylococcus, Streptococcus* and *Bacteroides* [28]. Small intestinal bacterial overgrowth (SIBO) is defined as the presence of excessive bacteria in the small intestine and has been implicated as a cause of chronic diarrhoea and nutrient malabsorption as well as being associated with many diseases [30]. SIBO is closely linked with IBS, and is thought to be associated with epithelial permeability, dysmotility, chronic inflammation, autoimmunity, decreased bile salt absorption and possibly neuronal activity. Several pathogenic strains of bacteria have been identified in the small intestines of individuals with SIBO, including *Enterococcus, Escherichia* and *Klebsiella* species [31].

The colon is by far the richest microbial region of the body, with on average 10^{12} bacteria/g content [23]. Numerous microbial niches are present within the colon, including luminal fluid and the epithelial mucus layer. In general, microbial diversity increases from the caecum and ascending colon to the descending colon and rectum [19]. Bacteria residing in the colon play an important role in the digestion of dietary fibre, which would be otherwise indigestible by the human host. This carbohydrate fermentation liberates short chain fatty acids (SCFAs) and lactic acid, which act as energy sources for resident microbiota and play a beneficial role in human health [32]. Aside from bacteria, the colon also houses a diverse array of fungi, viruses and protozoa. Fungal genera include *Aspergillus*, *Candida*, *Cryptococcus* and *Penicillium*, which have also been correlated with local concentrations of SCFAs [23]. Although minor in representation compared to bacteria, non-bacterial microbes do play significant roles in both intestinal and systemic physiology and pathology [33–35].

17.3   Key Causes of Microbiome Variability

Microbiome composition is as unique as individuals' fingerprints [36]. While the implications of this variability on health and biopharmaceutical considerations are still emerging, strain-level microbial variations have already been reported to have significant impacts on pharmacokinetics in a few circumstances [17, 18].

Early life microbiome composition has been associated with health later in life and has the propensity to affect infant pharmacokinetics. Several key factors lead to variability between infants' microbiomes, including whether they are born through natural delivery or caesarean section, whether they are breastfed, and when and how they are weaned [23, 37, 38].

Once adulthood is reached, the colonic microbiome enters into relative stability [39]. However, substantial changes to lifestyle, living arrangements and medicine use can lead to dramatic alterations in GI microbiota composition [40, 41]. Diet is a key driver of GI microbiome variability and can promote either health or dysbiosis. Drug intake is another important factor affecting GI microbial diversity and distribution [42, 43]. The effects of antimicrobials on the GI microbiome are well documented, with short courses of some oral antibiotics negatively impacting intestinal microbial diversity for several years after treatment cessation [15, 44, 45]. In addition to antimicrobials, many non-microbially targeted drugs are now recognised to affect gut microbiome composition. A recent *in vitro* study screened over 1000 marketed drugs for activity against 40 GI bacterial strains, and found that 24% of non-antibiotic drugs inhibit the growth of at least one strain [14]. A systematic review including 20 human studies have found proton pump inhibitors, metformin, non-steroidal anti-inflammatory drugs, opioids, statins and antipsychotics to all affect intestinal microbiota diversity or distribution [46]. It is important to note that not all drug effects on microbiome composition are negative. In the case of metformin, the drug has been found to promote the growth of beneficial species such as *Akkermansia muciniphila*, which may contribute towards its therapeutic effect in the treatment of type 2 diabetes mellitus [47, 48].

Age is another important feature governing individuals' microbiota; as humans age, so too do their microbiome compositions. The compositional and functional changes witnessed in an ageing microbiome are influenced by concurrent factors such as diet, multiple morbidities, pharmaceutical treatment and frailty status [49, 50]. Bacterial age-associated changes in the proximal and distal gut include an increased abundance of *Proteobacteria*, in conjunction with an decreased abundance of *Bacteroides* [13]. Both these bacterial populations are clinically relevant, from a disease and biopharmaceutical perspective. For example, *Proteobacteria* are known to affect the pharmacokinetics of several antineoplastic drugs, whilst *Bacteroides* activate the laxative lactulose in the colon [51, 52]. Microbiota shifts induced by ageing are also linked to immune function, which again may have biopharmaceutical implications [53]. Aside from bacteria, other microbiota in the GI tract are known to be responsive to their host's age. For example, gut virome diversity is thought to decline after the age of 65 [54]. Age should thus be a key consideration when considering the microbiome–biopharmaceutical relationship; an overview is presented in Figure 17.2.

17.4   Microbiome Influence on Key GI Parameters

17.4.1   pH

Metabolic activity of microbiota can have a significant effect on intestinal luminal pH. In the colon there resides a significant population of lactic acid-producing bacteria, such as *Lactobacilli*, *Bifidobacteria* and *Enterococci* [55]. This lactate is typically produced in the carbohydrate fermentation pathway and is utilised upon production by microbiota to produce SCFAs, particularly butyrate [32]. As luminal lactate is generally utilised quickly after production it is usually only present in small concentrations (<5 mM) in healthy humans' faeces [55]. However, in certain diseases such as IBD and short bowel syndrome, lactate can accumulate and lower luminal pH, leading to faecal lactate concentrations of up to 100 mM [56–58]. High concentrations of colonic lactate are correlated with the most severe cases of IBD, colonic lesions and low levels of faecal SCFAs [56]. In patients with SIBO, lactate accumulation in the small intestines has been linked to brain fogginess, bloating and acidosis [59]. This microbiome-driven alteration of luminal pH has the propensity to significantly alter the behaviour of medicines, and thus should be considered when delivering pharmaceuticals to dysbiotic intestinal regions. The administration of probiotic treatments, incorporating live microbiota, may also drive pH changes within the GI tract. Where lactic acid-producing probiotics are being administered, clinicians should consider how this could affect GI pH and thus the behaviour of co-administered medicines [60].

17.4.2   Bile Acid Concentration and Composition

Bile acid metabolism, which occurs primarily in the colon, is an important function carried out by microbiota. This is an example of a symbiotic relationship between the microbiome and its host; microbiota aid with the excretion of bile acids (which can become toxic if allowed to accumulate), and the antibacterial bile acids prevent bacterial overgrowth thus providing resident bacterial species with a homogenous habitat [61]. Numerous strains of gut bacteria produce hydrolases and dehydroxylases capable of deconjugating primary bile acids to secondary bile acids for faecal excretion [61, 62]. In certain cases of dysbiosis, bile acid metabolism by colonic bacteria may be altered. Bile acids are often responsible for the solubilisation of certain drugs in the intestinal lumen, namely lipid-soluble drugs. Thus, a change in bile acid composition or concentration can affect the absorption of drugs where solubility is the rate limiting step [63–65]. Microbiome-bile effects on pharmacokinetics is a phenomenon that warrants further research, and should be a consideration when administering lipid-soluble drugs in states of severe dysbiosis.

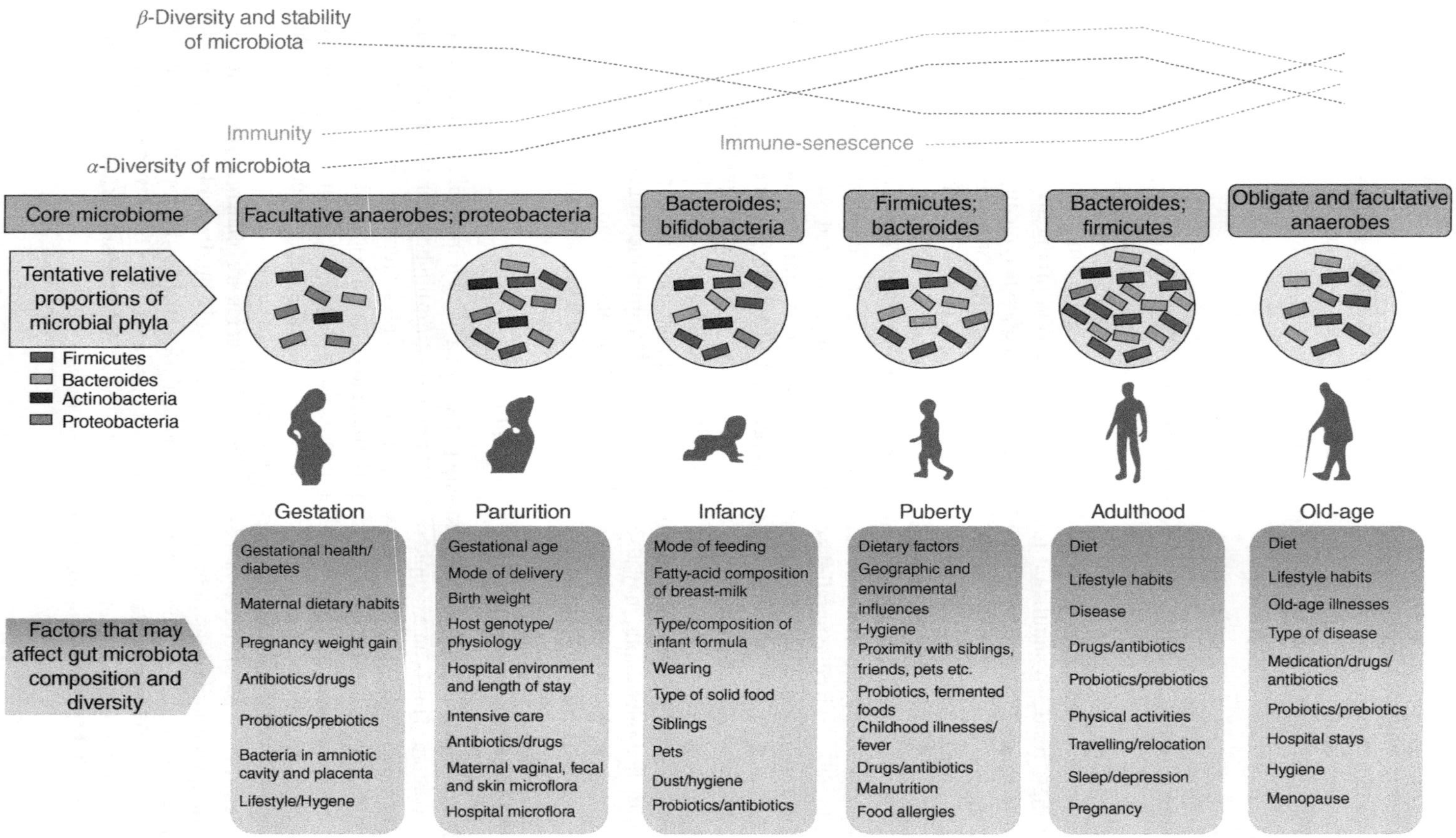

Figure 17.2 Age-related changes affecting gut microbiome composition, and thus the microbiome–biopharmaceutical relationship across the lifespan. Source: From Nagpal et al. [53] / with permission of IOS Press.

17.4.3   Drug Transporters

The microbiome has the potential to alter endogenous drug transporter expression. A study using germ free and antibiotic-treated mice as models of dysbiosis has shown that gut microbiome imbalances can lead to changes in hepatic transporter expression. Dysbiotic mice were found to have significantly lower levels of BCRP1/ABCG2 and BSEP/ABCB11 transporters on their liver plasma membranes than controls [66]. The liver is crucial for the metabolism of many drugs, thus changes to hepatic transporter expression could have important effects on pharmacokinetics. In the intestines, the presence of certain *Lactobacillus* species, specifically *Lactobacillus murinus*, has been found to downregulate the expression of efflux transporters, MDR1 and MRP2, on epithelial Caco-2 cells. When an efflux transporter substrate, glycyrrhizic acid, was orally administered to rats in conjunction with *L. murinus*, the substrate's bioavailability was significantly improved [67]. This highlights how subtle alterations in GI microbiota could affect the bioavailability of drugs that are substrates for endogenous transporters. There can also be competition for transporters where the microbiome or its derivatives have structures similar to those of ingested drugs; this has been reported for urolithin A (derived from ellagic acid by the gut microbiota) which was identified as a substrate for the breast cancer resistance protein, ABCG2/BCRP [68]. In addition, microbial enzymes can modify molecules that would otherwise be substrates for intestinal transporters. Bacteria sourced from human faeces has been shown to alter the structure of a number of pharmaceutical excipients that would otherwise inhibit the apical membrane OATP2B1 transporter, which mediates the absorption of drugs such as rosuvastatin and atenolol [69].

17.4.4   Motility

The GI microbiome's substantial metabolic activity also leads it to affect gut motility, a key physiological parameter within biopharmaceutics. Absence of GI microbiota, exemplified by germ free animals, results in slower small and large intestinal transit times – indicating that microbiota positively modulate myoelectric activity [70, 71]. Colonic transit time has been suggested as a marker for gut microbiome health. Research has demonstrated that longer colonic transit times are associated with increased microbial diversity and a shift from carbohydrate fermentation to protein catabolism [72]. Indeed, fibre-based interventions have long been employed as effective laxatives [73]. Patients with slow transit constipation have been seen to have significantly reduced concentrations of the SCFA acetate in their stools, in combination with differences in microbiome composition [74, 75]. Concentration of hydrogen-producing and hydrogen-utilising colonic mucosal microbiota is also positively correlated with constipation [76]. On the other hand, shorter colonic transit time is linked to increased epithelial cell turnover, indicated by the production of mucus-associated metabolites [72]. At the extremity of short transit, diarrhoea is often caused by states of dysbiosis. For example, severe diarrhoea is the principal feature of *Clostridium difficile* overgrowth in the GI tract, often arising from antibiotic-induced dysbiosis [77].

17.4.5   Hepatic Drug Metabolism

The gut–liver axis describes the relationship between GI microbiota and hepatic function. This relationship has recently received substantial attention, elucidating how microbial metabolites modulate the hepatic transcriptome and proteome via systemic circulation [78]. In this way, the GI microbiome regulates hepatic metabolism of drugs – whether they are orally administered or not. By comparing the hepatic gene expression of germ free and conventional mice, the GI microbiome has been identified as regulating a cluster of 112 genes predominantly involved in drug metabolism. This finding was consequently linked to germ free mice metabolising the anaesthetic pentobarbital more efficiently than their conventionally colonised counterparts [79]. Hepatic genes affected by gut microbiota typically encode for enzymes, transporters and mitochondrial proteins. One study has shown that mice treated with antibiotics have significantly decreased expression of hepatic CYP2B10, CYP3A11 and CYP51A1; all enzymes implicated in drug metabolism [66]. Elsewhere, it was confirmed that murine CYP3A expression is increased by enteric bacteria, supporting the conclusion that gut dysbiosis could significantly alter hepatic drug metabolism [80]. Recently, a sustained intervention of oral SCFA sodium butyrate and the prebiotic fructo-oligosaccharide-inulin has been shown to upregulate CYP2B10 and CYP3A13 whilst decreasing MDR1A expression in mice [81]. This highlights that the liver's metabolic functions are acutely sensitive to therapeutic interventions aimed at the intestinal microbiome.

17.4.6 Epithelial Permeability

Intestinal microbiota are known to play an important role in maintaining epithelial tight junction integrity. In IBD, epithelial permeability is often increased due to dysbiosis-led inflammation; treatment with probiotics has been shown to exert a protective effect by maintaining tight junction integrity [60]. Inflammation-associated dysbiosis is a feature of many diseases, and thus could affect the bioavailability of orally administered drugs in a significant patient population [82, 83]. The presence of several types of bacteria in faeces, including *Veillonella, Escherichia coli* and *Fusobacterium*, are positively associated with intestinal inflammation; possibly providing clinicians with a tool for predicting which patients may have impaired intestinal barrier function [84]. The mechanisms by which dysbiosis can lead to enhanced intestinal are manifold and complex. Serotonin, produced by intestinal enterochromaffin cells, is one molecule linked to the maintenance of epithelial integrity [85]. Drugs can also play a role; proton pump inhibitors have been shown to increase epithelial permeability in stressed mice via mechanisms involving vasoactive intestinal peptide and mast cells [86].

17.5 Enzymatic Degradation of Drugs by GI Microbiota

Orally administered drugs will pass through the GI tract and come into contact with microbial enzymes on the epithelium or luminal fluids; drugs that are poorly absorbed will be particularly exposed to microbiota, as they will remain in the GI tract for longer periods of time and reach the distal gut where microbiota concentration is the highest [23]. Parenteral drugs also have the propensity to be altered by GI microbiota; their systemic circulation may lead them into contact with epithelial cells, or excretion in bile could cause them to bypass microbiota in the colon.

Due to the broad substrate specificity of many microbiota enzymes, drugs are also vulnerable to chemical transformation [87, 88]. Over 180 drugs are now known to be susceptible to metabolism by intestinal bacteria, and there are likely many more cases that have not yet been characterised [89–94].

Enzymatic degradation of drugs can result in altered bioavailability and/or exposure to toxic metabolites. It can also be difficult to predict, due to the wide variability between patients' microbiome compositions. Whilst contributing microbiota are unknown in the majority of drug–microbiota interactions, a number of examples of drugs modulated by gut microbiota and their clinical implications are outlined in Table 17.1.

17.6 Exploitation of the GI Microbiome for Drug Delivery

As microbiota composition varies across the GI tract, with ascending bacterial concentration from the stomach to the descending colon, the metabolic potential of microbiota can be harnessed for the targeted delivery of drugs to the distal gut. The colon is an attractive site for delivery of many drugs; for example, those that are extensively degraded by small intestinal CYP450 or proteolytic enzymes, or those that exert their action locally on the colonic epithelium [102–104]. Prodrugs can also be designed for chemical activation by colonic microbiota enzymes. Tablet coatings formed from fermentable carbohydrates, such as chitosan, pectin, guar gum and amylose are popular types of microbiota-sensitive systems for colonic drug delivery [105]. Such formulations are intended to pass through the stomach and small intestines intact; the carbohydrate coating will only be degraded once it encounters the high concentrations of bacteria in the colon [106, 107]. Generally, microbiota-sensitive delivery systems are efficacious, as the broad metabolic capacity of the microbiome means that inter-individual species-level differences do not affect the degradation of carbohydrates [16]. Recently, dual-triggered colonic delivery coatings have been manufactured and marketed, composed of pH-sensitive and microbiota-sensitive materials (Figure 17.3). Such systems offer a fail-safe method of colonic drug delivery as the coating will still be degraded if either the pH- or microbiota-sensitive element fails [108, 109].

17.7 Models of the GI Microbiome

The utilisation of biorelevant models that reflect the microbiome is important where microbiota affect the release, stability, or metabolism of drugs. As the relevance of the microbiota–biopharmaceutical relationship emerges, it is likely that such models will become a routine step in the development of new medicines. Traditionally, animal models, typically murine, have been used to characterise the microbiome. However, the differences between microbiome compositions between laboratory animals and humans can be striking – potentially leading to invalid conclusions [110]. Moreover, as the pharmaceutical industry strives to move away from animal models as far as possible due to ethical and economic considerations, the development of accurate *in vitro* and *in silico* methods are becoming more and more sought after.

17.7.1 *In Vitro* Models

The vast majority of *in vitro* microbiome models to date use faeces as proxies for colonic microbiota compositions [89, 94, 103]. Human faeces can be incubated with drugs or

Table 17.1 Changes in pharmacokinetics mediated by gut microbiota with clinical implications.

Drug	PK change	Pharmacological/ toxicological consequences	Involvement of bacteria or enzymes	Clinical implications
Amiodarone	↑ AUC$_{0-30}$, ↑ absorption	↑ Bioavailability	*Escherichia coli*	Narrow T.I. range, ↓ dose and ↓ activity
L-Dopa	↑ Metabolism	↓ Bioavailability	*Lactobacilli* and *Enterococci* via tyrosine decarboxylase	↓ Activity in brain and dose variability
Calcitonin	Colonic absorption	↓ Bioavailability and activity	Protease	↓ Blood glucose
Insulin	Colonic absorption	↑ Bioavailability and activity	Protease	↑ Toxicity
Digoxin	↑ Metabolism	↓ Bioavailability	*Eggerthella lenta* via a flavoprotein reductase	Cardiac response lessened
5-ASA	↑ Metabolism	↓ Bioavailability	*N*-acetyltransferase	↑ Activity
Lovastatin	↑ Metabolism	↑ Bioavailability	Not reported	↑ Activity
Simvastatin	↑ Metabolism	↑ Bioavailability	*Lactobacillus*	↑ Activity
Loperamide oxide	↑ Metabolism	↓ Bioavailability	Not reported	↓ Activity
Deleobuvir	↓ C$_{max}$, ↓ AUC	↓ Bioavailability	Not reported	↓ Activity
Metronidazole	↓ AUC	↓ Bioavailability	*Clostridium perfringens*	↓ Activity
Clonazepam	↑ Metabolism and nitro reduction	↓ Bioavailability	Not reported	↑ Toxicity
Epacadostat	↑ Metabolism	↓ Bioavailability	Not reported	↓ Activity
Chloramphenicol	↑ Metabolism	↓ Bioavailability	Not reported	↑ Toxicity
Ranitidine	↑ Metabolism	↓ Bioavailability	Not reported	↓ Activity
Nizatidine	↑ Metabolism	↓ Bioavailability	Not reported	↓ Activity
Morphine	↑ Duration	Increases drug action	Not reported	↑ Toxicity
Lactulose	↑ Metabolism and hydrolysis	↓ Bioavailability	*Bacteroides*, *Lactobacillus* and *Clostridium*	↓ Activity
Loperamide oxide	Activation to loperamide	Prodrug activation	N-oxide reductases (caecal bacteria)	↑ Activity
Levamisole	↓ Metabolism (thiazole ring opening)	↑ Bioavailability	*Bacteroides* and *Clostridium*	↑ Activity

Drug	Effect	Bioavailability	Bacteria	Activity
Risperidone	↑ Metabolism	↓ Bioavailability	Not reported	↓ Activity
Zonisamide	Reduction to 2-sulphamolycetyphenol	↑ Bioavailability	Clostridium sporogenes	↑ Activity
Succinylsulfathiazole	Removal of succinate group	↑ Bioavailability	Not reported	↑ Activity
Paracetamol	↓ Metabolism	↓ Bioavailability	Clostridium difficile	↓ Activity
Aspirin	↑ AUC	↑ Bioavailability	Enterococci, Enterobacteria and Lactobacilli	↑ Therapeutic potency
Amlodipine	↓ C_{max}, ↓ AUC	↑ Bioavailability	Not reported	↑ Activity
Tacrolimus	↑ Metabolism	↓ Bioavailability	Faecalibacterium prausnitzii via keto-reductases	↓ Activity

Source: Adapted from Zhang et al. [95] with additional information from Refs. [17, 96–101].

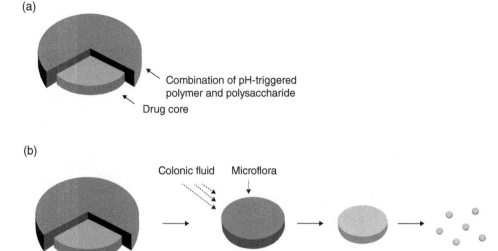

(a)

Combination of pH-triggered
polymer and polysaccharide

Drug core

(b)

Colonic fluid   Microflora

Acid resistance in the stomach   pH- and enzyme-triggered removal of coating layer

Figure 17.3 *Example of a dual-triggered tablet coating incorporating pH- and microbiota enzyme-sensitive materials. Source: From Lee et al. [106] / MDPI / CC BY 4.0*

pharmaceutical formulations in anaerobic conditions at 37 °C, thus replicating physiological conditions. The use of human faeces as microbiome models has several advantages, such as accessibility, cost-effectiveness and non-invasiveness. Moreover, faeces contain a population of microbiota, therefore allowing the study of microbiome–pharmaceutical interactions in which microbiota can interact as a community. Though useful work has been completed using faecal models, the methods do have their limitations [111]. Firstly, faeces cannot represent any single niche of the colon or broader GI tract, as microbial composition varies greatly between niches. Secondly, the storage, transport and sampling of faecal material can affect microbiota viability and representation, especially if exposed to oxygen. For example, faecal microbiota are not distributed uniformly in faeces, thus only using a portion of faecal material can affect the validity of results [112].

To address the limitations of simple batch fermentation models using human faeces, complex multi-compartmental systems modelling multiple niches within the GI tract have been developed. One such model is the Simulator of the Human Intestinal Microbial Ecosystem (SHIME®), an *in vitro* model mimicking the GI tract incorporating the stomach, small intestine and colon [60]. A further model, known as the M-SHIME® system, includes a mucosal niche in addition to the luminal environment, hence including surface-attached mucosal microbiota in experiments [113]. Inoculated with human faeces, the SHIME® and M-SHIME® models are dynamic systems in which fluid is pumped from the stomach through to the colon at physiological transit times and relevant changes in composition as it moves. These advanced systems can be considered as more physiologically relevant than traditional faecal studies, and allow the simulation of orally administered medicines along the length of the gut.

In situations where direct interactions between specific strains of microbiota and pharmaceuticals are intended to be studied, methods such as isothermal microcalorimetry

(IMC) can be utilised [114]. IMC allows the study of drug effects on bacterial growth; an important investigation when studying potential dysbiosis-inducing properties of new formulations. IMC methods involve the inoculation of bacteria and drug in growth medium, and continuous monitoring of bacterial growth via heat produced in the closed system. Using IMC, researchers can investigate precision-level effects of drugs on particular strains of bacteria, which can be useful adjuncts to investigations using models of multiple microbiota populations [115].

17.7.2 *In Silico* Models

More than ever, digital models of the microbiome are being applied to predict microbiota behaviour in specific scenarios [116]. *In silico* microbiome models have several advantages over *in vivo* and *in vitro* tools: they are fast, reproducible, unbiased, often cost effective, and do not require large quantities of consumables [117–119]. Artificial intelligence (AI) is a specific part of the *in silico* toolkit that shows great potential in microbiome modelling. AI can use existing experimental data to learn patterns, and output predictions based on new data [120]. For example, the postprandial glycaemic responses of patients have been accurately predicted based on numerous patient characteristics, including their GI microbiome profile [121]. Elsewhere, the chemical features of drugs that make them vulnerable to gut bacteria metabolism have been elucidated using a machine learning algorithm known as principal component analysis [89]. AI has also been used to predict how drugs affect the growth of gut bacteria, allowing *in silico* prediction of drug-promoted dysbiosis [122, 123]. The use of *in silico* tools and AI within microbiome modelling is still in its infancy, however it does promise a powerful way to predict drug–biopharmaceutical interactions, and will likely become widespread in future years [124].

17.8 Conclusion

The impact of the microbiome on biopharmaceutics is yet to be fully elucidated, however emerging research is highlighting the importance of the GI microbiota for drugs' bioavailability, metabolism and excretion. The entire GI tract is colonised with trillions of bacteria, fungi, viruses and archaea that substantially contribute towards key physiological parameters; from luminal pH, to motility, to epithelial permeability. Thus, the microbiome should not be overlooked when considering biopharmaceutical behaviour. It is now known that over 180 drugs are substrates for GI bacterial enzymes. These drug–bacteria interactions often lead to chemical reactions that change therapeutics' bioavailability and clinical performance. The metabolic potential of the microbiome can also be harnessed for the targeted delivery of drugs, exemplified by the use of fermentable tablet coatings designed for colonic release.

As well as microbiota affecting the performance of drugs, the effects of drugs on GI microbiota is also a key consideration. Dysbiosis within the GI tract has been closely associated with many diseases, hence maintaining a balanced microbiome is important for patients' health. It has been found that a significant proportion of non-antibiotic drugs alter the growth of gut bacteria. Whilst this may not always be a negative feature, it is important that such interactions are documented, as they may have clinical implications.

Due to the extensive variability between individuals' GI microbiome compositions, it can be difficult to predict microbiome–biopharmaceutical interactions. For this reason, the development of accurate models that allow the prediction of treatment-microbiota relationships is important. Though historically the majority of this work has been performed in rodents, new *in vitro* and *in silico* tools are coming to the fore. An industry move towards routinely predicting and testing treatments' relationship with the GI microbiome may soon become a reality; thus there is a need to standardise and validate the biopharmaceutical models utilised.

References

[1] Huttenhower, C., Gevers, D., Knight, R. et al. (2012). Structure, function and diversity of the healthy human microbiome. *Nature* **486** (7402): 207–214.

[2] Qin, J., Li, R., Raes, J. et al. (2010). A human gut microbial gene catalogue established by metagenomic sequencing. *Nature* **464** (7285): 59–65.

[3] Proctor, L.M., Creasy, H.H., Fettweis, J.M. et al. (2019). The integrative human microbiome project. *Nature* **569** (7758): 641–648.

[4] Rosa, B.A., Mihindukulasuriya, K., Hallsworth-Pepin, K. et al. (2020). Improving characterization of understudied human microbiomes using targeted phylogenetics. *mSystems* **5** (1): e00096-20.

[5] Cundell, A.M. (2018). Microbial ecology of the human skin. *Microb. Ecol.* **76** (1): 113–120.

[6] Almeida, A., Nayfach, S., Boland, M. et al. (2020). A unified catalog of 204,938 reference genomes from the human gut microbiome. *Nat. Biotechnol.* **39**: 105–114.

[7] Thursby, E. and Juge, N. (2017). Introduction to the human gut microbiota. *Biochem. J.* **474** (11): 1823–1836.

[8] Mizrahi-Man, O., Davenport, E.R., and Gilad, Y. (2013). Taxonomic classification of bacterial 16S rRNA genes using short sequencing reads: evaluation of effective study designs. *PLoS One* **8** (1): e53608.

[9] Dutta, S.K., Verma, S., Jain, V. et al. (2019). Parkinson's disease: the emerging role of gut dysbiosis, antibiotics, probiotics, and fecal microbiota transplantation. *J. Neurogastroenterol. Motil.* **25** (3): 363–376.

[10] Jostins, L., Ripke, S., Weersma, R.K. et al. (2012). Host–microbe interactions have shaped the genetic architecture of inflammatory bowel disease. *Nature* **491** (7422): 119–124.

[11] Marques, F.Z., Nelson, E., Chu, P.Y. et al. (2017). High-fiber diet and acetate supplementation change the gut microbiota and prevent the development of hypertension and heart failure in hypertensive mice. *Circulation* **135** (10): 964–977.

[12] Aslam, H., Marx, W., Rocks, T. et al. (2020). The effects of dairy and dairy derivatives on the gut microbiota: a systematic literature review. *Gut Microb.* **12** (1): 1799533.

[13] Chaudhari, D.S., Dhotre, D.P., Agarwal, D.M. et al. (2020). Gut, oral and skin microbiome of Indian patrilineal families reveal perceptible association with age. *Sci. Rep.* **10** (1): 5685.

[14] Maier, L., Pruteanu, M., Kuhn, M. et al. (2018). Extensive impact of non-antibiotic drugs on human gut bacteria. *Nature* **555** (7698): 623–628.

[15] Willing, B.P., Russell, S.L., and Finlay, B.B. (2011). Shifting the balance: antibiotic effects on host-microbiota mutualism. *Nat. Rev. Microbiol.* **9** (4): 233–243.

[16] Tian, L., Wang, X.W., Wu, A.K. et al. (2020). Deciphering functional redundancy in the human microbiome. *Nat. Commun.* **11** (1): 6217.

[17] Guo, Y., Crnkovic, C.M., Won, K.-J. et al. (2019). Commensal gut bacteria convert the immunosuppressant tacrolimus to less potent metabolites. *Drug Metab. Dispos.* **47** (3): 194.

[18] Haiser, H.J., Gootenberg, D.B., Chatman, K. et al. (2013). Predicting and manipulating cardiac drug inactivation by the human gut bacterium *Eggerthella lenta*. *Science (New York, N.Y.)* **341** (6143): 295–298.

[19] James, K.R., Gomes, T., Elmentaite, R. et al. (2020). Distinct microbial and immune niches of the human colon. *Nat. Immunol.* **21** (3): 343–353.

[20] Krom, B.P., Kidwai, S., and ten Cate, J.M. (2014). *Candida* and other fungal species: forgotten players of healthy oral microbiota. *J. Dent. Res.* **93** (5): 445–451.

[21] Mira, A. (2018). Oral microbiome studies: potential diagnostic and therapeutic implications. *Adv. Dent. Res.* **29** (1): 71–77.

[22] Abbayya, K., Puthanakar, N.Y., Naduwinmani, S., and Chidambar, Y.S. (2015). Association between periodontitis and Alzheimer's disease. *N. Am. J. Med. Sci.* **7** (6): 241–246.

[23] Dieterich, W., Schink, M., and Zopf, Y. (2018). Microbiota in the gastrointestinal tract. *Med. Sci. (Basel, Switzerland)* **6** (4): 116.

[24] Dunn, B.E., Cohen, H., and Blaser, M.J. (1997). *Helicobacter pylori*. *Clin. Microbiol. Rev.* **10** (4): 720–741.

[25] Angelakis, E., Armougom, F., Carrière, F. et al. (2015). A metagenomic investigation of the duodenal microbiota reveals links with obesity. *PLoS One* **10** (9): e0137784.

[26] Sundin, O.H., Mendoza-Ladd, A., Zeng, M. et al. (2017). The human jejunum has an endogenous microbiota that differs from those in the oral cavity and colon. *BMC Microbiol.* **17** (1): 160.

[27] Li, N., Zuo, B., Huang, S. et al. (2020). Spatial heterogeneity of bacterial colonization across different gut segments following inter-species microbiota transplantation. *Microbiome* **8** (1): 161.

[28] Kastl, A.J. Jr., Terry, N.A., Wu, G.D., and Albenberg, L.G. (2020). The structure and function of the human small intestinal microbiota: current understanding and future directions. *Cell. Mol. Gastroenterol. Hepatol.* **9** (1): 33–45.

[29] Booijink, C.C., El-Aidy, S., Rajilić-Stojanović, M. et al. (2010). High temporal and inter-individual variation detected in the human ileal microbiota. *Environ. Microbiol.* **12** (12): 3213–3227.

[30] Dukowicz, A.C., Lacy, B.E., and Levine, G.M. (2007). Small intestinal bacterial overgrowth: a comprehensive review. *Gastroenterol. Hepatol.* **3** (2): 112–122.

[31] Takakura, W. and Pimentel, M. (2020). Small intestinal bacterial overgrowth and irritable bowel syndrome – an update. *Front. Psychiatry* **11**: 664.

[32] Moens, F., Van den Abbeele, P., Basit, A.W. et al. (2019). A four-strain probiotic exerts positive immunomodulatory effects by enhancing colonic butyrate production *in vitro*. *Int. J. Pharm.* **555**: 1–10.

[33] Sokol, H., Leducq, V., Aschard, H. et al. (2017). Fungal microbiota dysbiosis in IBD. *Gut* **66** (6): 1039.

[34] Campisciano, G., de Manzini, N., Delbue, S. et al. (2020). The obesity-related gut bacterial and viral dysbiosis can impact the risk of colon cancer development. *Microorganisms* **8** (3): 431.

[35] Wang, W., Jovel, J., Halloran, B. et al. (2015). Metagenomic analysis of microbiome in colon tissue from subjects with inflammatory bowel diseases reveals interplay of viruses and bacteria. *Inflamm. Bowel Dis.* **21** (6): 1419–1427.

[36] Franzosa, E.A., Huang, K., Meadow, J.F. et al. (2015). Identifying personal microbiomes using metagenomic codes. *Proc. Natl. Acad. Sci.* **112**: E2930–E2938.

[37] Song, S.J., Dominguez-Bello, M.G., and Knight, R. (2013). How delivery mode and feeding can shape the bacterial community in the infant gut. *CMAJ Can. Med. Assoc. J. J. l'Assoc. Med. Canad.* **185** (5): 373–374.

[38] Scheiwiller, J., Arrigoni, E., Brouns, F., and Amadò, R. (2006). Human faecal microbiota develops the ability to degrade type 3 resistant starch during weaning. *J. Pediatr. Gastroenterol. Nutr.* **43** (5): 584–591.

[39] Coyte, K.Z., Schluter, J., and Foster, K.R. (2015). The ecology of the microbiome: networks, competition, and stability. *Science* **350** (6261): 663.

[40] Keohane, D.M., Ghosh, T.S., Jeffery, I.B. et al. (2020). Microbiome and health implications for ethnic minorities after enforced lifestyle changes. *Nat. Med.* **26** (7): 1089–1095.

[41] Merchant, H.A., Liu, F., Orlu Gul, M., and Basit, A.W. (2016). Age-mediated changes in the gastrointestinal tract. *Int. J. Pharm.* **512** (2): 382–395.

[42] Hatton, G.B., Madla, C.M., Rabbie, S.C., and Basit, A.W. (2018). All disease begins in the gut: influence of gastrointestinal disorders and surgery on oral drug performance. *Int. J. Pharm.* **548** (1): 408–422.

[43] Hatton, G.B., Madla, C.M., Rabbie, S.C., and Basit, A.W. (2019). Gut reaction: impact of systemic diseases on gastrointestinal physiology and drug absorption. *Drug Discov. Today* **24** (2): 417–427.

[44] Jernberg, C., Löfmark, S., Edlund, C., and Jansson, J.K. (2010). Long-term impacts of antibiotic exposure on the human intestinal microbiota. *Microbiology (Reading)* **156** (Pt 11): 3216–3223.

[45] Mulder, M., Radjabzadeh, D., Kiefte-de Jong, J.C. et al. (2020). Long-term effects of antimicrobial drugs on the composition of the human gut microbiota. *Gut Microb.* **12** (1): 1795492.

[46] Le Bastard, Q., Al-Ghalith, G.A., Grégoire, M. et al. (2018). Systematic review: human gut dysbiosis induced by non-antibiotic prescription medications. *Aliment. Pharmacol. Ther.* **47** (3): 332–345.

[47] Rodriguez, J., Hiel, S., and Delzenne, N.M. (2018). Metformin: old friend, new ways of action-implication of the gut microbiome? *Curr. Opin. Clin. Nutr. Metab. Care* **21** (4): 294–301.

[48] Wu, H., Esteve, E., Tremaroli, V. et al. (2017). Metformin alters the gut microbiome of individuals with treatment-naive type 2 diabetes, contributing to the therapeutic effects of the drug. *Nat. Med.* **23** (7): 850–858.

[49] Ghosh, T.S., Rampelli, S., Jeffery, I.B. et al. (2020). Mediterranean diet intervention alters the gut microbiome in older people reducing frailty and improving health status: the NU-AGE 1-year dietary intervention across five European countries. *Gut* **69** (7): 1218.

[50] An, R., Wilms, E., Masclee, A.A.M. et al. (2018). Age-dependent changes in GI physiology and microbiota: time to reconsider? *Gut* **67** (12): 2213–2222.

[51] Weber, J.F.L. (1996). Lactulose and combination therapy of hepatic encephalopathy: the role of the intestinal microflora. *Dig. Dis.* **14** (Suppl. 1): 53–63.

[52] Chankhamjon, P., Javdan, B., Lopez, J. et al. (2019). Systematic mapping of drug metabolism by the human gut microbiome. *bioRxiv*: 538215.

[53] Nagpal, R., Mainali, R., Ahmadi, S. et al. (2018). Gut microbiome and aging: physiological and mechanistic insights. *Nutr. Healthy Aging* **4** (4): 267–285.

[54] Gregory, A.C., Zablocki, O., Zayed, A.A. et al. (2020). The gut virome database reveals age-dependent patterns of virome diversity in the human gut. *Cell Host Microbe* **28** (5): 724–740.e8.

[55] Duncan, S.H., Louis, P., and Flint, H.J. (2004). Lactate-utilizing bacteria, isolated from human feces, that produce butyrate as a major fermentation product. *Appl. Environ. Microbiol.* **70** (10): 5810.

[56] Vernia, P., Caprilli, R., Latella, G. et al. (1988). Fecal lactate and ulcerative colitis. *Gastroenterology* **95** (6): 1564–1568.

[57] Hove, H., Nordgaard-Andersen, I., and Mortensen, P.B. (1994). Faecal dl-lactate concentration in 100 gastrointestinal patients. *Scand. J. Gastroenterol.* **29** (3): 255–259.

[58] Mayeur, C., Gratadoux, J.-J., Bridonneau, C. et al. (2013). Faecal D/L lactate ratio is a metabolic signature of microbiota imbalance in patients with short bowel syndrome. *PLoS One* **8** (1): e54335–e54335.

[59] Rao, S.S.C., Rehman, A., Yu, S., and de Andino, N.M. (2018). Brain fogginess, gas and bloating: a link between SIBO, probiotics and metabolic acidosis. *Clin. Transl. Gastroenterol.* **9** (6): 162.

[60] Ghyselinck, J., Verstrepen, L., Moens, F. et al. (2020). A 4-strain probiotic supplement influences gut microbiota composition and gut wall function in patients with ulcerative colitis. *Int. J. Pharm.* **587**: 119648.

[61] Chiang, J.Y.L. and Ferrell, J.M. (2018). Bile acid metabolism in liver pathobiology. *Gene Expr.* **18** (2): 71–87.

[62] Enright, E.F., Joyce, S.A., Gahan, C.G.M., and Griffin, B.T. (2017). Impact of gut microbiota-mediated bile acid metabolism on the solubilization capacity of bile salt micelles and drug solubility. *Mol. Pharm.* **14** (4): 1251–1263.

[63] Pavlović, N., Goločorbin-Kon, S., Danić, M. et al. (2018). Bile acids and their derivatives as potential modifiers of drug release and pharmacokinetic profiles. *Front. Pharmacol.* **9**: 1283.

[64] Enright, E.F., Griffin, B.T., Gahan, C.G.M., and Joyce, S.A. (2018). Microbiome-mediated bile acid modification: Role in intestinal drug absorption and metabolism. *Pharmacol. Res.* **133**: 170–186.

[65] Enright, E.F., Gahan, C.G.M., Joyce, S.A., and Griffin, B.T. (2016). The impact of the gut microbiota on drug metabolism and clinical outcome. *Yale J. Biol. Med.* **89** (3): 375–382.

[66] Kuno, T., Hirayama-Kurogi, M., Ito, S., and Ohtsuki, S. (2016). Effect of intestinal flora on protein expression of drug-metabolizing enzymes and transporters in the liver and kidney of germ-free *and* antibiotics-treated mice. *Mol. Pharm.* **13** (8): 2691–2701.

[67] Yuan, T., Wang, J., Chen, L. et al. (2020). *Lactobacillus murinus* improved the bioavailability of orally administered glycyrrhizic acid in rats. *Front. Microbiol.* **11**: 597.

[68] González-Sarrías, A., Miguel, V., Merino, G. et al. (2013). The gut microbiota ellagic acid-derived metabolite urolithin A and its sulfate conjugate are substrates for the drug efflux transporter breast cancer resistance protein (ABCG2/BCRP). *J. Agric. Food Chem.* **61** (18): 4352–4359.

[69] Zou, L., Spanogiannopoulos, P., Pieper, L.M. et al. (2020). Bacterial metabolism rescues the inhibition of intestinal drug absorption by food and drug additives. *Proc. Natl. Acad. Sci.* **117** (27): 16009.

[70] Husebye, E., Hellström, P.M., and Midtvedt, T. (1994). Intestinal microflora stimulates myoelectric activity of rat small intestine by promoting cyclic initiation and aboral propagation of migrating myoelectric complex. *Dig. Dis. Sci.* **39** (5): 946–956.

[71] Caenepeel, P., Janssens, J., Vantrappen, G. et al. (1989). Interdigestive myoelectric complex in germ-free rats. *Dig. Dis. Sci.* **34** (8): 1180–1184.

[72] Roager, H.M., Hansen, L.B., Bahl, M.I. et al. (2016). Colonic transit time is related to bacterial metabolism and mucosal turnover in the gut. *Nat. Microbiol.* **1** (9): 16093.

[73] Lever, E., Scott, S.M., Louis, P. et al. (2019). The effect of prunes on stool output, gut transit time and gastrointestinal microbiota: A randomised controlled trial. *Clin. Nutr.* **38** (1): 165–173.

[74] Tian, H., Chen, Q., Yang, B. et al. (2020). Analysis of gut microbiome and metabolite characteristics in patients with slow transit constipation. *Dig. Dis. Sci.* https://link.springer.com/article/10.1007/s10620-020-06500-2#citeas.

[75] Müller, M., Hermes, G.D.A., Canfora, E.E. et al. (2020). Distal colonic transit is linked to gut microbiota diversity and microbial fermentation in humans with slow colonic transit. *Am. J. Physiol. Gastrointest. Liver Physiol.* **318** (2): G361–G369.

[76] Wolf, P.G., Parthasarathy, G., Chen, J. et al. (2017). Assessing the colonic microbiome, hydrogenogenic and hydrogenotrophic genes, transit and breath methane in constipation. *Neurogastroenterol. Motil.* **29** (10): e13056.

[77] Fadda, H.M. (2020). The route to palatable fecal microbiota transplantation. *AAPS PharmSciTech* **21** (3): 114.

[78] Fu, Z.D. and Cui, J.Y. (2017). Remote sensing between liver and intestine: importance of microbial metabolites. *Curr. Pharmacol. Rep.* **3** (3): 101–113.

[79] Björkholm, B., Bok, C.M., Lundin, A. et al. (2009). Intestinal microbiota regulate xenobiotic metabolism in the liver. *PLoS One* **4** (9): e6958.

[80] Ishii, M., Toda, T., Ikarashi, N. et al. (2012). Effects of intestinal flora on the expression of cytochrome P450 3A in the liver. *Yakugaku Zasshi* **132** (3): 301–310.

[81] Walsh, J., Gheorghe, C.E., Lyte, J.M. et al. (2020). Gut microbiome-mediated modulation of hepatic cytochrome P450 and P-glycoprotein: impact of butyrate and fructo-oligosaccharide-inulin. *J. Pharm. Pharmacol.* **72** (8): 1072–1081.

[82] Silveira-Nunes, G., Durso, D.F., de Oliveira, L.R.A. Jr. et al. (2020). Hypertension is associated with intestinal microbiota dysbiosis and inflammation in a Brazilian population. *Front. Pharmacol.* **11**: 1–14.

[83] Gogokhia, L., Taur, Y., Juluru, K. et al. (2020). Intestinal dysbiosis and markers of systemic inflammation in viscerally and generally obese persons living with HIV. *J. Acquir. Immune Defic. Syndr.* **83** (1): 81–89.

[84] Chen, L., Reynolds, C., David, R., and Peace Brewer, A. (2020). Development of an index score for intestinal inflammation-associated dysbiosis using real-world stool test results. *Dig. Dis. Sci.* **65** (4): 1111–1124.

[85] Szoke, H., Kovács, Z., Bókkon, I. et al. (2020). Gut dysbiosis and serotonin: intestinal 5-HT as a ubiquitous membrane permeability regulator in host tissues, organs, and the brain. *Rev. Neurosci.* **31** (4): 415–425.

[86] Takashima, S., Tanaka, F., Kawaguchi, Y. et al. (2020). Proton pump inhibitors enhance intestinal permeability via dysbiosis of gut microbiota under stressed conditions in mice. *Neurogastroenterol. Motil.* **32** (7): e13841.

[87] Koppel, N., Maini Rekdal, V., and Balskus, E.P. (2017). Chemical transformation of xenobiotics by the human gut microbiota. *Science (New York, N.Y.)* **356** (6344): eaag2770.

[88] Crouwel, F., Buiter, H.J.C., and de Boer, N.K. (2020). Gut microbiota-driven drug metabolism in inflammatory bowel disease. *J Crohns Colitis* **15**: 307–315.

[89] Zimmermann, M., Zimmermann-Kogadeeva, M., Wegmann, R., and Goodman, A.L. (2019). Mapping human microbiome drug metabolism by gut bacteria and their genes. *Nature* **570** (7762): 462–467.

[90] Basit, A.W., Newton, J.M., and Lacey, L.F. (2002). Susceptibility of the H2-receptor antagonists cimetidine, famotidine and nizatidine, to metabolism by the gastrointestinal microflora. *Int. J. Pharm.* **237** (1): 23–33.

[91] Sousa, T., Paterson, R., Moore, V. et al. (2008). The gastrointestinal microbiota as a site for the biotransformation of drugs. *Int. J. Pharm.* **363** (1–2): 1–25.

[92] Yadav, V., Varum, F., Bravo, R. et al. (2016). Gastrointestinal stability of therapeutic anti-TNF alpha IgG1 monoclonal antibodies. *Int. J. Pharm.* **502** (1–2): 181–187.

[93] Yadav, V., Gaisford, S., Merchant, H.A., and Basit, A.W. (2013). Colonic bacterial metabolism of corticosteroids. *Int. J. Pharm.* **457** (1): 268–274.

[94] Wang, J., Yadav, V., Smart, A.L. et al. (2015). Stability of peptide drugs in the colon. *Eur. J. Pharm. Sci.* **78**: 31–36.

[95] Zhang, J., Zhang, J., and Wang, R. (2018). Gut microbiota modulates drug pharmacokinetics. *Drug Metab. Rev.* **50** (3): 357–368.

[96] Koppel, N., Bisanz, J.E., Pandelia, M.E. et al. (2018). Discovery and characterization of a prevalent human gut bacterial enzyme sufficient for the inactivation of a family of plant toxins. *elife* **7**: e33953.

[97] Lavrijsen, K., van Dyck, D., van Houdt, J. et al. (1995). Reduction of the prodrug loperamide oxide to its active drug loperamide in the gut of rats, dogs, and humans. *Drug Metab. Dispos.* **23** (3): 354.

[98] van Kessel, S.P., Frye, A.K., El-Gendy, A.O. et al. (2019). Gut bacterial tyrosine decarboxylases restrict levels of levodopa in the treatment of Parkinson's disease. *Nat. Commun.* **10** (1): 310–310.

[99] Zhu, H., Xu, G., Zhang, K. et al. (2016). Crystal structure of tyrosine decarboxylase and identification of key residues involved in conformational swing and substrate binding. *Sci. Rep.* **6**: 27779–27779.

[100] Lee, J.R., Muthukumar, T., Dadhania, D. et al. (2015). Gut microbiota and tacrolimus dosing in kidney transplantation. *PLoS One* **10** (3): e0122399.

[101] Sousa, T., Yadav, V., Zann, V. et al. (2014). On the colonic bacterial metabolism of azo-bonded prodrugs of 5-aminosalicylic acid. *J. Pharm. Sci.* **103** (10): 3171–3175.

[102] Tubic-Grozdanis, M., Hilfinger, J.M., Amidon, G.L. et al. (2008). Pharmacokinetics of the CYP 3A substrate simvastatin following administration of delayed versus immediate release oral dosage forms. *Pharm. Res.* **25** (7): 1591–1600.

[103] Coombes, Z., Yadav, V., McCoubrey, L.E. et al. (2020). Progestogens are metabolized by the gut microbiota: implications for colonic drug delivery. *Pharmaceutics* **12** (8): 760.

[104] Allegretti, J.R., Fischer, M., Sagi, S.V. et al. (2019). Fecal microbiota transplantation capsules with targeted colonic versus gastric delivery in recurrent clostridium difficile infection: a comparative cohort analysis of high and lose dose. *Dig. Dis. Sci.* **64** (6): 1672–1678.

[105] Basit, A. (2005). Advances in colonic delivery. *Drugs* **65** (14): 1991–2007.

[106] Lee, S.H., Bajracharya, R., Min, J.Y. et al. (2020). Strategic approaches for colon targeted drug delivery: an overview of recent advancements. *Pharmaceutics* **12** (1): 68.

[107] Capel, A.J., Rimington, R.P., Lewis, M.P., and Christie, S.D.R. (2018). 3D printing for chemical, pharmaceutical and biological applications. *Nat. Rev. Chem.* **2** (12): 422–436.

[108] Varum, F., Freire, A.C., Fadda, H.M., Bravo, R., and Basit, A.W., (2020). A dual pH and microbiota-triggered coating (PhloralTM) for fail-safe colonic drug release. *Int. J. Pharm.* **583**: 119372. doi: 10.1016/j.ijpharm.2020.119379.

[109] Ibekwe, V.C., Khela, M.K., Evans, D.F., and Basit, A.W. A new concept in colonic drug targeting: a combined pH-responsive and bacterially-triggered drug delivery technology. *Alimentary Pharmacology & Therapeutics* **2008**, *28,* 911–916, doi:10.1111/j.1365-2036.2008.03810.x.

[110] Hatton, G.B., Yadav, V., Basit, A.W., and Merchant, H.A. (2015). Animal farm: considerations in animal gastrointestinal physiology and relevance to drug delivery in humans. *J. Pharm. Sci.* **104** (9): 2747–2776.

[111] Tang, Q., Jin, G., Wang, G. et al. (2020). Current sampling methods for gut microbiota: a call for more precise devices. *Front. Cell. Infect. Microbiol.* **10**: 151.

[112] Swidsinski, A., Loening-Baucke, V., Verstraelen, H. et al. (2008). Biostructure of fecal micro-biota in healthy subjects and patients with chronic idiopathic diarrhea. *Gastroenterology* **135** (2): 568–579.

[113] Venema, K. and van den Abbeele, P. (2013). Experimental models of the gut microbiome. *Best Pract. Res. Clin. Gastroenterol.* **27** (1): 115–126.

[114] Braissant, O., Wirz, D., Gopfert, B., and Daniels, A.U. (2010). Use of isothermal microcalo-rimetry to monitor microbial activities. *FEMS Microbiol. Lett.* **303** (1): 1–8.

[115] Said, J., Dodoo, C.C., Walker, M. et al. (2014). An *in vitro* test of the efficacy of silver-containing wound dressings against Staphylococcus aureus and Pseudomonas aeruginosa in simulated wound fluid. *Int. J. Pharm.* **462** (1–2): 123–128.

[116] Namkung, J. (2020). Machine learning methods for microbiome studies. *J. Microbiol.* **58** (3): 206–216.

[117] Elbadawi, M., Gaisford, S., Basit, A.W. Advanced machine-learning techniques in drug discovery. *Drug Discovery Today* **2021**, *26*, 769–777, doi:https://doi.org/10.1016/j.drudis.2020.12.003.

[118] Elbadawi, M., Muniz Castro, B., Gavins, F.K.H. et al. (2020). M3DISEEN: a novel machine learning approach for predicting the 3D printability of medicines. *Int. J. Pharm.* **590**: 119837.

[119] Elbadawi, M., Gustaffson, T., Gaisford, S., and Basit, A.W. (2020). 3D printing tablets: pre-dicting printability and drug dissolution from rheological data. *Int. J. Pharm.* **590**: 119868.

[120] Cammarota, G., Ianiro, G., Ahern, A. et al. (2020). Gut microbiome, big data and machine learning to promote precision medicine for cancer. *Nat. Rev. Gastroenterol. Hepatol.* **17** (10): 635–648.

[121] Zeevi, D., Korem, T., Zmora, N. et al. (2015). Personalized nutrition by prediction of glycemic responses. *Cell* **163** (5): 1079–1094.

[122] Nava Lara, R.A., Aguilera-Mendoza, L., Brizuela, C.A. et al. (2019). Heterologous machine learning for the identification of antimicrobial activity in human-targeted drugs. *Molecules* **24** (7): 1258.

[123] Zheng, S., Chang, W., Liu, W. et al. (2019). Computational prediction of a new ADMET end-point for small molecules: anticommensal effect on human gut microbiota. *J. Chem. Inf. Model.* **59** (3): 1215–1220.

[124] McCoubrey, L.E., Elbadawi, M., Orlu, M., Gaisford, S., and Basit, A.W. Harnessing machine learning for development of microbiome therapeutics. *Gut Microbes* **2021**, *13*, 1–20, doi:10.1080/19490976.2021.1872323.

Index
